Lithic Analysis at the Millennium

Lithic Analysis at the Millennium

Edited by Norah Moloney and Michael J. Shott

Routledge
Taylor & Francis Group
LONDON AND NEW YORK

First published 2003 by Left Coast Press,Inc.

Published 2016 by Routledge
2 Park Square, Milton Park, Abingdon, Oxon OX14 4RN
711 Third Avenue, New York, NY 10017, USA

Routledge is an imprint of the Taylor & Francis Group, an informa business

British Library Cataloguing in Publication Data
A catalogue record for this book is available from the British Library

Production by Archetype Publications
Typeset by Kris Lockyear, Institute of Archaeology, UCL, using LaTeX2ε

ISBN 13: 978-0-90585-339-0 (pbk)

Contents

List of Figures

List of Tables

Lithic Analysis at the Millennium: Introduction

Norah Moloney[1] and Michael J. Shott[2]

[1]*Institute of Archaeology, University College London;* [2]*Department. of Sociology, Anthropology and Criminology, University of Northern Iowa*

Archaeologists study stone tools for all of the textbook reasons: stone is abundant because imperishable, and its use spans a range of space and time vastly greater than any other material that archaeologists encounter. Yet in post-Pleistocene context, lithic analysis is often ghettoised as a narrow domain of arguable importance to archaeology at large. However, precisely because of its qualities, lithic analysis can address issues from the evolution of modern human cognition to pragmatic action in the past to labour organisation to symbolic manipulation. In a field that encompasses everything from physical-science instrumentation to post-modern hermeneutics, nothing is better suited than stone to unify archaeology, never in outlook of course but at least in subject. If we wish to place on the broadest canvas our study of prehistory from pragmatics to symbol to cognition, we might well argue that we can do it best in stone.

The present volume had its genesis in a conference held at the Institute of Archaeology, London, in March 1998. The aim of the conference was to review the scope of lithic analysis at a number of levels: inevitably in relation to Stone Age societies, where it has always served a crucial interpretative rôle, possibly because of the paucity of other types of evidence (particularly in the Palaeolithic); also the rôle it may have in later periods, including the present. It was hoped that by emphasising a temporal panorama, we lithic analysts and aficionados would be united primarily in the versatility of stone evidence more so than in our own particular period of interest. Indeed, it was evident that many questions which arose concerning practical as well as interpretative aspects of lithic analysis, cross-cut periods and one could say that we were united through the medium of stone. While most of the papers in this volume are conference contributions, the editors invited other contributions which they felt might add breadth to the coverage (Bradbury and Carr, De Loecker *et al.*, Pettitt, Wilhemsen and Feathers).

No collection can faithfully register the whole range of approaches possible in stone; this one necessarily is uneven in its treatment. Collectively, the papers show less interest in tool typology than an equivalent group might have as recently as a decade ago, which reflects larger trends in the field. But the chapters demonstrate the breadth that lithics can accommodate and illustrate the range of new subjects which lithic studies has recently addressed. When viewed from a more generalised perspective, the papers can been loosely grouped under two broad topics, those relating primarily with interpretation and those discussing analytical techniques.

In some ways this collection remains faithful to the traditional preoccupations of lithic analysis. Understandably, the volume has a strongly European cast. Two of the North American contributions concentrate on bifacial projectile points, reflecting the comparative abundance and diversity of these bifaces as finished tools there. Most papers by far concern stone in the context of simpler societies, whether of modern humans or of our hominid ancestors. The only archaeological study of civilisations (Shaw and Bunbury's) concerns gemstones, not stone tools. Yet Martingell's remarkable contribution reminds us that stone continues in use today, and in what we certainly would consider civilised society. This point assimilated, only our imaginations prevent us from extending lithic analysis to any prehistoric context that we please.

Of course, there are lithic studies of prehistoric civilisations but they are lost among the concern for pottery, architecture, metal, practically everything except for stone tools. But even from a setting comparatively rich in other materials and more complex culturally than the subjects of most lithic analyses, Conolly's examination of obsidian hoards in Neolithic Turkey (Çatalhöyük) at once suggests the symbolic treatment of obsidian tools and demonstrates the value of lithic analysis in uncommon contexts. Even for pragmatic purposes, Humphrey and Young's paper, on flint use during the Bronze and Iron Ages in England, shows that stone was not abandoned instantly once metallurgy arose, indeed for certain purposes was not abandoned at all. We can understand the process of (partial) replacement only by abandoning naive ideas about the universal superiority of metal to stone. Theirs is a cautionary tale for those who excavate later period sites: if we look for stone tools we may find them; if we don't look we surely won't.

On a continent that has a long recorded history and that abounds in standing (if sometimes teetering) architecture of great age, sparse scatters of stone tools and debris can seem very thin gruel by comparison. Yet Barrowman and Stuart emphasise the value of the modest lithic scatter as a way to both appreciate and empathise with the past. We would stress that they also possess great analytical value in understanding the past. Aboriginal North America's traditions of settlement and architecture are foreign to European sensibilities. As a result, European North Americans largely missed the ancient cities that were hidden in plain sight (Kennedy 1994), but today North American archaeologists appreciate the analytical value of equivalent scatters there, which are not so much destroyed as transformed.

In methodological terms, Shaw and Bunbury's paper demonstrate in gems what is equally true in stone: that 'source' is no primitive concept but one that engages many assumptions, and that methods cannot be better than the questions that impel their use. The subject is gems, but this chapter registers the most current and reasoned perspectives in provenance studies. Use-wear analysis has survived the giddy claims of some original advocates and the criticism and anguish that inevitably followed. Ibáñez and González chronicle this 'progression from optimism to realism'; their chapter is perhaps the best and most current expression of use-wear's maturity as a field within lithic studies, not above it. Polish formation remains controversial and poorly understood, but use-wear otherwise is better grounded, if somewhat diminished in prospect, than it was a decade ago. Thermoluminescence (TL) has improved Middle Palaeolithic chronology, but not all of its properties are appreciated. For instance, TL moots the association problem that plagues radiocarbon dating. Widely regarded as a ratio measure, TL can be an interval one, trading information for precision (Feathers 1997). Wilhelmsen and Feathers' test case illustrates such use. At the assemblage level, it also models the relationship in time between manufacture and use histories. Because the frequency and duration of use after last manufacture of specimens can differ greatly between types, only direct dating methods like TL can disentangle the complex histories of type use and, in the process, improve the resolution of typological cross-dating. TL applies only to heated specimens, destroys them in the process of dating and, like radiocarbon dating, is capital-intensive. But it can be cheaper and works on tools that have been long exposed to sunlight. Here is one example of corollary benefits to other domains; direct TL dating of constituents can increase the analytical value of the lithic scatters whose study Barrowman and Stuart champion.

Typology and technology are central to several chapters. The papers by Dmochowski and Copeland and Moloney discuss materials excavated some time ago and are successful in demonstrating that useful information can be gained from past excavated materials. Dmochowski uses refitting to reconstruct Upper Palaeolithic technological sequences and combines this information with artefact typology in an attempt to identify the origins and associations of his assemblage. While a number of unavoidable factors led to the interrupted study of the Mousterian assemblage of Ras el-Kelb (Lebanon), the paper provides an example of how older and newer techniques of lithic analysis can be usefully employed, in this case the Bordean 'method', operational chains and micro wear analysis. Pettitt addresses in detail the nature of technological variability in the Middle Palaeolithic with particular reference to Levallois and Quina technology and their relationship to climatic and environmental changes. As a result of his investigations into French assemblages he proposes a model to account for the interaction between Neanderthal mobility and Mousterian lithic technology.

Two contributions raise doubts about the integrity of stone-tool types. Using computer methods to measure orthogonal and polar dimensions, Bradbury and Carr show how techniques influence our perceptions of tool size and form. They resolve a celebrated sequence of early Holocene North American discrete biface types into a continuum of changing tool size and form through time. Types become artefacts of our assumptions, not empirical norms or templates characteristic of discrete time intervals. North Americanists and perhaps others will be inspired by Bradbury and Carr's method to seek time-dependent metric trends in stone tools that can resolve discrete time intervals into continua. Shott questions the integrity of assemblage types, whether signifying tool kit or normative facies, and suggests an extension of the Middle Palaeolithic reduction thesis — that types are arbitrary subdivisions of continua of metric reduction in flake blanks — to understanding variation in 'type' proportions between assemblages.

The study of raw material choice, acquisition and use has been profitably employed as an avenue to elucidating hominid behaviour patterns. McNabb, Sinclair and Quinney's survey in the Makapansgat environs has uncovered differential exploitation of stone sources by Late Acheulean peoples which suggests that the presence of good raw material does not necessarily determine exploitation. Their survey highlights the importance of landscape studies in our effort to understand hominid behaviour patterns on a larger scale and in relation to larger sites, in this case the Cave of Hearths, reminiscent of the 'scatters among the patches' advocated by Glynn Isaac (1981).

Although it has a long history, the refitting of stone artefacts has come into its own since the 1960s with the pioneering work at Pincevent (Leroi-Gourhan & Brézillon 1966). When it is feasible to do so, the refitting of stone artefacts has the potential to address a variety of issues as comprehensively explained and exemplified in the paper by de Loecker *et al.*, detailing the refitting programme undertaken at the sites of Maastricht-Bélvédere, (the Netherlands). Along with micro-wear studies, the possibility of refitting forms a part of most modern studies. Difficult as it is in modern excavations, it is, therefore, all the more incredible that Dmochowski was able to refit artefacts from an excavation which took place 80 years ago.

Stone forms a fundamental part of the earth. For more than three million years stone has been knapped, sometimes shaped and used for a variety of activities. Its durability has ensured the preservation of many of these artefacts and human curiosity has led to their discovery which in turn raised questions concerning their manufacture and use. From the earliest forays into lithic analysis, experimental replication was used to answer such questions and continues to be important in our present-day studies. In order to test Leroi-Gourhan's (1964) claim that cutting edge-to-mass ratios of stone tools increased from the Lower to the Upper Palaeolithic, Tactikos had 23 assemblages replicated which she then compares with 22 archaeological assemblages. In doing so, she makes a forceful argument on the importance of carefully prepared experimentation and analysis of experimental material as an aid to lithic studies.

There remain many issues on which lithic analysts do not agree, some of which are evident in the pages of this volume. For instance, the French concept of *chaîne opératoire* has grown very fashionable, especially in parts of Western Europe. Certainly it extends systematic treatment to the entire reduction process, not just finished products and arguably it contextualises, rather than merely describes, that process. Nevertheless, to many, *chaîne opératoire* and reduction sequence are one and the same. One future challenge is to identify the legitimate differences between *chaîne opératoire* and reduction sequence and to dispassionately weight their comparative merits.

Several contributors use the concept of curation to understand their lithic subjects. Curation, of course, has been contested since the concept arose amidst debates about the meaning of Middle Palaeolithic assemblage variation. Broadly, in Europe curation seems to possess a single meaning, the practice of carrying tools about the landscape. In North America, curation has meant many different things, leading some archaeologists to doubt its value at all (*e.g.,* Nash 1996). Others have sought to refashion curation by integrating some of its earlier meanings and yet preserving its analytical value (*e.g.,* Shott 1996). Perhaps it is time for analysts to re-examine the assumptions they engage by invoking curation.

Across the range that lithic studies encompass, there is no area or approach that cannot be improved. But more than any other, we must attend to the symbolic properties that reside in stone tools, their manufacture and use. Post-modernists have contributed unevenly but most to symbolic interpretation in recent years, yet there is no reason that analytical approaches cannot do the same, differently but no less well. Stones were tools but at the same time symbols, no matter their antiquity or context. To a large extent, ardent sectarian advocacy has divided archaeologists into fortified camps. The tragedy in this situation is our failure to synthesise the best in every approach while abandoning the rest. Archaeology arose in substantial measure as the disciplined study of stone tools. Perhaps it can advance in the same way by arriving where it began, lithic studies leading the détente that reunifies archaeology as the grounded study of our common past.

Acknowledgements

The editors would like to thank Professor Peter Ucko, Director of the Institute of Archaeology, London, who suggested hosting the conference which forms the basis of this volume. Without his encouragement and funding from the Institute of Archaeology the project would never have reached fruition. Thanks too, to Diane Rees, Norena Shopland and Steve Townend who worked tirelessly during the conference, and in particular to Jo Dullagan for her constant help and interest throughout. We are truly indebted to Kris Lockyear who undertook the difficult task of preparing the manuscript for publication. Finally, our thanks to those who participated in the conference making it an enjoyable and fruitful event, and to the contributors of this volume.

References

FEATHERS, J. K. 1997. 'The application of luminescence dating in American archaeology.' *Journal of Archaeological Method and Theory* 4: 1–66.

ISAAC, G. 1981. 'Stone age visiting cards; approaches to the study of early land use patterns.' *In* I. Hodder, G. Isaac & N. Hammond (eds.), *Pattern of the Past: Studies in Honour of David Clarke*, pp. 131–55. Cambridge University Press, Cambridge.

KENNEDY, R. C. 1994. *Hidden Cities: The Discovery and Loss of Ancient North American Civilization.* Penguin, New York.

LEROI-GOURHAN, A. 1964. *Le Geste et la Parole 1. Techniques et langage.* Albin Michel, Paris.

LEROI-GOURHAN, A. & A. BRÉZILLON 1966. 'L'habitation magdalénienne n. 1 de Pincevent près Montrau.' *Gallia Préhistoire* 9: 263–385.

NASH, S. E. 1996. 'The confusion on 'curation': A framework for evaluation of concepts and terms.' *In* G. H. Odell (ed.), *Stone Tools: Theoretical Insights into Human Prehistory*, pp. 81–99. Plenum Press, New York.

SHOTT, M. J. 1996. 'An exegesis of the curation concept.' *Journal of Anthropological Research* 52: 259–80.

Part I

Interpretative Approaches to Lithic Analysis

1

Recent Investigations into the Later Acheulean of the Makapansgat Region, Northern Province, South Africa

John McNabb[1], Anthony Sinclair[2], and Patrick Quinney[3]

[1]*Department of Archaeology, The University of Southampton;* [2]*Department of Archaeology, University of Liverpool;* [3]*Department of Archaeology, University of the Witwatersrand*

Abstract

Extensive fieldwalking in the Swartkrans and Umuritzi (colloquially also known as the Makapan) Valleys in the Northern Province of South Africa, near Potgeitersrus, has revealed a rich archaeological landscape. The pattern suggests a differential appreciation of affordances within the environment. The importance of this new understanding is that it contextualises the archaeological assemblages from the important Cave of Hearths (Makapansgat) site in the Umuritzi Valley.

1.1 Introduction

There is little doubt that there were significant changes in hominid behaviour that occurred with the appearance of anatomically modern humans. Evidence for symbolic communication using material artefacts, extended social networks, the development of long-term planning in both subsistence and technological economies, and the appearance of long-range transport of material resources (500km or more) has been identified in the archaeological record of anatomically modern humans (see papers in Gamble 1996, 1998; Mellars 1990; Mellars & Stringer 1989; Steele 1996; Trinkaus 1989). The question still remains as to whether these developments had greater chronological depth, or whether they were an abrupt innovation dependent on the evolution of our own biological species (see Mithen 1996 for an argument along this line). It can be argued that it is in the Middle-Pleistocene that the first evidence for complex behaviour, among archaic-grade hominids can be seen, and that it is this new behaviour that represents the background context for the later developments that have been so emphasised of late. For instance, the development of prepared core technologies — such as Levallois technique and some blade production, an extension of habitats and geographical areas occupied (Deacon & Deacon 1999), an increase in hominid group sizes (Dunbar 1993; Steele 1996) and possibly even language (Aiello & Dunbar 1993) appear at various times in the Middle Pleistocene, coincident with the evidence for archaic-grade hominids (archaic *Homo sapiens*) and predate the appearance of anatomically modern humans.

This is particularly relevant to an examination of the Middle Pleistocene archaeological and palaeontological record of southern Africa. Intense debate has surrounded the origin and spread of anatomically modern humans, and the subsequent patterns of replacement of indigenous archaic populations in other regions of the world. The current and generally accepted models suggest that anatomically modern humans originated in sub-Saharan Africa and then radiated into the rest of the world from there in the relatively recent past (290,000–140,000). However, whilst the specific relationship between current palaeontological specimens (such as Florisbad, Klasies River Mouth, Jebel Irhoud, Border Cave, Omo Kibish 1) is currently unresolved, what is clear is that archaic-grade hominids, at least in Africa, are the ancestral genetic resource that gave rise to anatomically modern humans in the rest of the world. They therefore represent the hominid context in which these crucial behavioural developments must have occurred.

From an archaeological perspective, the understanding of these archaic-grade hominids and their archaeology poses fundamental interpretative challenges (Sinclair 1998). In trying to interpret their behaviour we cannot easily draw upon the modern analogues of either contemporary human gatherer-hunter societies or modern great ape populations (Gowlett 1997). The former are clearly not only too modern, but they are also participants within a world system of state societies and world economic structures. The latter display quite different mental and physical characteristics from archaic *Homo sapiens*. Instead, analytical techniques have to be developed which can identify intelligent / strategic behaviour in its own right from 'first principles' and, first and foremost, these techniques must be applicable to interpreting the archaeological and palaeontological evidence

At a practical level the investigation of behavioural change in archaic-grade hominids presents a number of avenues of investigation for fieldwork at a landscape level:

1. Prepared core technologies such as Levallois technology indicate that hominids were able to perceive of managing the volume of raw material resource through time and through an intended successive sequence of actions (see discussion in van Peer 1992). This is different from the production of other complex technologies such as handaxe production or even Victoria West cores in a South African context where volume is managed to an end point. From this perspective prepared core technologies represent the first instance of economising in the sense that we would understand it. If this is present in a single production sequence, we need to determine whether it can be observed at the scale of the landscape.

2. Deacon & Deacon (1999) have argued that Middle Stone Age (MSA) hominids in South Africa display an extension in habitats: sites are no longer constrained by having to be close to permanent water sources.

3. The development of hominid settlement systems is one of the major changes identified between anatomically modern hominids and archaic-grade hominids (*e.g.*, Lieberman & Shea 1994). Attempts to reconstruct settlement systems are vital.

It is through these three areas of analysis that the Makapansgat Middle-Pleistocene Project is attempting to relate ideas of broader changes in archaic-grade hominid behaviour to the reality of the archaeological record in a landscape.

1.2 Investigating Middle-Pleistocene behaviour in the Makapansgat environs

The Makapansgat environs lie approximately 280km north of Johannesburg, and 16km east-northeast from the town of Potgietersrus, Northern Province, South Africa (Fig. 1.1). The Makapansgat and Swartkrans Valleys, which have been the focus of work to date, are located within a mountainous landscape. The principal substrate is the Malmani Dolomite, which then overlays the Pre-Cambrian Black Reef quartzite series. This in turn is backed by quartzites, tuffaceous shales, mudstones, and feldspars of the Wolkberg group. The Malmani series is inter-bedded with chert breccias, cherts, sandstones and shales. Pronounced continental weathering (*c.* 3 million years ago) with accompanying local solution processes have roughened the topography creating jagged, micro-karstic slopes, and opened up several large and complex doline cave systems (Butzer 1984). These have subsequently been in-filled with late Pliocene and Pleistocene sediments containing extensive accumulations of faunal and artefactual remains.

Exploration during the last 60 years has recorded a number of fossil and artefactual deposits within the Malmani formation of exceptional note (Dart 1948a; Mason 1962; van Riet Lowe 1938, 1943, 1948). Though the region drew the attention of amateur and professional palaeontologists and archaeologists from the 1920s onwards (Dart 1925; Eitzman 1958) systematic research in the valley was only initiated in 1937 (van Riet Lowe 1938). The region is probably best known for the Plio-Pleistocene deposits at the Limeworks site that have produced numerous specimens of ***Australopithecus africanus***, and where research has recently begun again conducted by Kevin Kuykendal of the University of the Witwatersrand and Kaye Reed of the Institute of Human origins.

The principal archaeological site in the valley is the Cave of Hearths which presents the most important, deeply-stratified archaeological site in the region. The site itself was discovered by van Riet Lowe in 1937 when he was asked to assess the neighbouring Historic Cave as a potential National Historic Monument. (This cave was the site of a major conflict and massacre between a Boer Commando and members of the Kekana Tribe led by their chief Makapane — after whom the Makapansgat Valley is named (Maguire 1998; Mason 1988)). Van Riet Lowe chanced upon the Cave of Hearths when he explored a neighbouring lime workers' cutting which exposed breccia containing bones and stone tools. The basal units were noted to contain lenses of ash and carbon, and it was thought that these were traces of hominid fireplaces — hence the name. Following its discovery, excavations began under the aegis of van Riet Lowe: first in the late 1940s by James Kitchen, and then two major campaigns of work undertaken by Revil Mason in 1953 and 1954. The results of Mason's excavations were recently published in a monograph (Mason 1988). The Cave of Hearths contains what may be one of the most complete archaeological records anywhere in Africa — over 250,000 years of almost continuous habitation (Rayner *et al.* 1993). The deposits have yielded fossil remains of archaic-grade *Homo sapiens* (*Homo* cf. *rhodesiensis*; Dart 1948b; Tobias 1971) and archaeological assemblages of Early, Middle and Later Stone Age industries. Historically, the site was deemed important for yielding the first recorded series of stratified Acheulean artefacts, hominids, and faunas from a sub-Saharan African site (Mason 1988, 183). It is still one of just three cave sites with Acheulean industries (the others being Montagu Cave in the southern Cape and Wonderwerk Cave near Kimberley).

Since 1954, little archaeological fieldwork has been undertaken within the Makapansgat environs, with

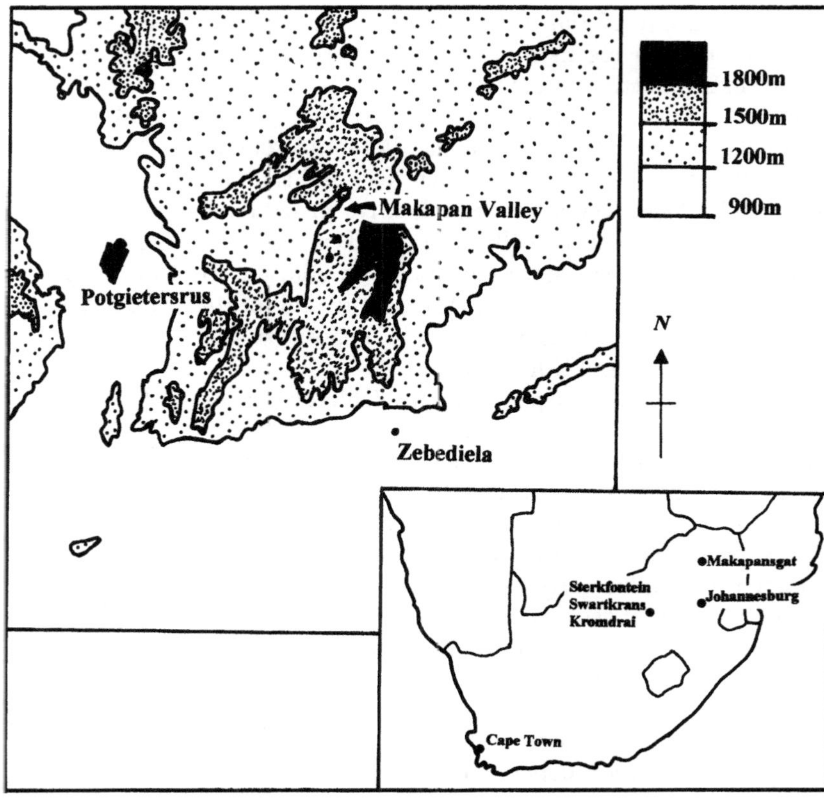

Figure 1.1: Makapansgat and other important early hominid sites in South Africa

the exception of sporadic work on local Iron Age communities (*e.g.*, Hoffmeyr 1988; Moore 1981) and at the nearby Middle Stone Age site of Mwulu's Cave just outside of the present area of study to date (Tobias 1949).

Butzer (1984) and Maguire (1998) have both noted that the Makapansgat landscape has not been subject to major modification in the last 3 my. Uplift during the Early Miocene and Late Pliocene gave rise to the development of local karst features, but these have been imposed on the basic pre-Cambrian structure without obliterating it. In terms of the geomorphology, therefore, the Makapansgat Valley is a young valley written onto a mature, structurally controlling landscape. This structural control has resulted in relative stability in terms of drainage patterns and upland landscape form, and thus, whilst stream channels running into the wide Swartkrans Valley display shallow palaeo-channels and terraces, many other features of the landscape remain largely unchanged. The Makapansgat environs, therefore, provide an ideal area in which to investigate landscape-scale behaviour in the Middle Pleistocene.

The Makapansgat Middle-Pleistocene Research Project is an inter-disciplinary undertaking, directed by Quinney and Sinclair. The project has aimed to investigate the changing nature of human use of environmental resources through the Middle-Pleistocene taking the Makapansgat environs (defined as including the Makapansgat Valley proper, the immediately abutting Swartkrans Valley and neighbouring upland and valley areas) as our focus of investigation. The main work involves a programme of intensive landscape survey to locate and identify new archaeological sites, as well as a programme of small-scale research excavation of these sites to understand their composition and formation history. The project also includes a re-appraisal of the formation history, dating and archaeological assemblages from previous excavations at the Cave of Hearths. Surrounding these field-based activities is a range of other scientific post-survey and post-excavation studies. Fieldwork started in 1997 and has continued in 1998 and 1999 (McNabb *et al.* 1998; Quinney *et al.* 1998).

What follows below is a discussion of the landscape survey and the re-assessment of the Early Stone Age lithic assemblages from the Cave of Hearths. Our aim is to show how localised fieldwork may begin to contribute to the broader questions of the development of complex behaviour in the Middle-Pleistocene. Work is still ongoing and the results below are necessarily provisional.

1.3 Landscape survey in the Makapansgat environs

Landscape survey has followed a standard fieldwalking methodology. In simple terms, small crews numbering from 6 to 12 individuals (average 9) as appropriate fieldwalk predetermined areas of terrain along the valley bottoms, valley sides or uplands. They maintain as closely as possible, a straight line with approximately 5 m distance between walkers. Of necessity, the steep nature of many of the valley sides, and their dense vegetation cover means that it is often impossible for fieldwalking parties to arrange themselves in a perfect line. Finds (lithic, ceramic or breccia deposits) are collected and later described. Potential sources of lithic raw material are also identified and samples collected. From 2000, a comprehensive programme of thin-section analysis will be undertaken for these possible raw material source locations. In addition to lithic materials, survey teams also identify and locate breccia outcrops, which are potential depositories of preserved faunal and lithic remains.

For operational purposes in the field, two types of find grouping — individual find spots and 'localities' — have been defined. In simple terms, localities are defined as places where lithic artefacts cluster. When encountered, detailed report forms are used to describe, amongst other aspects, the local terrain, major topographic features, local vegetation, ground visibility, natural geology, *etc.* All localities are examined in detail with a note made of the artefacts visible and the apparent extent of the spread. A sample of artefacts is then collected for examination; the collected sample size of these locality assemblages varies from 5 to 600 pieces (depending on permit provisions) allowing, in many cases, a reasonable understanding of the nature of the assemblage to be determined. All individual findspots are collected.

All find spots and localities are located by use of a hand-held Ground Positioning Satellite device (Magellan 2000). Each GPS-derived location is noted on a fieldwalking daysheet and subsequently on a (1:50,000) topographic map and a larger-scale (1:10,000) aerial-photographic map. Experience, based on the GPS data for localities, indicates that the GPS-derived data is accurate to within 30m within the horizontal dimension (Eastings and Southings), but quite inaccurate vertically for height above sea level (up to 150m at times). The use of large-scale aerial photographic maps has largely overcome this particular problem and allows real co-ordinates to be entered in the major artefact catalogue. All sites and localities are also being entered into a card-based site database (a sort of regional sites and monuments record based on the local South African ordnance survey maps at 1:50,000 scale) located in the Department of Archaeology at the University of the Witwatersrand. The artefact assemblages will also be permanently stored here at the end of the Project.

As a check on the accuracy and/or potential bias of fieldwalking observations that may exist between field crews and times of day walked, a number of areas walked in 1997 and 1998 were re-examined in later years, using different crews and at different times of day. On these occasions field survey teams have not been told of the previous survey results from these areas. Observations have usually confirmed the presence/absence of archaeological findspots and localities. On just one occasion have fieldwalkers missed the presence of an albeit small archaeological locality. This form of double checking suggests that the presence or absence of survey findings can be relied upon and the practice will be continued in following years. Re-walking of areas through the three years has also made it possible to re-appraise localities previously identified in light of the greater experience of what is to be found in the Makapansgat environs. As a result of this process, a number of localities (6 in total) have been downgraded from locality status.

1.4 Survey results

In the three seasons of work to date, the Project has conducted 57 days of field survey, walking on average 10km each day, with 4km being field-surveyed. In total an area of approximately 15km^2 has been intensively surveyed. Before the Project began, the database held at Witwatersrand recorded the presence of 4 Early and Middle Stone Age 'sites' in the Makapansgat environs: the Cave of Hearths, Rainbow Cave, Magazine Cave and Rufus Cave. With the exception of Rufus Cave, the three other sites are in fact part of the same infill system. This pattern suggests a concentration of activity in the Makapansgat Valley. Field survey has now increased this number of localities to 37 (Fig. 1.2). In addition, 13 other palaeontological localities, and 2 Iron Age localities have been discovered. More than 330 individual lithic artefacts have also been recovered. Just as importantly, systematic field survey has also revealed large areas of landscape where findspots are either absent or very rare.

In comparison with the areas commonly utilised by studies of modern gatherer-hunter populations in a year and especially over a lifetime, the 15km^2 of area surveyed by the Project is not large. It includes, however, a wide range of topographies and lithologies, and realistically represents a coverage of all areas that were accessible from the Cave of Hearths within a day. The number of localities discovered and the variation in presence of findspots suggests that the survey data collected so far provides a good basis upon which to begin to understand local geographical planning decisions by Middle Pleistocene hominids. And this addresses the three major aims of the fieldwork.

From a survey perspective all of the larger localities were encountered in the first season of work. The second season produced 17 localities, and the final season 6 localities. The number of findspots corresponds closely to the duration spent in the field and the number of personnel on survey (Table 1.1). We anticipate that we have now located all major localities and most other localities in the Makapansgat environs.

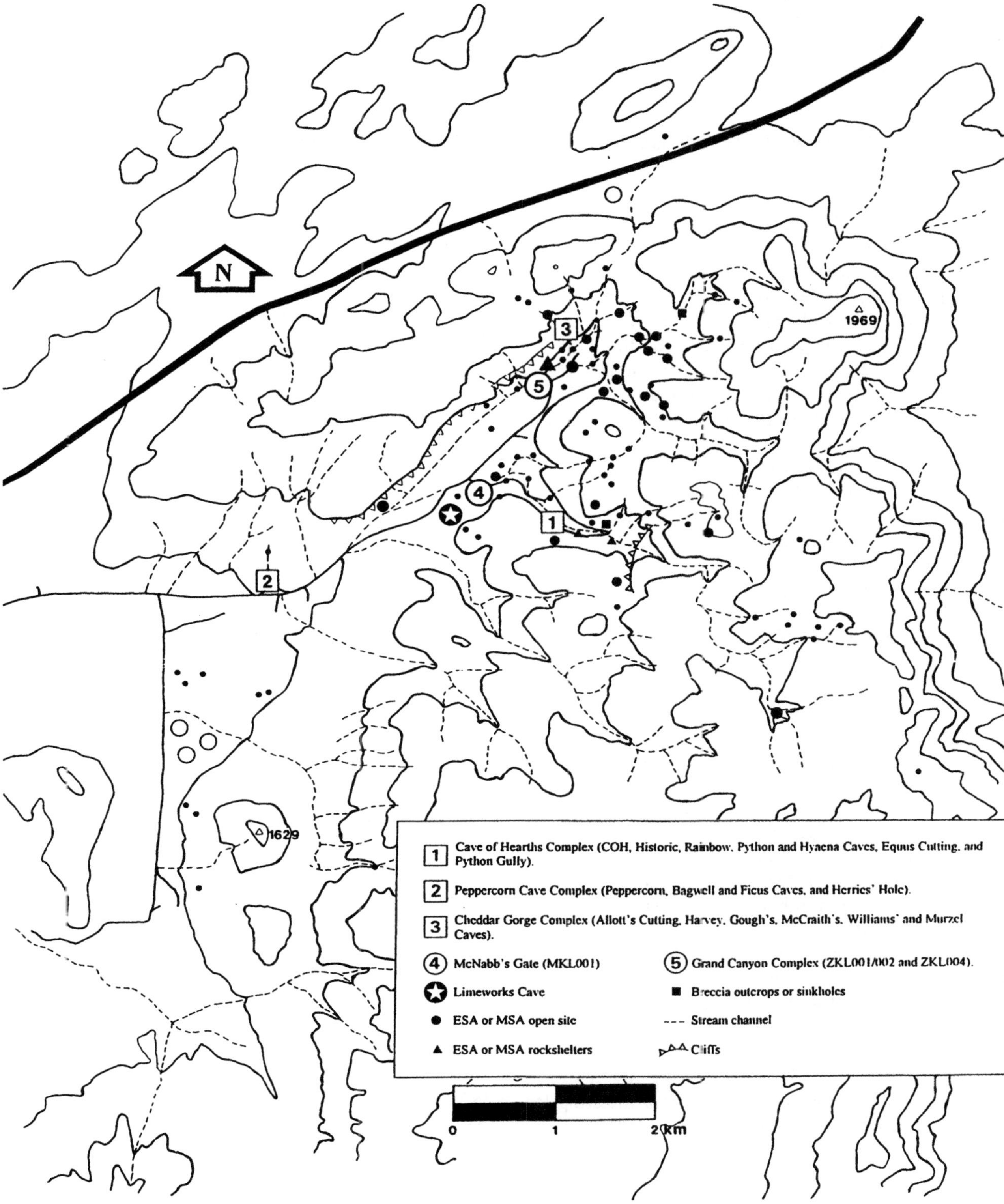

Figure 1.2: The landscape survey areas for the Makapansgat Middle-Pleistocene Research Project.

Field Seasons	No. of major localities	No. of other localities	No. of find spots
1997	5	5	90
1998	0	17	150
1999	0	6	92

Table 1.1: The frequencies of major localities, other localities and find spots discovered by field survey. Major localities possess more than 80 pieces. Four other localities are known from previous work.

1.5 Locality variation and find spot density

Among the 37 localities identified in the Makapansgat environs there is considerable variability in the number / density of artefacts. Excluding the excavated assemblage from the Cave of Hearths, this varies from approximately 15 artefacts visible to 600.

Localities identified so far are to be found on all lithologies and in all topographic situations; upland areas, valley bottom and side locations, as well as a number of localities that are some distance from a water source. Across most of the Makapansgat environs it is possible to encounter individual findspots, discarded evidence of a 'background noise' of hominid activity. There are two areas that are notable exceptions, both located on dolomite bedrock. The first includes the north west side of the Swartkrans Valley (an area of approximately 3km^2), and the second includes the south-east side of the Makapansgat Valley (an area of approximately 2km^2). There are localities in both areas, but the background noise of general hominid activity is absent despite good visibility for field survey and repeated survey work. Interestingly, background noise of activity is usually to be found on quartzite lithology (indeed it obviously appears when crossing from dolomite to quartzite), but this is not the case with all areas of quartzite. A large area of quartzite lithology, again regularly surveyed and with good visibility, is barren of artefacts.

The specific reasons for the placing of these localities and the discard of a background noise of artefacts are, therefore, likely to involve a number of factors, among them raw material availability, local topography (including such factors as local shelter and good vantage points) and access to / distance from water. In the sections that follow we shall look at the Early Stone Age (ESA) lithic assemblages from the Cave of Hearths and other localities considering raw material provenance and use as a potential factor affecting locality. Interpretation of the MSA levels remains to come.

1.6 Technological behaviour at the local scale

The quality of Mason's excavations at the Cave of Hearths and his analysis of the lithic assemblage still makes the archaeological assemblages from the Cave of Hearths immensely valuable, and pivotal to an understanding of chronological change in the region. Despite the heavily consolidated and breccified nature of the sediments, especially in the lower part of the sequence, Mason excavated by 12 inch spits and with a fixed horizontal grid so as to maximise the stratigraphic information that might be assigned to each piece. Using controlled blasting techniques, Mason related his faunal and artefact assemblages, where it was practical, to a three dimensional grid. It is still possible to reconstruct his excavation and the position of artefacts, other than those in the debitage category, from the three dimensional co-ordinates preserved on each artefact, and from the excavation notes contained in the site monograph. The facsimile publication of his weekly excavation reports, submitted to van Riet Lowe in 1953 and 1954, also makes it possible to understand his excavation strategy through both seasons of work.

According to Mason (1988), the earliest occupants of the cave were bats who left a substantial deposit of guano on a travertine floor. The bats were succeeded by owls, whose pellets litter the guano surface. There follows three levels of occupation by ESA biface making hominids (beds 1–3). On the basis of artefact typology these were identified as belonging to a late Acheulean complex. A provisional date of *c.* 250,000 was assigned to these beds. This is followed by an hiatus during which a considerable talus of éboulis from roof collapse built up in the front of the cave. Beds 4–9 represent MSA occupation. Bifaces no longer appear in the archaeological levels and the most significant tools are now points and flake tools, many of which are made by Levallois prepared core techniques. Beds 4 and 5 represent a manifestation of the earlier phases of the Pietersburg complex which is the regionally dominant MSA variant in the northern Transvaal. Beds 6–9 contain a later manifestation of the Pietersburg. Bed 10 contains the cultural debris of Late Stone Age (LSA) gatherer-hunter peoples; at the time of excavation this was allied to the Fauresmith culture, later renamed the Albany. In the uppermost level, bed 11, Mason recovered remains of occupation by indigenous Iron Age farming and pastoral communities. In addition to the extensive Acheulean assemblage, the site has also yielded an extensive faunal record (Cooke 1988) and this is currently under study as part of the Project's re-appraisal of the site.

Whilst it is now clear that Mason's account of the formation history of the site will have to be significantly revised, the considerable depth of archaeologically rich sediments spanning the ESA and MSA continue to make the Cave of Hearths a pivotal site for any understanding of changes in hominid activity in the area through the Middle-Pleistocene. Although the original excavations focused on the establishment of developmental sequences within local cultural historical frameworks, it is a tribute to Mason's work that his insistence on an exacting recovery strategy still allows modern investigators to use the Cave of Hearths assemblages in the context of new research programmes and very different research questions.

1.7 Raw material use at the Cave of Hearths

The diversity of raw materials present in the Cave of Hearths assemblage covers the full range of materials available in the Makapansgat environs (Table 1.2). The majority of stone tools from the Cave of Hearths are quartzite. It is reasonable to suppose that, from the point of view of ESA knappers, rocks known to be suitable for flaking and readily accessible would be the primary targets for raw material acquisition. Visual homogeneity and a non-granular appearance would be the triggers for further investigation, much as with modern knappers. Table 1.2 gives the proportions of occurrence of a selection of the larger flaked artefacts, spheroids, bifaces and cleavers from the cave. The dominance of quartzite and diabase is clear.

The major question concerning the link between landscape and raw material usage is the sources of the major locations for raw material procurement and their relationship to the archaeological assemblages excavated by Mason from within the cave. This directly links the field walking programme to the analysis of the cave data.

1.8 The geology of raw materials around the Cave of Hearths

A number of major lithological divisions are present in the landscape in the immediate vicinity of the Cave of Hearths (Fig. 1.3). Most of the raw material types that appear in the archaeological record — quartzite, diabase, shale and chert — are present within a day's walk from the cave (verified by fieldwalking in 1999).

In the Makapansgat Valley, along which runs the Mwaridzi River, the Cave of Hearths (Fig. 1.3, a), (and Magazine and Rufus Caves) is cut into dolomitic limestone. Upstream of the Cave, the valley head is formed by a dramatic cliff face of quartzite, known locally as the red cliffs (Fig. 1.3, b). These have been formed by the erosion of the overlying Malmani dolomites, exposing the underlying Black Reef Quartzite series. The country to the east of the cave is therefore dominated by quartzite. The plateaux and highlands (c), whose western limit in this vicinity is the red cliffs, are quartzite. The quartzite here represents a mixture of fine grained homogeneous rock, suitable for flaking, and the less suitable porphyritic quartzites where bubbles of air have been infilled with silica to form a more heterogeneous rock. This quartzite plateaux and upland landscape is one of slopes and hills dissected by steep valleys and ravines. Erosion produces a litter of quartzite scree over the slopes (especially at Fig. 1.3, c).

Quartzite is also present closer to the Cave. An exposure of quartzite (Fig. 1.3, d) runs from south-east to north-west across the southern slopes of the Makapansgat Valley, less than 500m from the Cave of Hearths. On the southern slope of the valley it appears on the surface as a narrow band partially exposed through erosion. Just downstream from the Cave, the Mwaridzi has cut through this quartzite exposure creating deep sumps. But at the mouth of the Mwaridzi (Fig. 1.3, e) the northern slopes of the valley have experienced greater lateral erosion, and more of the quartzite has been exposed. On these northern slopes erosion has also produced carpets of scree (Fig. 1.3, e) composed of fine grained, homogeneous, quartzite slabs. At various points upstream, between the cave and the red cliffs, the river has exposed large quartzite outcrops. So it is apparent that within easy distance of the Cave high quality tool making raw material was present. The bed of Mwaridzi river stream in its valley is also a major modern source of quartzite and also contains quantities of quartzite boulders brought down from upstream by the summer rains and replenishing the clast budget of the stream when the input of water is at a sufficient velocity to initiate entrainment. There is little reason to doubt that this situation would not have persisted during the Middle Pleistocene.

Away from the immediate vicinity of the cave, quartzite is present in the dissected landscape to the north of the present site of the Swartkrans farmhouse (Fig. 1.3, f). The western side of the Swartkrans valley is composed of vertical dolomite cliffs, with dolomite plateaux to the west of the cliff line (Fig. 1.3, g). At the north-eastern end of the valley the cliff walls are replaced by a deeply dissected landscape of steep hill slopes. Within this area there are outcrops of quartzite, now forming the tops of the hills, which are exposed as erosion removes the blanket of dolomitic cover and begins to sculpt the hills themselves. These have become the source for colluvial fans of quartzite scree which have built up at the level of the valley floor (Fig. 1.3, f).

Diabase is also present within easy walking distance of the Cave. An extensive outcrop is present to the south of the Makapansgat Valley. It runs east-

Artefact	Diabase	Quartzite	Quartz	Shale	Felsite	Chert	Others
Bifaces Beds 1-3	15	63	4	4	4	2	0
Cleavers Beds 1-3	33	228	0	3	2	1	0
Discoids Bed 3	6	50	0	1	0	0	0
Chopper Bed 3	7	57	0	1	0	0	0
Selected Flaked Pieces Bed 3	70	710	10	5	8	7	8
Spheroids Beds 1–3	166	32	25	0	0	3	1
Total	297	1140	39	14	14	13	9

Table 1.2: Raw material usage at the Cave of Hearths. Data from Mason (1988, 192ff.). Selected flaked pieces are Mason classes — irregular artefacts, bifacial quads, cuboids, ellipsoid cores, cuboid cores. A smaller percentage of artefacts classed as quartzite may in fact be amphibolite. The examples of this rock type which was used as blanks for stone tools, are dark in colour, homogeneous and fine grained. In terms of its suitability for flaking and visual appearance, to the geologically untrained eye amphibolite is visually identical to quartzite. For the purposes of this paper the two rock types are grouped under the umbrella term quartzite.

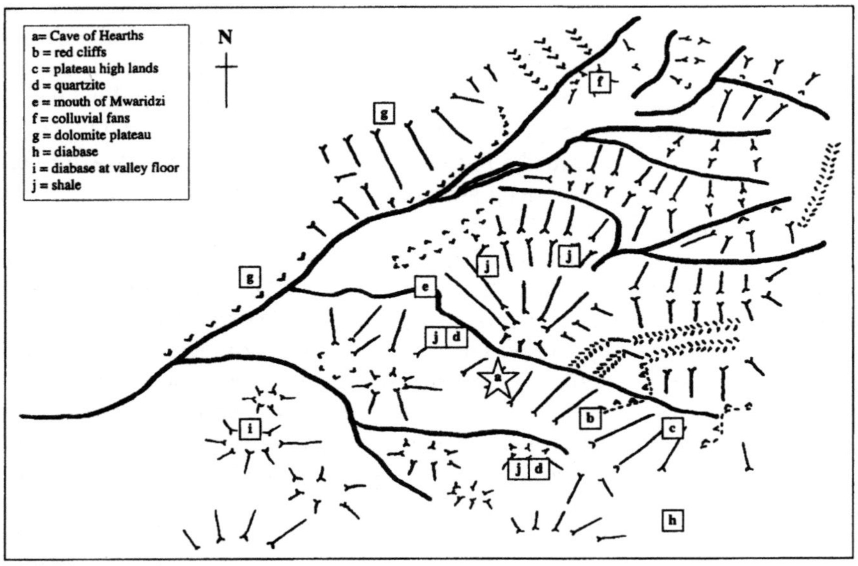

Figure 1.3: Sketch map of the Swartkrans and Mwaridzi Valleys showing raw material sources and major topographic features in the Makapansgat Environs.

wards in a narrow band from the Swatrkrans Valley floor (Fig. 1.3, i) to a ridge overlooking the red cliffs (Fig. 1.3, h). At both extremities it is readily accessible. Shale is present in a series of exposures that form a band encircling the cave (Fig. 1.3, j). Shales form a part of the Malmani complex and are exposed as the limestone is eroded down to the level of the underlying Black Reef Quartzites. The southernmost part of this encircling band parallels the band of quartzite (Fig. 1.3, d). Other exposures of shale are directly across the Makapansgat Valley from the entrance of the Cave of Hearths. Chert is present in the dolomites throughout the Valley as a constituent part of the limestone, and quartz is also present to the south of the Makapansgat Valley, associated with the Black Reef Quartzites.

1.9 Hominid use of these landscapes

As noted above, the field survey has encountered a complex landscape of plateaux, uplands and hill tops, slopes, deep ravines and gullies and valley bottoms, some steep and narrow others wide and flat. This topographical diversity cross cuts a wide range of lithologies in a relatively small area. It is very apparent that the exploitation of these landscapes during the ESA (as evidenced by the presence of diagnostic artefacts, primarily bifaces and cleavers and large factory size flakes) was very fragmented, and there does not appear to have been an extensive exploitation of all areas where suitable raw materials were to be found.

This is very evident on the plateaux lands to the east of the red cliffs. Here the Mwaridzi stream has cut a shallow valley through quartzite. As noted above, the valley's flanks to the north and south are quartzites, erosion having produced a cover of quartzite scree. Extensive field walking revealed only a light density of artefacts at this location, little more than background noise. The quartzite here is a light coloured fine grained and homogeneous rock, similar examples were used in the Cave of Hearths for making bifaces, cleavers and cores. Yet at this location there is little evidence of any extensive exploitation of the raw material potential (either ESA or otherwise).

Artefact type	Dark/black Quartzites	White/light quartzites	Non-quartzites	Indeterminate
Bifaces		8	1	
Cores		12		
Other artefacts	5	54	4	1

Table 1.3: Raw material composition of artefacts from colluvial fans behind the old Swartkrans farm house. Other artefacts include debitage, Levallois, flake tools. Bifaces include cleavers, rough-outs and broken pieces.

This is despite the obvious attractions of this plateau valley with all the affordances a stream would provide (even during modern dry, winter conditions there is still standing water in the pools at the cliff edge). What makes this absence of activity even more puzzling is the definite presence of hominids in this area. Towards the top of the southern slopes of this stretch of the valley the quartzite is replaced by shales and then diabase (Fig. 1.3, h). The field walking recovered a number of artefacts — cores, bifaces and a number of possible large 'factory' sized flakes in diabase. Furthermore, weathering and erosion at this locale was noted to produce large exfoliation flakes of diabase which would have made ideal blanks for bifaces and cleavers. Spheroids of diabase were also present on the ground at this location. Following the ridge top, field walkers noted that when they passed off the diabase onto the quartzite the presence of human activity persisted. Bifaces and cores were recovered. In summary, activity was located near the top of the southern slopes associated with the line of a ridge. But further down slope where good quality raw material was located there is little evidence of any activity. The quartzites of the northern slope and the northern summits only showed a background noise.

Elsewhere this pattern of prime locations being ignored is repeated. At the de Hoop farm where the diabase outcrops near the level of the valley floor (Fig. 1.3, i), extensive field walking has not revealed anything other than the background noise of ESA exploitation; yet this, like the quartzite plateau is a prime raw material location for a rock type noted to have been targeted for manufacture.

Is it possible that the archaeology has been missed, buried by sedimentation or obscured by undergrowth? In these two cases this is unlikely. Both areas have been extensively walked, using different teams of people. The quartzite plateau is open country and grass cover low in the winter when it was walked. Although the de Hoop diabase exposures may have been covered by modern colluvial and alluvial process the area was walked with this in mind and especial attention was paid to drainage ditches, foundation diggings *etc.* No evidence of the presence of intensive archaeology, other than the normal light background scatter was identified. However it was noted that large diabase exfoliation flakes were present here as well.

Other locations within a day's walking distance of the cave present a completely different picture. At the northern end of the main Swartkrans Valley where the colluvial fans are located (Fig. 1.3, f), considerable accumulations of archaeological material are present. One, immediately behind the old Swartkrans farmhouse, directly at the head of the valley, was field walked over two seasons, and a considerable quantity of material was recovered from the surface of the fans. The data for this is presented in Table 1.3.

Given the low density of manufacture, discard or use, represented by surface finds elsewhere in the Makapansgat environs, this accumulation of artefacts stands out. Raw material characteristics can not be the only determining factor since the rock type utilised here is the same light coloured, fine grained, homogeneous quartzite as was encountered on the plateau adjacent to the red cliffs. The majority of the artefacts from the fans reflect the use of the scree as a ready made source of suitable blanks.

In addition to a concentration of artefacts, the nature of the assemblage merits attention. The background noise of hominid activity in the landscape is usually a light presence of flakes, cores and the occasional biface or cleaver. At the scree fans, bifaces and cleavers were present, debitage was present as were cores and Levallois pieces. In addition, there were scrapers and other flake tools and also a small number of biface rough-outs in the local light coloured quartzite. One of these had been broken during manufacture. Rough-outs and flake tools are very rare in the landscape. The only other location where they occur in any numbers is the Cave of Hearths. The assemblage from the fans suggests specific activities carried out at this location. So why here? One possible answer represents other affordances offered by this location. From only a short way up the hill slopes adjacent to the fans there is a commanding view of the whole northern end of the broad flat Swartkrans valley. It is possible, then, that this fact, combined with a ready made source of raw material and ease of access to the valley floor, decided this location as a focus for persistent activity by hominids. In addition, the journey back to the Cave of Hearths would be over gentle valley bottom terrain. Perhaps the considerable hike back to the cave from the plateau, especially if carrying food and/or blanks for knapping, represented a higher degree in energy investment. At the very least the above makes it clear that the presence of a good source of raw material was not the principle reason for choosing a spot as an activity location.

Artefact	Dark quartzite	Other quartzite	Non-quartzite	Totals
Bifaces	121 (44.49%)	58 (21.32%)	93 (34.19%)	272
Core element	583 (65.95%)	129 (14.59%)	172 (16.86%)	884
Totals	704 (60.90%)	129 (14.59%)	265 (19.46%)	1156

Table 1.4: Raw material usage for cores and bifaces in the ESA levels of the Cave of Hearths. Bifaces include cleavers, rough-outs, indeterminate large cutting tools.

One way of organising this information is to suggest that the landscape was apportioned up into parcels of differing perceived importance on the basis of a number of different but interrelated factors.

1.10 The relationship between the landscape and the Cave of Hearths

This idea of a landscape of localised significance is reinforced when the character of the stone tool assemblages recovered in Mason's excavations in the Cave of Hearths is compared to those from the colluvial fans. Both have bifaces and cleavers, rough-outs, debitage, scrapers and flake tools. But the character of the material usage is very different. In Table 1.4 the totals are presented for the different varieties of quartzite used in the knapping of cores and bifaces noted during the recent analysis of the cave assemblages.

From Table 1.4 it is evident that the use of dark quartzite predominates, marking a strong contrast with the use of the light quartzites at the head of the Swartkrans valley. This dark quartzite is the predominant facies in the immediate vicinity of the cave. It is present in the outcrop on the south side of the valley to the west of the cave entrance (Fig. 1.3, d); it is exposed through lateral erosion at the mouth of the valley as bedrock and as scree covered slopes (Fig. 1.3, e); it is also found in exposures along the margins of the stream, and it represents the main constituent of the Mwaridzi's bedload drawn from the extensive scree accumulations at the red cliffs. The presence of light coloured quartzite and other hues does not imply transport of raw materials from other locations since there are facies variations in terms of colour (and homogeneity and crystal coarseness) within all the outcrops of quartzite at Makapansgat. An outcrop of predominantly dark coloured quartzite will contain a small proportion of lighter coloured material, and *vice versa.* The data conforms to the working hypothesis of a local usage of suitable raw material at or close to the source of use/activity.

The next question is a simple 'so what?'. Should we be surprised at an extensive usage of localised resources, and the answer of course is 'no'. It makes sense for hominids to use what is available to them and involves the least amount of effort, particularly when lugging big chunks of rock up and down hills, which is what resource transport amounts to. But Table 1.4 does highlight two particular points of interest. Where exactly in the vicinity does the dark quartzite come from? With so much excellent quality raw material available on the doorstep, why bother to transport the other, non-quartzite raw materials at all?

1.11 Quartzite and the transport web

In studies of faunal assemblages other agencies of bone accumulation, apart from hominids, have to be disentangled. These include carnivores, scavengers and other animals. Mercifully animals are not known to transport knapped lithics into caves. Moreover, the suite of rock types that occur in the Cave of Hearths are not a natural component of limestone, apart from chert (however the exposures are thin bands and not geologically massive, *i.e.,* they are not sufficiently homogeneous and are lacking in structure sufficient to make them suitable for knapping). Consequently all stone tools have been introduced into the cave by hominids, either made within the confines of the cave, or brought in as finished or near finished items.

It is clear from the nature of the artefact content that material was brought into the cave in both states. Anvils, spheroids and hammer stones were transported in to be used as they were and not modified by flaking in any other way (as far as received wisdom suggests). Manuports and blocks of raw material were brought in presumably to be flaked. The presence of cores, hammer stones and large quantities of small flake debitage (< 50mm) indicate the working of raw material within the confines of the cave. In addition, however, large flakes were also being introduced into the cave having first been made outside. This was a point originally noted by Mason (1988) who demonstrated that in terms of the arithmetic median the bifaces were larger than any of the cores in the ESA assemblages. This is supported by the current analysis. From a total of 230 bifaces, cleavers and rough outs for which length could be measured (mean

	Flake size	Dark quartzite	Other quartzite	Other raw materials
Beds 1 and 2				
total flakes = 283				
	⩾ 100mm	13	5	2
	⩾ 150mm	5	0	0
Mean flake size for all flakes ⩾ 100mm = 127.28mm, sd 20.33				
Max flake size 175mm				
(from a total of 25, 16 are retouched)				
Bed 3				
total flakes = 1772				
	⩾ 100mm	48	5	4
	⩾ 150mm	3	0	0
Mean flake size for all flakes ⩾ 100mm = 117.88mm, sd 18.61				
Max flake size 193mm				
(from a total of 60, 36 are retouched)				

Table 1.5: Frequency and size ranges for flakes greater than 10cm in the Cove of Hearths ESA assemblages.

139.76mm, sd 29.59), and a total of 511 cores (mean 82.14mm, sd 21.90), a random sample of 100 artefacts was generated (SPSS 8.0; 70 cores, 30 bifaces). A Students t test for independent samples found a statistically significant difference between the two data sets ($t = 11.509, \text{df} = 60.659, p = 0.000$, therefore the null hypothesis could be rejected for a two tailed test at 5%).

In the case of both beds the numbers of these large pieces are few (Table 1.5), both represent less than 10% of their respective assemblages, assuming that the excavated sample is representative of the original proportions of occurrence. Nevertheless, they clearly imply transport into the cave. And, as with the biface frequency noted in Table 1.4, the dark quartzite clearly dominates the sample.

In fact the importance of the manufacture and subsequent transport of large quartzite flakes may have had significance for classes of artefact other than bifaces and cleavers. One of us (JMcN) believes that these flakes formed the principal blank for the production of discoids. This category of core represents 63.91% of the total number of cores studied from beds 1–3 of the Cave of Hearths. Given that these tool forms are intensely reduced by their very nature, this hypothesis is difficult to prove. Details of blank morphology for the 565 discoids identified in this analysis are presented in Table 1.6.

The flake and split piece category of blank type forms nearly a quarter of the total number of discoids, and the clast category less than 5%. One other piece of circumstantial data may be introduced in support of this contention. A test was conducted to determine whether there was a difference in the overall thickness of discoids and those bifaces and cleavers clearly made on flakes. A Student's t-test for independent samples demonstrates no statistically significant difference between the thickness of bifaces and cleavers made on flakes, and the thickness of discoids. A random sample of 100 pieces were selected by SPSS 8.0 (53 bifaces and 47 discoids) from a total of 103 bifaces on flakes from beds 1–3, and a total of 79 discoids from beds 1–3 which could be confidently identified as having been made on flakes. The results were as follows: $t = 1.423, \text{df} = 96.919, p = 0.158$, therefore the null hypothesis could not be rejected for a two tailed test at 5%. The test does not contradict the hypothesis that the large flake blanks were used for making large cutting tools as well as discoidally worked cores.

1.12 Factory localities in the Makapansgat environs?

In advance of the results of the dating programme initiated within the Cave of Hearths by the Project, the ESA beds are loosely interpreted as belonging to the latter end of the Acheulean time range; this is on the basis of Levallois, and the presence of a small but undoubted sample of Levallois blades and points associated with the bifaces and the cleavers. Similar biface assemblages from East and South Africa from this time range (*i.e.,* Middle Pleistocene) are often associated with large accumulations of lithic material, often interpreted as factory sites. At these locations either outcrops or boulder fields are exploited for the production of large flakes and/or blanks. South African examples are present in the Lower Vaal river basin (Helgren 1978), and the frequency of these kinds of site in the landscape and their richness is indicated by Sampson's work in the Seacow valley (Sampson 1985). Yet at Makapansgat, despite intensive field walking such sites have never been identified. Other researchers intimate with the valley over a period of many years (Maguire 1998; Mason 1988) have likewise never reported such sites in the immediate vicinity. Yet the archaeological record clearly indicates that the manufacture and transport of large flakes across the raw material spectrum was an integral component of the tool making and using continuum.

Core type	Flakes and split pieces	Cobbles and pebbles	Other blanks and indeterminate
Discoid (24.60%)	139 (4.78%)	27 (70.62%)	399

Table 1.6: Frequencies of different blank types for discoid cores from the ESA levels of the Cave of Hearths.

Our preliminary conclusions, based on the absence of such sites, is that they were never there. At Makapansgat, the focusing of large scale industrial activity at particular locations, factory sites, was not a part of the resource procurement strategy. Activity foci appear to be located at particular points in the landscape where a number of primary requirements would be met, not just raw material proximity.

1.13 The non-quartzite raw materials

The second question posed earlier was why, with so much excellent quality raw material so close to hand, did the Middle Pleistocene hominids of Makapansgat bother to transport large flakes of other raw material types from more distant locations into the cave? As with the quartzites no convincing evidence has been identified by the current field project, or reported by other investigators, for the presence of large working sites in other raw materials. The only location which remotely resembles a factory is the diabase exposure (Fig. 1.3, c). However, since diabase weathers so deeply it was difficult to be certain whether some of the large flakes found here were humanly struck or were exfoliation pieces, and the number of unequivocally struck flakes was very small. Mason's data presented in Table 1.1, and that of the current analysis (Table 1.4) make it clear that it would only be in diabase that we would expect there to be more than just casual or opportunistic collecting of raw materials / exfoliation flakes. Combining the data from Table 1.4, the percentages in Table 1.7 highlight two points: firstly, the use of non-quartzite raw materials is more pronounced in the biface / cleaver element of the assemblage, and secondly there is a concomitant increase in the frequency of cores on quartzite.

We might suggest that this latter feature is a result of the use of the very local quartzite to make cores, in particular the use of flakes as blanks for discoids may influence this pattern. Perhaps the majority of large flakes entering the cave were destined for the manufacture of discoids anyway. But that still does not explain why the large cutting tool element should show a greater frequency of the non-quartzites, which must represent a greater investment in expended effort than the quartzite; especially since there is nothing about the diabase that is inherently superior to quartzite. At the moment it is not possible to do anything more than speculate on the answer. Mason's data in table 1 shows that spheroids in diabase are the most common form of this artefact category encountered in the ESA levels. If hominids were targeting the diabases here (Fig. 1.3, c) where spheroids have been noted to occur naturally, then it may have been simple opportunism to bring home a few large exfoliation flakes at the same time. Serendipity may explain the low frequencies of other raw materials brought into the cave as well.

1.14 Conclusions

Field walking has revealed a patterning of occupation and use of the Makapansgat landscape, and a relationship between the outer world and the inhabitants of the Cave of Hearths, that was unexpected when the programme was initiated. In simple terms, there is less archaeological material than the presence of a famous site, such as the Cave of Hearths, might lead one to think. Moreover, most of the hominid activity is clearly outside of the Cave and, in fact, located in the nearby Swartkrans Valley. We have encountered a full range of localities across different lithologies and in different topographic locations. Whilst it is clear that some material is being brought into the Cave of Hearths, much of the artefactual record seemingly suggests local activity and a very localised use of resources.

Currently this pattern is being interpreted as a reflection of behaviour that perceived the landscape in terms of a series of situationally important behavioural foci. These locations did not require the landscape to be provisioned through caching, provisioning from factory sites, or through the use of mobile tool kits. If correct the pattern of resource exploitation may reflect a different strategy to that seen in other 'late Acheulean' localities. We postulate that the ease of access to a plentiful high quality resource strongly influenced the responses of the Makapan hominids to subsistence problems. Of course this model remains, for the moment, provisional.

Perhaps most intriguingly, the density of findspots, and localities leads us to think that the Makapansgat area is not the major focus of ESA hominid activity in the area. Research into other neighbouring areas and the possible density and frequency of occupation in the Cave of Hearths through into the MSA levels may have a considerable affect upon the hypotheses put forward here.

Artefact	Quartzite	Non-quartzite	Totals
Bifacial tools	65.81	34.19	272
Cores	80.54	19.46	884

Table 1.7: Percentage totals for raw material usage in bifaces and cores from Beds 1–3 at the Cave of Hearths, Makapansgat.

Acknowledgements

The Project would like to thank the National Monuments Council of South Africa, and in particular Dr Jeannette Deacon. We would also like to thank the British Academy and the Arts and Humanities Research Board for financial assistance in 1997, 1998 and 1999, and the School of Archaeology, Classics and Oriental Studies, University of Liverpool for financial assistance from the Vacation Field Studies Fund. PSQ gratefully acknowledges the support of the Natural Environment Research Council (NERC, UK), the Boise Fund, Oxford, and the Palaeoanthropological Heritage Fund of the Department of Arts, Culture, Science and Technology (DACST), South Africa. We would also like to thank Lyn Wadley, Tom Huffman, Kathy Kuman, Simon Hall, John and Molly Calabrese in the Department of Archaeology; Philip Tobias, Ron Clarke and Kevin Kuykendall at the Department of Anatomy; and Bruce Rubidge and Judy Maguire at the Bernard Price Institute of Palaeontology at The University of the Witwatersrand. We would also like to thank the field supervisors, Susan Andressen, Phil Rye, Jim Williams, Emma Nelson, Chris Duke and Jason Hall for assistance in the field; and last but not least all the students who have trekked their way over the uplands and valleys of the Makapansgat landscape during the last field seasons. The Project would especially like to thank the Murzel family from the Swartkrans Farm, the Knight family from Buffalshoek Farm and the University of Witwatersrand for permission to work on their land in the Makapansgat and Swartkrans Valleys.

References

AIELLO, L. & R. DUNBAR 1993. 'Neocortex size, group size and the evolution of language.' *Current Anthropology* 34: 184–93.

BUTZER, K. W. 1984. 'Archaeology and Quaternary environment in the interior of southern Africa.' *In* R. G. Klein (ed.), *Southern African Prehistory and Palaeoenvironments*, pp. 1–64. Balkema, Rotterdam.

COOKE, H. S. B. 1988. 'The larger mammals from the Cave of Hearths.' *In* Mason (1988), pp. 507–34.

DART, R. A. 1925. 'A note on Makapansgat: a site of early human occupation.' *South African Journal of Science* 22: 454.

DART, R. A. 1948a. 'An adolescent promethean australopithecine mandible from Makapansgat.' *South African Journal of Science* 45: 73–5.

DART, R. A. 1948b. 'The Makapansgat proto-human Australopithecus prometheus.' *American Journal of Physical Anthropology* 6: 391–412.

DEACON, H. J. & J. DEACON 1999. *Human Beginnings in South Africa: Uncovering the Secrets of the Stone Age.* David Philip, Cape Town.

DUNBAR, R. 1993. 'Coevolution of neocortical size, group size and language in humans.' *Behavioral and Brain Sciences* 16: 681–735.

EITZMAN, W. I. 1958. 'Reminiscences of Makapansgat Limeworks and its bone-breccia layers.' *South African Journal of Science* 54: 177–82.

GAMBLE, C. 1996. 'Making tracks: hominid networks and the evolution of the social landscape.' *In* Steele & Shennan (1996), pp. 253–77.

GAMBLE, C. 1998. 'Palaeolithic society and the release from proximity: a network approach to social relations.' *World Archaeology* 29(3): 426–49.

GOWLETT, J. A. J. 1997. 'Why the muddle in the middle matters: the language of comparative and direct in homanid evolution.' *In* C. M. Barton & G. A. Clark (eds.), *Rediscovering Darwin: Evolutionary Theory and Archaeological Explanation*, Archaeological Papers of the American Anthropological Association No. 7, pp. 49–65. American Anthropological Association, Arlington.

HELGREN, D. M. 1978. 'Acheulean settlement along the lower Vaal River, South Africa.' *Journal of Archaeological Science* 5: 39–60.

HOFFMEYR, I. 1988. 'Oral and written versions of the Makapansgat Siege.' *In* Mason (1988), pp. 417–26.

LIEBERMAN, D. E. & J. J. SHEA 1994. 'Behavioral differences between Archaic and Modern Humans in the Levantine Mousterian.' *American Anthropologist* 96(2): 300–32.

MAGUIRE, J. 1998. *Makapansgat: a guide to the Palaeontological and archaeological sites of the Makapansgat Valley.* Desktop Creations, Pretoria. Dual Congress of the International Association for the Study of Human Palaeontology and the International Association of Human Biologists.

MASON, R. J. 1962. *The Prehistory of the Transvaal.* Witwatersrand University Press, Johannesburg.

Mason, R. J. (ed.) 1988. *Cave of Hearths, Makapansgat, Transvaal.* Archaeology Research Unit Occasional Paper No. 21. University of the Witwatersrand Press, Johannesburg.

McNabb, J., A. Sinclair, A. Andresen & P. Quinney 1998. *Artefactual raw material economy in the Cave of Hearths and the Makapansgat Valley.* Paper presented at the Dual Congress of the International Association for the study of Human Palaeontology and the International Association of Human Biologists, Sun City, South Africa, July 1998.

Mellars, P. (ed.) 1990. *The Emergence of Modern Humans: an Archaeological Pespective.* Cornell University Press, Ithaca.

Mellars, P. A. & C. Stringer (eds.) 1989. *The Human Revolution: Behavioral and Biological Perspectives in the Origins of Modern Humans.* Edinburgh University Press, Edinburgh.

Mithen, S. J. 1996. *The Prehistory of the Mind.* Thames and Hudson, London.

Moore, M. P. J. 1981. *The Iron Age of the Makapan Valley.* Master's thesis, Department of Archaeology, University of the Witwatersrand, South Africa.

Quinney, P., A. Sinclair, A. Andresen, A. Latham, J. McNabb & K. Kuykendall 1998. *Cultural hominids in the landscape: archaeological fieldwork at Makapansgat, South Africa.* Paper presented at the Dual Congress of the International Association for the study of Human Palaeontology and the International Association of Human Biologists, Sun City, South Africa, July 1998.

Rayner, R. J., B. P. Moon & J. C. Masters 1993. 'The Makapansgat australopithecine environment.' *Journal of Human Evolution* 24: 219–31.

Sampson, C. G. 1985. *Atlas of Stone Age settlement in the Central and Upper Seacow Valley.* Memoirs of the National Museum, No 20, Bloemfontein.

Sinclair, A. 1998. 'Footnotes to Plato.' *Antiquity* 72: 686–91.

Steele, J. 1996. 'On predicting hominid group sizes.' *In* Steele & Shennan (1996), pp. 231–52.

Steele, J. & S. J. Shennan (eds.) 1996. *The archaeology of human ancestry: power, sex and tradition.* Routledge, London.

Tobias, P. V. 1949. 'The excavation of Mwulu's Cave, Potgietersrus District.' *South African Archaeological Bulletin* 4: 2–13.

Tobias, P. V. 1971. 'Human skeletal remains from the Cave of Hearths, Makapansgat.' *American Journal of Physical Anthropology* 34: 335–67.

Trinkaus, E. (ed.) 1989. *The Emergence of Modern Humans: Biocultural Adaptations in the Later Pleistocene.* Cambridge University Press, Cambridge.

van Peer, P. 1992. *The Levallois reduction strategy.* Monographs in World Archaeology 13. Prehistory Press, Madison, Wisconsin.

van Riet Lowe, C. 1938. 'The Makapan caves — an archaeological note.' *South African Journal of Science* 35: 371–81.

van Riet Lowe, C. 1943. 'Further notes on the Makapan caves.' *South African Journal of Science* 40: 289–95.

van Riet Lowe, C. 1948. 'Cave breccias in the Makapan Valley.' *In* A. L. Du Toit (ed.), *Special Publication of the Royal Society of South Africa: Robert Broom Commemorative Volume*, pp. 127–31. Royal Society of South Africa, Cape Town.

John McNabb
Department of Archaeology
University of Southampton
Southampton SO17 1BJ
England

Anthony Sinclair
Department of Archaeology
School of Archaeology, Classics, and Oriental Studies
P.O. Box 147
University of Liverpool
Liverpool, L69 3BX
England
`A.G.M.Sinclair@liv.ac.uk`

Patrick Quinney
Department of Archaeology
University of the Witwatersrand
C. van Riet Lowe Building, Private Bag 3
Wits 2050
Johannesburg, South Africa

2
The Mousterian Lithic Assemblages from Ras El-Kelb Cave, Lebanon

Lorraine Copeland[1] and Norah Moloney[2]
[1]*Chateau de Marouette, France;* [2]*Institute of Archaeology, University College London*

Abstract

A rescue excavation at the Middle Palaeolithic cave site of Ras el-Kelb, Lebanon revealed a long sequence of occupation levels with a rich, stable Levallois lithic tradition throughout. Unfortunate circumstances prevented full publication of the excavation for decades. However, this period saw the emergence of important approaches to stone tool studies which have dominated the second half of the 20th century and which were incorporated into the lithic analysis of the Ras el-Kelb industry. The results of the Ras el-Kelb investigations contribute towards the Near Eastern Middle Palaeolithic in general and the Tabun C facies, which has up until now received relatively little attention, in particular.

2.1 Setting and excavation

Ras el-Kelb cave is located about 15km north of Beirut at the foot of a limestone promontory that juts out into the Mediterranean Sea at the mouth of the Nahr el-Kelb (Dog River) (Fig. 2.1). Behind the cave are the forested ravines and foothills that rise to the summits of Mount Lebanon. The cave became available for occupation during the Late Pleistocene, when the fluctuating Quaternary sea-levels had regressed to below the level of its floor; this lay at an altitude above present sea-level of 5.8m. The original cave mouth was destroyed by road and rail building around the headland, and the middle parts of the deposits can now only be seen in the side of the cliff facing the sea. The cave was known to contain a rich Mousterian industry having typological links with the industry of Level C at Tabun Cave in Mount Carmel. It was excavated in 1959 as a rescue operation by Professor Dorothy Garrod and G. Henri-Martin, since it was scheduled to be further destroyed by the blasting of a road tunnel through the promontory. Although two preliminary accounts were published shortly after the excavation (Garrod & Henri Martin 1959, 1961), for various reasons a full report was not published until 40 years later (Copeland & Moloney 1998).

When excavations began in 1959 it was possible to access the prehistoric deposits from two directions: from the rear of the cave outwards, where the blasting had breached the back wall (this became known as the Tunnel Trench) and from the remains of the mouth area beside the railway line (the Rail Trench — Fig. 2.2). After many difficulties, not the least of which was the extreme hardness of the brecciated deposits, the excavators succeeded in coming down onto what seemed to be the original cave floor in the Rail Trench exposure (Fig. 2.3). This was seen to be covered by beach pebbles and shells to a depth of *c.* 50cm, over which began the archaeological material, consisting of flints, animal bones, hearth deposits *etc.* This result was exactly what Garrod had hoped for; it allowed the marine chronology of raised beaches along the Levant coast to be tied in with the prehistory of the region (Garrod 1962). In the Tunnel Trench, bedrock was encountered at *c.* 7m above sea level. A lens of marine material, regarded as a storm beach, was thought to link the two exposures through the intervening unexcavated deposit (Fig. 2.3). The levels continued upward from this floor for *c.* 4m, and were overlain by an alluvial soil eroding from chinks in the roof and without artefacts in place. The remains of the twice-truncated cave are no longer accessible behind the tunnel wall inside, and below tarmac outside.

2.2 The industry

The Mousterian lithics recovered from the 19 distinguishable levels in the two small trenches at Ras el-Kelb numbered more than 30,000 artefacts. An initial inventory of the artefacts was undertaken during the original rescue excavation. Later a sample of 300 pieces was sent to Europe to be analysed and drawn by Copeland (1978) who also began a detailed attribute recording of the material stored in Beirut in the 1970s (Copeland 1998). She had analysed the industry from the lowest 11 of the 19 levels when she was forced to interrupt work by the outbreak of Civil War. Conse-

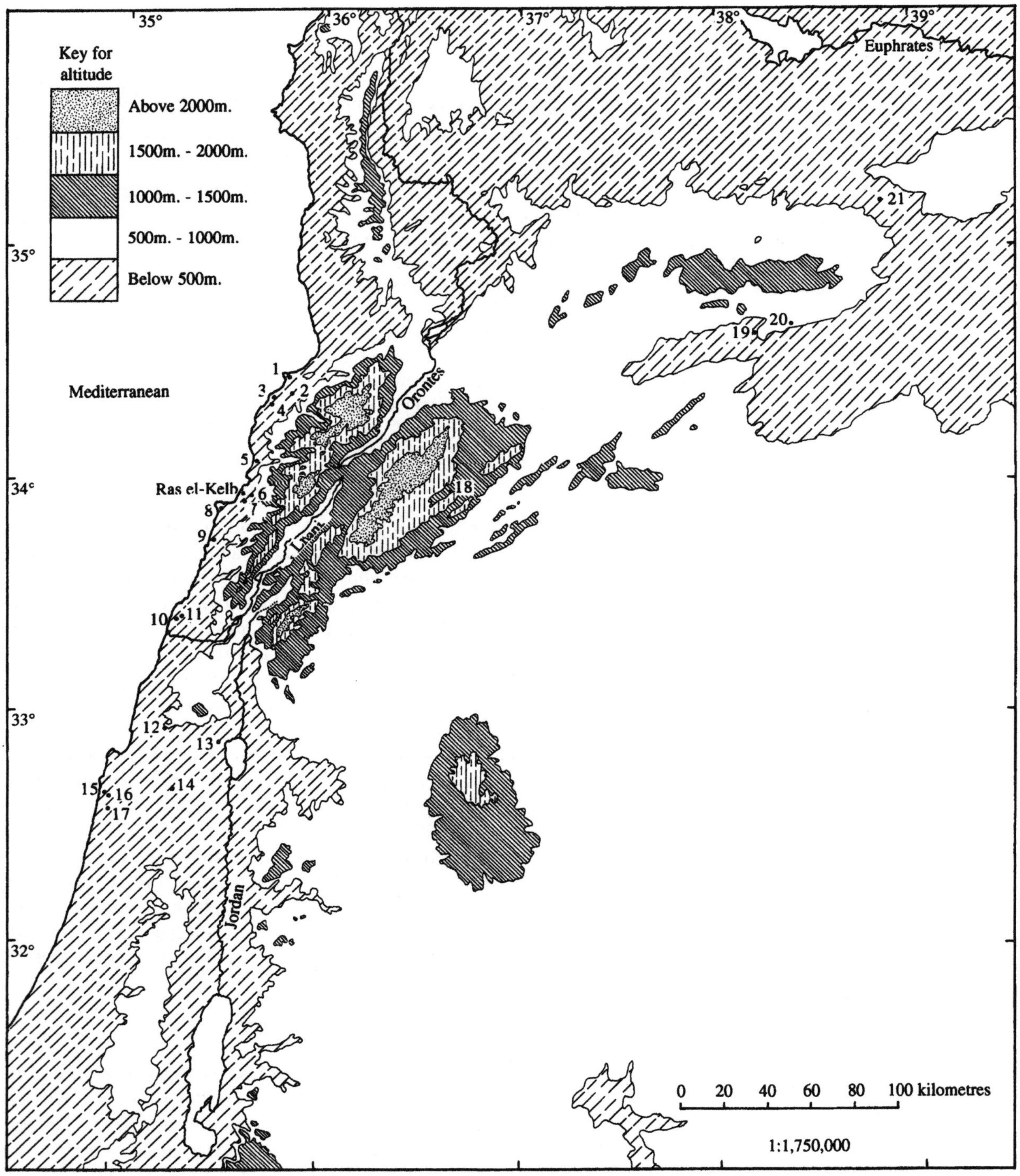

Figure 2.1: The location of Ras-el-Kelb, Lebanon.

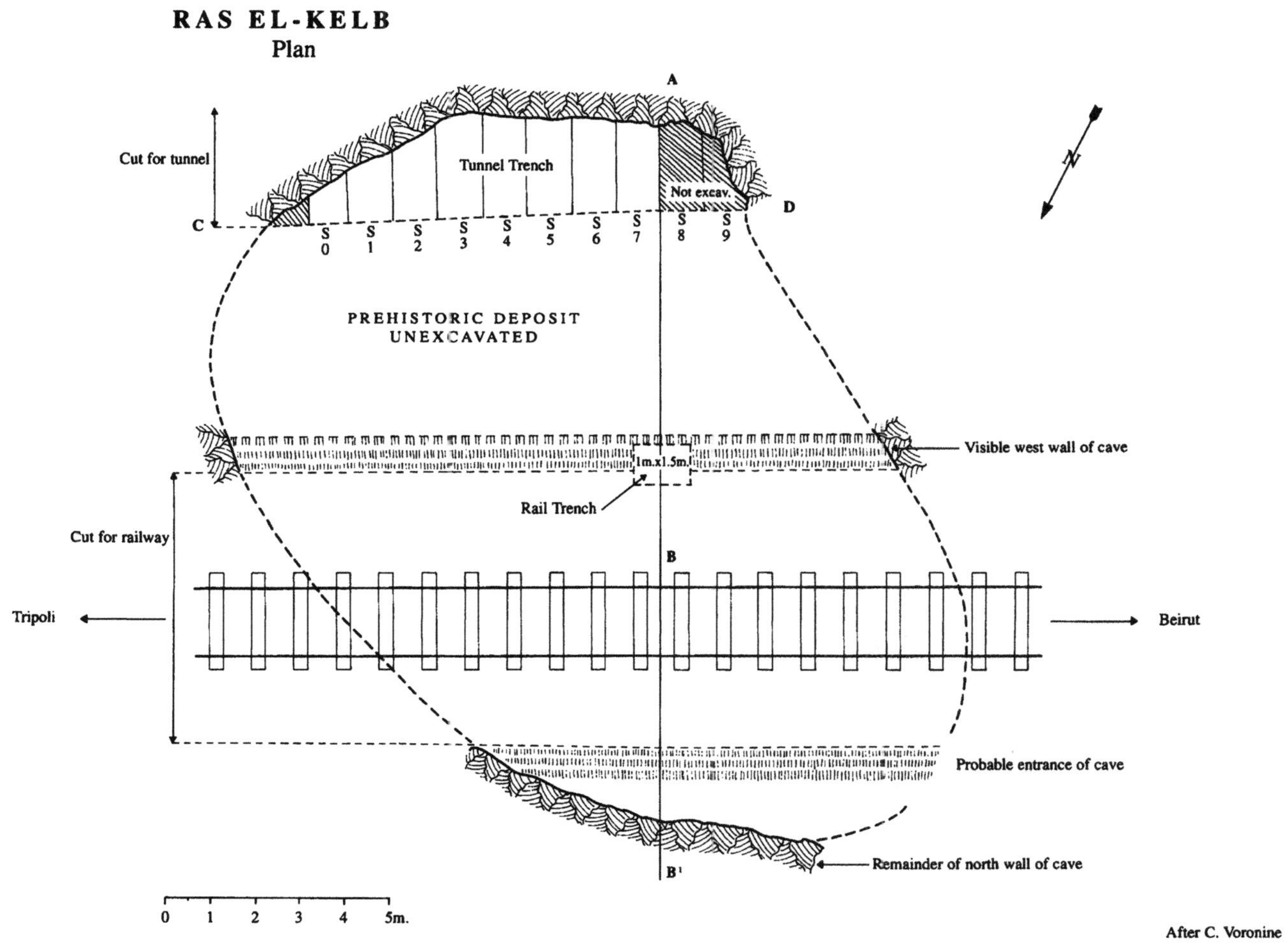

Figure 2.2: Plan of Ras-el-Kelb.

quently, the material from the remaining 8 levels has yet to be studied.

The lengthy period between excavation and publication enabled Copeland to study the lithic industry from Ras el-Kelb in two phases which appropriately reflect the broadening spectrum of approaches to lithic analysis in the last decades of the 20th century: the first phase followed the Bordean techno-typological system while the second (and later study) was based on reduction sequences and the concept of the *chaîne opératoire*. Although the Bordean system has been criticised more recently for its emphasis on 'tools', it was normal practise in the 1970s and still plays an important role in Middle Palaeolithic Levantine studies (Bar-Yosef & Meignen 1992). The later analysis was undertaken in the light of research by Tixier *et al.* (1980), Geneste (1985) and Boëda (1988, 1995), whose work has enabled analysts to recognise the various stages of the reduction sequence within the global aspect of the *chaîne opératoire*. Understanding the stages in, and approaches to Levallois production has been particularly helpful in the context of the Levantine Middle Palaeolithic (Bar-Yosef & Meignen *e.g.*, at Kebara 1992; Nishiaki & Copeland and at Keoue 1992). Additionally, in recent years a sample of the Ras el-Kelb industry was examined microscopically for use-wear traces, an area of lithic studies which was not available in the 1950s.

2.3 Raw material

The industry is exclusively in flint, of which there is no shortage in Lebanon; it outcrops in Jurassic, Cretaceous and Tertiary seams in the sedimentary limestone rocks of the country. It is present in valley terraces as flat pebbles, smoothed by fluviatile transportation downstream from outcrop sources, and finally as rounded beach pebbles along the coast (Emery & George 1963). The flint is varied in quality and colour but is mainly medium to fine-grained and in beige to dark brown shades. Some flint has mottled, speckled and banded areas indicating internal impurities which led Yamada, in his use-wear study of a sample collection to consider the material as relatively coarse-grained (Yamada 1998). However, very coarse chert and siliceous limestone (which occurs nearby) are virtually absent.

The amount and nature of the cortex remaining on artefacts can indicate the initial blank type: thick chalky cortex is indicative of fresh, bedrock nodules;

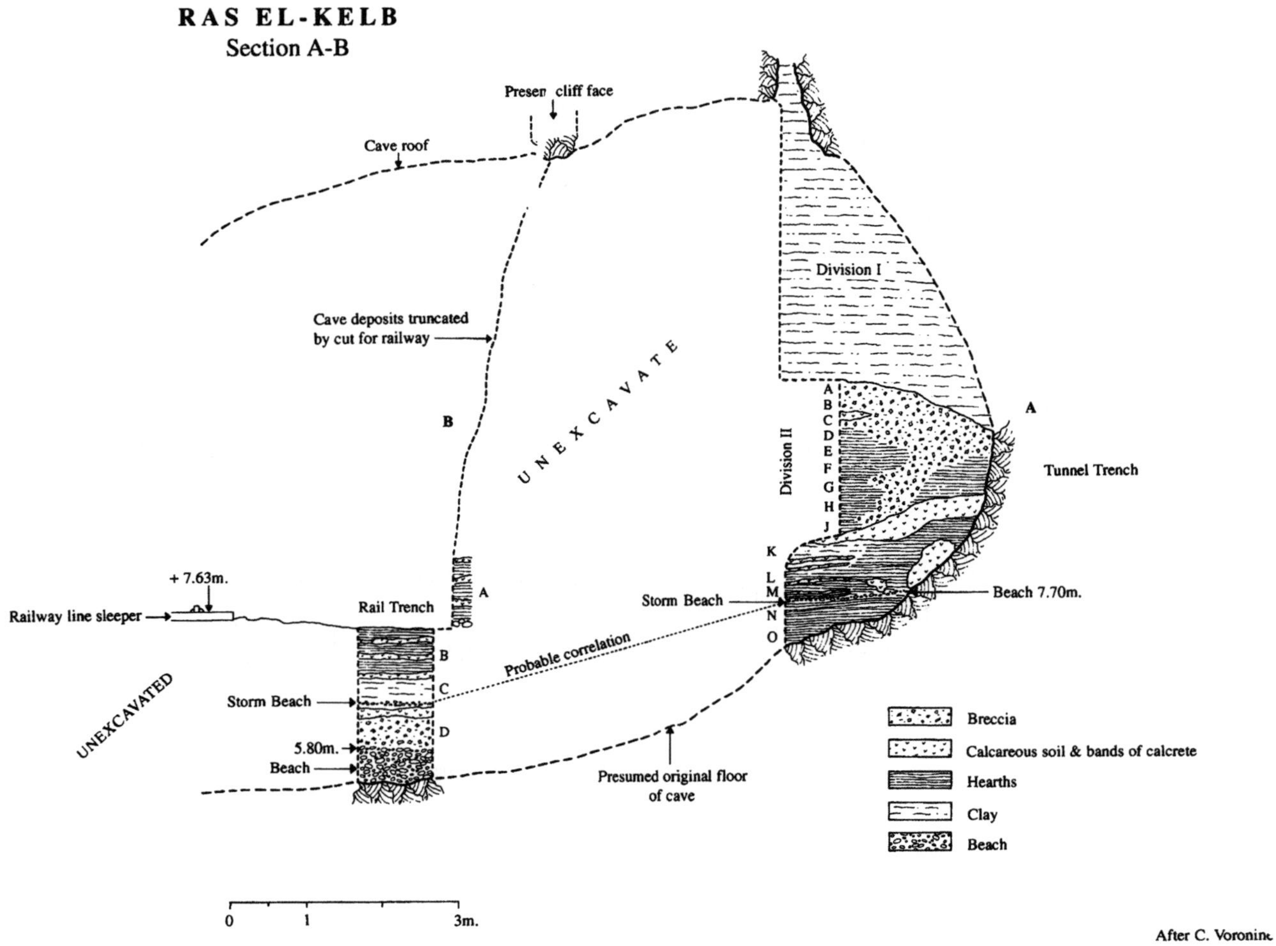

Figure 2.3: Section A–B of Ras-el-Kelb (see Fig. 2.2).

thin flat brown cortex suggests tabular-like material deriving from bedrock seams, while the original cortex on rounded, smooth riverine and beach pebbles tends to be replaced by a battered-looking outer surface. Most pieces were generally non-cortical; however, in those cases where the type and location of cortex had been recorded, (and, admittedly, the type of cortex was recorded on only a small number of pieces) it would seem that nodules were preferred over tabular-like flint and that pebbles were rarely chosen for reduction. In other words, the material seems to have been preferentially retrieved from flint seams in the limestone bedrock of the neighbouring hills, instead of, as would have been expected, from the river terraces and seashore beaches 'on the doorstep'.

2.4 Technological and typological considerations

The reduction sequences, as seen from the numerous cores, preparation flakes and other flake forms, were strongly Levallois; especially prevalent were traces of the 'Classic' bun-shaped or tortoise cores from which the typical oval flake had been removed and the core then discarded or reshaped in order to produce further, often narrower flakes and the occasional blade. This is the process named 'lineal' and 'recurrent' by Boëda (1995) but termed 'radial' by Copeland (1998). The method produces oval flakes with multi-convergent dorsal facet margins. Among other results of the core reshaping are core edge flakes, *enlèvements* II *etc.*, all of which are regarded by Boëda as part of the Levallois system of reduction. These types of flakes are prevalent at Mount Carmel in Tabun Level C, the 'Middle' Mousterian level there (Garrod & Bate 1937). The flakes, besides having 60–70% radial facet margin patterns, depending on level, as well as broad oval outlines, had finely faceted butts (IF between 40 and 67% of the butts), often of *chapeau de gendarme* type. The most numerous flake form is the small preparation flake only *c.* 2–3cm long or wide. Blades are scarce and seem to be slightly more numerous in the lower Rail Trench levels where they can reach 20–18%, while the ILam of the Tunnel Trench units range between 4 and 12%.

Large numbers of the flakes (and some of the cores) are retouched into tools. These amounted to *c.* 15–33% of the flakes (compare with Kebara Cave, with only 2–3%). Levallois blanks run to about 37–67% de-

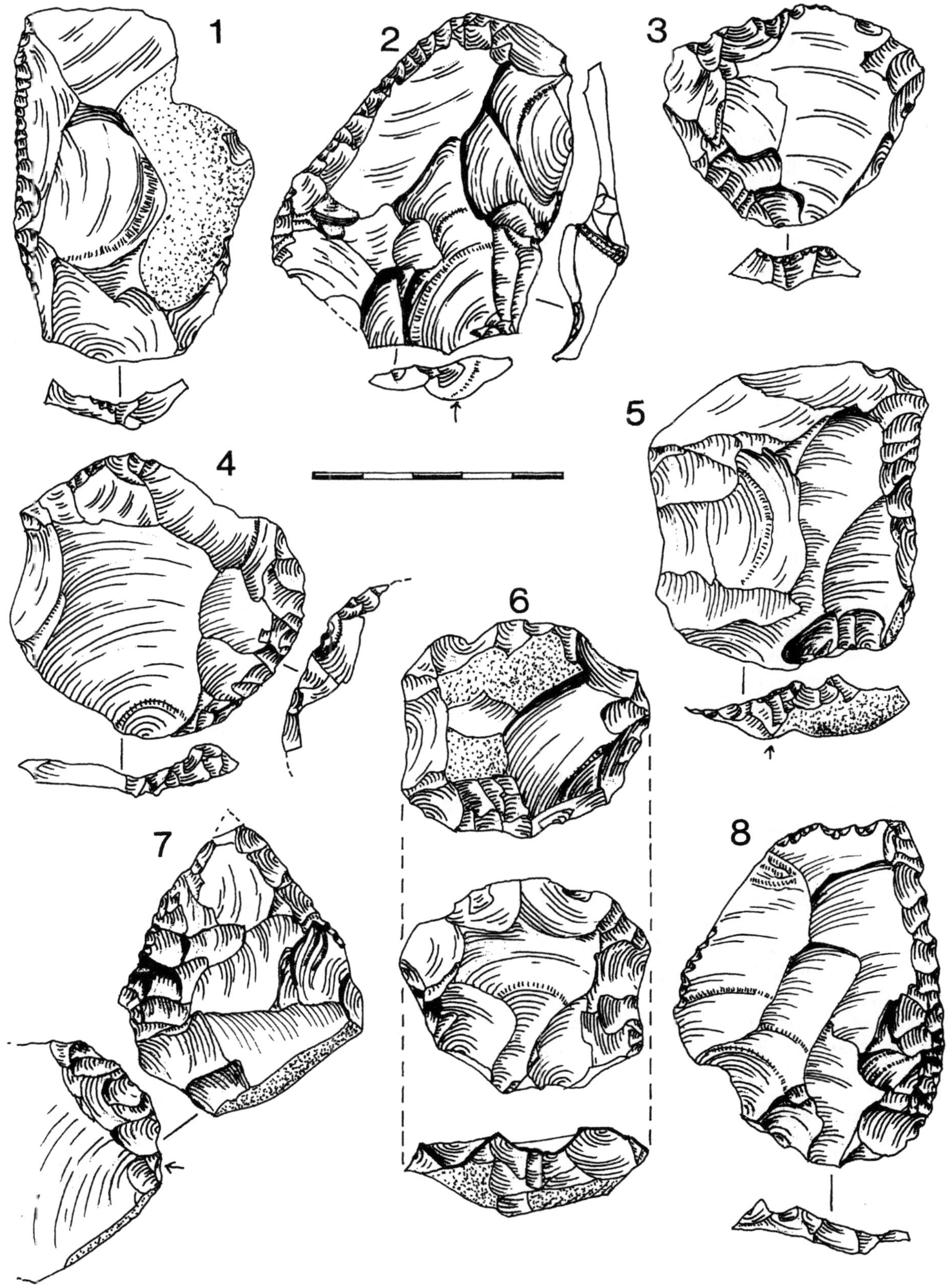

Figure 2.4: Tunnel Units F (nos 1–4), Unit B (nos. 5 and 6) and Unit A (nos. 7 and 8). **1:** Single straight *racloir* of brown, translucent flint. **2:** F4 nn Offset *racloir* of banded brown flint. **3:** F5 nn Denticulate on a broad Levallois flake of olive-grey flint. **4:** F6 nn Double *racloir* biconvex, of olive brown banded flint. **5:** B7 nn Denticulate or single straight *racloir* on a Levallois 'first flake' of dark brown flint. **6:** B7 nn Discoid core, very worked down and not rejuvenated. **7:** A4 nn Double convex straight *racloir* of brown flint. **8:** A2 nn Single straight *racloir*, slightly denticulated, of banded, olive-grey flint on a radial Levallois flake.

pending on level, although in no particular chronostratigraphic pattern. The Index of Levallois Points fluctuates between 3% and zero, again depending on level and again in no particular pattern except that they are very slightly more frequent in the lower levels of the trenches. The main retouched tool type is the side-scraper (Fig. 2.4). These occur in all shapes and sizes, such as lateral, transverse, offset, single or double, bifacial, inverse etc. The IR runs from 40–60%. Mousterian Points are distinctive but rare, and there are a few end-scrapers, burins, notches and denticulates in each level as well as naturally-backed knives or knives on core-edge flakes (where the perpendicular core edge makes a useful finger rest). The few discs present probably represent re-used cores. As is now seen to be usual in Mousterian assemblages, the 'truncated faceted flake' is much in evidence. Its meaning has often been discussed in the literature; it is regarded by some as representing a thinning technique, by others as a core on a flake (Solecki & Solecki 1970). In Europe it has been called the Kostienki thinning technique (Geneste 1985). The retouch type is just as consistent as the other attributes (except possibly for the ILam): abrupt and Quina styles are rare or absent, and the prevailing retouch type is a delicate scalar or squamous flaking that skims the surface of the blank.

The Bordean indices (IL, ILty) show that the industry, which seems to represent the same facies from base to top, is 'of Levallois facies'. However, although the technical aspects are so similar, the percentages of the various tool types differ slightly from level to level; for example the IR in level B of the Rail Trench was 31.57% (in 133 tools essential), and the index of Levallois blanks was only 27% in Unit O of the Tunnel Trench (both values markedly lower than in the other levels). Nevertheless one could not argue for a change of facies. Only at the base of the Rail Trench was the more 'bladey' ambience there able to suggest something different from the rest of the material. The uppermost layers in the Tunnel Trench were not studied in as much Bordean detail as were the lower assemblages but a careful examination of samples does not suggest that there is any change here either.

2.5 The *chaîne opératoire* at Ras el-Kelb

The reconstruction of the *chaîne opératoire* has to be tentative given the nature of the excavation (Fig. 2.5). It begins with the collection of flint nodules from non-local, but perhaps not too distant, sources. Primary working may have taken place outside the cave or at the raw material source. At Ras el-Kelb itself, cores were shaped radially to produce the 'preferred' large, ovoid, Levallois flakes. The presence of core-edge flakes and other left- or right-sided flakes suggests that some of the tortoise cores were reshaped to produce a number of other Levallois products including some blade forms but virtually no points. Many of the products were retouched into recognisable forms with fine scalar flaking. Sometimes cores were used as tools while some Levallois flakes were used as cores and there are examples of other by-products and used tools being exploited as blanks for further artefacts. Pieces which were no longer useful were discarded on the cave floor, many ending up in hearth deposits.

2.6 Faunal and lithic evidence for subsistence

The Ras el-Kelb Mousterians hunted the moderate-sized mammals and other small game which lived on the coastal plain and among the foothills of the Lebanese mountains. Their staple diet seems to have been fallow deer, to judge by the high percentages of the remains of the latter in the deposits (Garrard 1998). Garrard believes that the larger mammal bones which are also present (rhino, cattle) may have been obtained by scavenging. Carnivores, such as bear or hyena, and non-mammalian species (tortoises) also contributed to the diet but evidence of exploitation of seashore resources is missing.

As mentioned above, there are very few points in the Ras el-Kelb industry, which distinguishes it from many other Near Eastern Middle Palaeolithic assemblages. However, the abundance of scrapers in the industry indicates their constant utility, most likely in food processing activities. If this is the case, then other materials, possibly wood, may have been used for hunting or scavenging activities. Use-wear studies have proved unsuccessful in determining specific activities; no use-wear polish was found on a sample of the industry studied by Yamada (1998). Yamada proposes four possible factors to account for the lack of polish:

1. the flint is too coarse-grained for polish to develop
2. the polish has been obliterated by post depositional surface modification
3. the tools were not used intensively
4. a combination of these factors.

2.7 Engravings on flint: a note of caution

Of some interest is the single, incised flake which was found during study of the lithic assemblage (Fig. 2.6). Viewed under the binocular microscope, the dorsal, cortical surface of this flake is heavily scored with clear lines which form no particular pattern. As a result of an earlier study of the markings on the flake, Moloney (1998) concluded that they had not been caused by

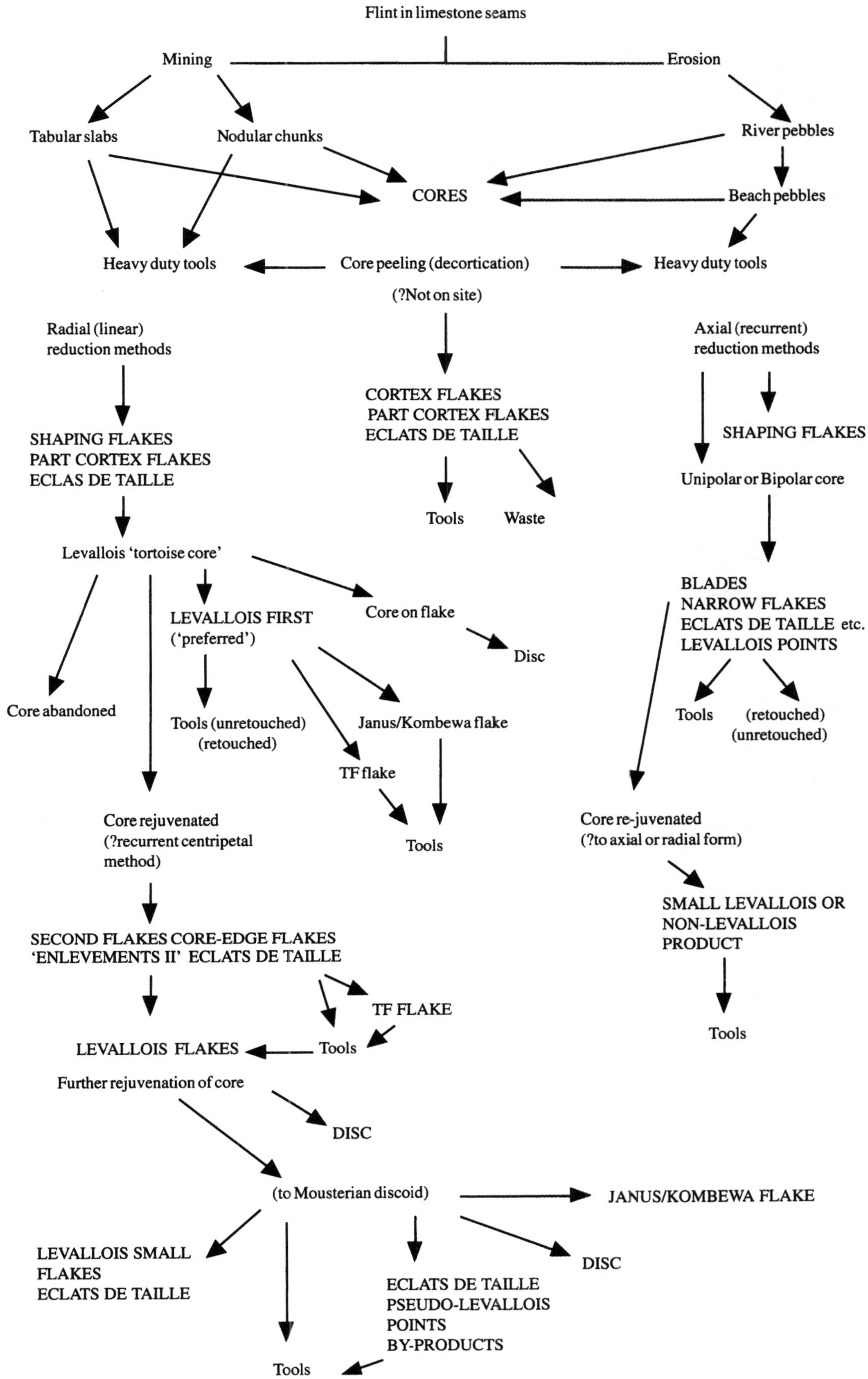

Figure 2.5: The *chaînes opératoires* determined for the Ras el-Kelb lithic industry.

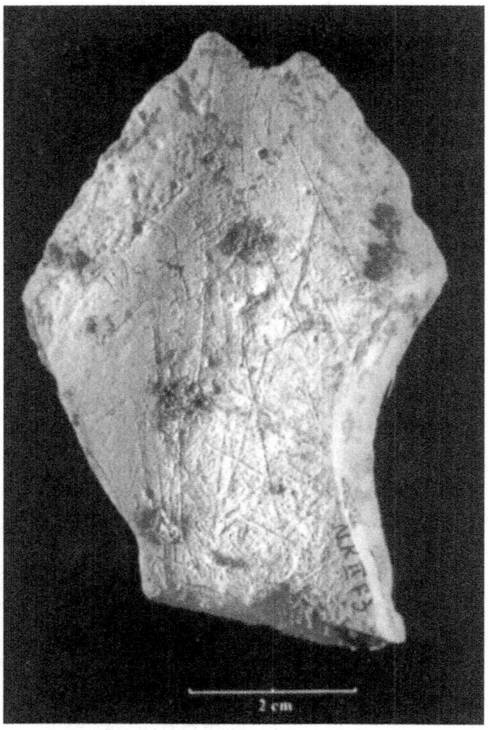

Figure 2.6: The original photograph of the Ras el-Kelb flint with incisions.

Figure 2.7: Experimental incising into hard cortex ×40.

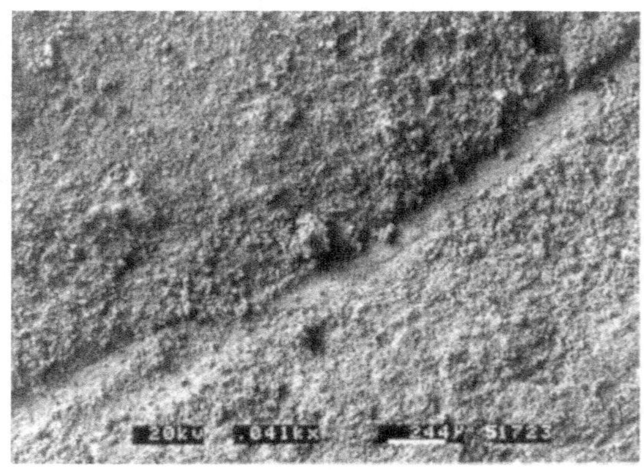

Figure 2.8: Experimental incising into soft cortex ×40.

Figure 2.9: One of the deeper incisions on the Ras el-Kelb flint ×40.

natural agents (*e.g.*, root marks, glacial action, trampling) and suggested that they were, in fact, the result of human action. Additionally, as there is no difference in weathering between the flake cortex and incised lines, and given that concretions cover part of the flake including the incisions, it was felt likely that the latter date to a period during the Mousterian occupation. As claims for engraving objects by Middle Palaeolithic humans can be contentious, it was recently decided to have the flint surface analysed under a Scanning Electron Microscope (SEM), a task undertaken in November 1999 by Dr Fernandez-Jalvo at the Natural History Museum, London. The results were compared with two flint flakes (one with a hard cortex and one with a soft cortex) which had been experimentally incised with a sharp-edged flint flake. The experimental incisions made into the hard cortex flake are shallow and narrow and some have clear V-shaped striations typical of those made by the edge of a flint tool (Fig. 2.7). Incisions into the soft cortex flake are much deeper and wider than in the hard cortex and are also clearly V-shaped (Fig. 2.8). When compared with the experimental pieces under the SEM at ×40, the incisions on the Ras el-Kelb artefact were seen to be wider than the experimental soft cortex piece but shallower than the hard cortex piece and with a fully rounded rather than V-shape profile (Fig. 2.9). Fernandez-Jalvo (*pers. comm.*) suggests that these marks are the result of friction against sediments. It is possible, too, that the flint nodule was abraded before it had been exploited by the occupants of Ras el-Kelb because of the homogenous surface weathering and because some of the incisions extend beyond the edge of the flake. Similar scratches have been found on the cortex of some of the refitted Boxgrove artefacts which have been attributed to natural agents (de Loecker *pers. comm.*).

The SEM results have been given in some detail here as an example of the relative ease with which we may confuse natural with artificial modification, particularly in reference to Middle Palaeolithic engravings and symbolism. While well-lit photographs and good drawings may emphasise engravings, perhaps we need to be cautious in our interpretation of those engravings which are less well-defined.

Although this particular Ras el-Kelb flake does not in itself add to the growing body of evidence for engraving and possible symbolism in a Middle Palaeolithic context (*e.g.*, Goren 1990; Hovers *et al.* 1997; Kozlowski 1992; Leroi-Gourhan 1961; Marshack 1996), three pieces of ochre, which do not appear to have originated in the cave, were recovered and an entry in the dig diary mentions the presence of ochre in one area.

In the earlier paper Moloney (1998) included the 'incised' Ras el-Kelb flake as one of the similarities with the site and Qafzeh which has an incised Levallois core fragment (Hovers *et al.* 1997). However, these two sites do share other similarities: ochre was found on both sites, their lithic industries are similar in some respects (Copeland 1998) and they both fall in an age range of *c.* 100,000 years. Additionally, human remains have been found at both sites; *Homo sapiens sapiens* at Qafzeh, but the four human teeth at Ras el-Kelb cannot yet be positively identified to taxa. It is hoped that there may be more success with identification in a future analysis of the teeth.

2.8 Dating

The date of the occupation of Ras el-Kelb is, in the opinion of Sanlaville (1998), dependent on the interpretation of the chronology of the Eastern Mediterranean marine transgressions and regressions during the Quaternary. This is today understood in terms of the Oxygen Isotope Stages of Shackleton & Opdyke (1976) and especially the warm sea transgressive Stage 5. The sea-level of this Stage is marked by the presence of lithophage worms (Vermets) on the rock faces which live at the exact sea-level as well as by the shells in the beach conglomerates of the mollusc *Strombus bubonius*, which form part of the Senegalese warm-sea fauna.

At Ras el-Kelb traces of these Last Interglacial markers were seen in the immediate environs of the cave. Recent research into the evolution of the Levant shorelines has generally resulted in the relative dates of such archaeological sites as were associated with marine features (marine benches or raised beaches *etc.*), being put back in time to an earlier transgressive or regressive phase. It appears that the sea-level variations already observed were indeed induced by regional tectonic movement and glacio-eustatic oscillations in relatively uniform upthrust of the region. However, today, climatic change such as an increase in rainfall is seen as an additional and major factor; it affects the dynamics of the Levant littoral by causing the streams to erode the mountain soils and to deposit quantities of debris onto the seashore. Once consolidated, these can give a researcher a false impression of a raised beach, especially when the material includes marine detritus (shells and sand brought in by storm action at sea). Sanlaville sees these marine effects as resulting from a series of three-phase climato-sedimentary cycles, occurring as follows:

- Phase 1: Storms; erosion (rhexistasy) which covers the shore with terrestrial debris; this migrates northward propelled by the prevailing wind and waves.

- Phase 2: Amelioration of the climate; discharge decreases and the coastline recedes; beachrock forms; equilibrium.

- Phase 3: Biochemical processes in calm, warm water favour the formation of wave-cut notches, corrosion benches, settlement of Vermets at sea-level and Strombus. Finally a new wave of violent storms upsets the balance and marks the onset of another cycle.

At Ras el-Kelb the features present are regarded by Sanlaville (1998, 39) as referring to the 8–12 m above present sea-level transgression attributed to Stage 5 and associated with a rich Senegalese fauna. For him, the pebble beach on the cave floor represents Stage 5e, the earliest transgressive oscillation in stage 5, dated in the region to *c.* 150–125,000 by UTh and ESR dating methods. The following short-lived regression (5d) to below 6.40m above sea-level would date to *c.* 125–100,000 and would be the time of the earliest Mousterian occupation of the cave. It is possible that the uppermost units in the Tunnel Trench could date to the later stages of stage 5 (5c–a) or even continue into the regressive, glacial Stage 4, since the sea never returned to the 6m level.

Although the above reading places Ras el-Kelb occupation in the Stage 5 (specifically in Stage 5d) on marine evidence, an archaeological problem arises. Other dating methods have also advanced during the last decade, and Tabun Cave levels have been put back in time considerably. For example, the industry of Level C, which is closely similar to that of Ras el-Kelb, is dated on TL evidence to 170,000 by Mercier *et al.* (1995). Also in the Southern Levant, the recently reported C-type Mousterian industry at Hayonim Upper E (Meignen 1998, 168) has been dated to *c.* 161,000 by ESR and U-series (Bar-Yosef 1998, Table 1). This is in spite of the fact that the industry of Qafzeh Cave, thought to contain a relatively similar Tabun C-like industry, has been dated to *c.* 96,000 by ESR (Bar-Yosef & Meignen 1992). These discrepancies suggest, against all reason and other evidence such as that at Naame, that if both Sanlaville's interpretations and those of Mercier *et al.* (1995) are valid, perhaps the Tabun C-like industry made by the Ras el-Kelb Mousterians was not contemporary with that at Tabun C and Hayonim and that the two groups lived *c.* 50,000 apart! If so, a similar group lived much later still at Qafzeh.

It is possible that coming dates from Nahr lbrahim Cave will help to resolve these difficulties. Nahr Ibrahim is a seashore cave some 15km to the north of Ras el-Kelb. Like Ras el-Kelb it was occupied when the sea-level was 12m higher than that of today; it was occupied first by a Mousterian group with a Tabun C-type industry closely comparable to that of Ras el-Kelb, followed by (as also at Tabun) a Tabun B-type industry. It was excavated in 1974 but has not so far been finally published (Solecki 1975).

Another site with a Tabun C-like Mousterian industry, this time located about 30 km south of Ras el-Kelb, is the open air site of Naame, excavated by Fleisch (1970). Although it has dates of *c.* 90,000 and is associated with the same interglacial fauna (Vermets, *Strombus*), these dates are thought to be too young (Farrand 1994).

2.9 Conclusion

The evidence from Ras el-Kelb has so far given us a glimpse into the life of a group with a particularly stable lithic tradition, lasting long enough to produce at least 4m of archaeological deposit in a region (the Northern Levant) where relatively few Mousterian sites have been examined, and fewer still published. Although many years elapsed between excavation and publication of the report, the thoroughness of the excavation and detailed analyses of material will hopefully have allowed the site to take its place within the Levantine Middle Palaeolithic. The work has enabled us to present, for the first time (Copeland & Moloney 1998), an extensively described and fully illustrated account of a Tabun C-type Levantine Mousterian lithic industry. Further research on the as yet unstudied material is strongly desired and can be expected to further augment our knowledge of the Levantine Mousterian, hopefully sooner rather than later in the future.

Acknowledgements

We would like to thank the many people without whose assistance the final Ras el-Kelb report could not have been completed: the individual contributors to the final report, Mme Haikimian of the Lebanese Department of Antiquities, the staff at the Musée des Antiquités Nationales, St. Germain-en-Laye, Paris, members of the Institute of Archaeology who helped with aspects of production and Andrew Garrard for his constant encouragement and advice, and to all of those who are too numerous to name here, but whose help was invaluable. Our thanks, also, to Yolanda Fernandez-Jalvo for her SEM work and subsequent discussion. Above all, we acknowledge the superb work of Dorothy Garrod and Germaine Henri-Martin.

References

AKAZAWA, T., K. AOKI & O. BAR-YOSEF (eds.) 1998. *Neandertals and Modern Humans in Western Asia.* Plenum Press, New York and London.

BAR-YOSEF, O. 1998. 'The chronology of the Middle Paleolithic of the Levant.' *In* Akazawa *et al.* (1998), pp. 39–56.

BAR-YOSEF, O. & L. MEIGNEN 1992. 'Insights into Levantine Middle Paleolithic variability.' *In* H. L. Dibble & P. Mellars (eds.), *The Middle Paleolithic: Adaptations, Behaviour and Variability,* pp. 68–182. University Museum of Philadelphia Museum Monograph 78, Philadelphia.

BOËDA, E. 1988. 'Le concept Levallois et évaluation de son champ d'application.' *In* M. Otte (ed.), *L'Homme de Néanderthal, 4. La technique,*

Etudes et Recherches Archéologiques de l'Université de Liège 31, pp. 13–26. Université de Liège, Liège.

Boëda, E. 1995. 'Levallois; un concept volumétrique, des methods, une technique.' *In* H. L. Dibble & O. Bar-Yosef (eds.), *The Definition and Interpretation of Levallois Technology*, Monographs in World Archaeology 23, pp. 13–16. Prehistory Press, Madison.

Copeland, L. 1978. 'The Middle Palaeolithic of Adlun and Ras el-Kelb (Lebanon): the first results from a study of the flint industries.' *Paléorient* 9: 15–27.

Copeland, L. 1998. 'The Middle Paleolithic flint industry of Ras el-Kelb.' *In* Copeland & Moloney (1998), pp. 73–175.

Copeland, L. & N. Moloney (eds.) 1998. *The Mousterian Site of Ras el-Kelb, Lebanon*. British Archaeological Reports International Series 706, Oxford.

Emery, K. C. & C. J. George 1963. *The Shores of Lebanon*. American University of Beirut, Misc. Papers in Natural Sciences No. 1, Contribution no. 1385. Woods Hole Oceanographic Institution, Woods Hole.

Farrand, W. R. 1994. 'Confrontation of geological stratigraphy and radiometric dates from Upper Pleistocene sites in the Levant.' *In* O. Bar-Yosef & R. S. Kra (eds.), *Late Quaternary Chronology and Paleoclimates of the Eastern Mediterranean*, pp. 33–53. Radiocarbon, and the American School of Prehistoric Research, Tucson, Arizona and Cambridge, Massachusetts.

Fleisch, H. 1970. 'Les habitats du Paléolithique Moyen à Naame (Liban).' *Bulletin du Musée de Beyrouth* 23: 25–96.

Garrard, A. 1998. 'Food procurement by Middle Palaeolithic hominids at Ras el-Kelb Cave in Lebanon.' *In* Copeland & Moloney (1998), pp. 45–65.

Garrod, D. 1962. 'The Middle Paleolithic of the Near East and the problem of Mount Carmel.' *Man. Journal of the Royal Anthropological Institute* 92: 232–59.

Garrod, D. & D. Bate 1937. *The Stone Age of Mount Carmel*. Oxford University Press, Oxford.

Garrod, D. & G. Henri Martin 1959. 'Fouilles à Ras el-Kelb, Liban, 1959.' In *Actes du 16ᵉ Congrès Préhistorique de France*. Monaco.

Garrod, D. & G. Henri Martin 1961. 'Rapport préliminaire sur la fouille d'un grotte à Ras el-Kelb, Liban, 1959.' *Bulletin du Musée de Beyrouth* 16: 61–8.

Geneste, J. M. 1985. *Analyse lithique d'industries Moustériennes du Périgord; une approche téchnologique de comportement des groupes humaines au Paléolithique Moyen.* Ph.D. thesis, Université de Bordeaux I.

Goren, N. 1990. *Quneitra: a Mousterian site on the Golan Heights*. QEDEM, 31. Hebrew University of Jerusalem, Jerusalem.

Hovers, E., B. Vandermeersch & O. Bar-Yosef 1997. 'A Middle Palaeolithic engraved artefact from Qafzeh Cave, Israel.' *Rock Art Research* 14(2): 79–87.

Kozlowski, J. K. 1992. *L'Art de la Préhistoire in Europe Orientale*. CNRS Editions.

Leroi-Gourhan, A. 1961. 'Les fouilles d'Arcy-sur-Cure.' *Gallia Préhistoire* 4: 1–16.

Marshack, A. 1996. 'A Middle Paleolithic symbolic composition from the Golan Heights: the earliest known depictive image.' *Current Anthropology* 37(2): 357–65.

Meignen, L. 1998. 'Hayonim lithic assemblages in the context of the Near Eastern Palaeolithic.' *In* Akazawa *et al.* (1998), pp. 165–80.

Mercier, N., G. Valladas, H. Valladas, A. Jelinek, L. Meignen, J. L. Joron & J. L. Reys 1995. 'TL dates of burned flint from Jelinek's excavations at Tabun and their implications.' *Journal of Archaeological Science* 22: 496–509.

Moloney, N. 1998. 'Ochre and incised flint from Ras el-Kelb.' *In* Copeland & Moloney (1998), pp. 181–5.

Nishiaki, Y. & L. Copeland 1992. 'Keoue Cave, Northern Lebanon and its place in the context of the Levantine Mousterian.' *In* T. Akazawa, K. Aoki & T. Kimura (eds.), *The Evolution and Dispersal of Modern Humans in Asia*, pp. 107–27. Hokusen-Sha, Tokyo.

Sanlaville, P. 1998. 'The deposits of Ras el-Kelb in the framework of the regional chronostratigraphy.' *In* Copeland & Moloney (1998), pp. 37–44.

Shackleton, N. J. & N. D. Opdyke 1976. 'Oxygen-Isotope and Paleomagmetic Stratigraphy of Pacific Core V.28–239. Late Pliocene to Latest Pleistocene.' *In* R. M. Cline & J. D. Hays (eds.), *Investigation of Late Quaternary Paleoceanography and Paleoclimatology*, Geological Society of America Memoir 145, pp. 449–64. Geological Society of America, Boulder, Colorado.

Solecki, R. 1975. 'The Middle Paleolithic site of Nahr Ibrahim (Asfouriyeh Cave) in Lebanon.' *In* F. Wendorf & A. Marks (eds.), *Problems in Prehistory, North Africa and the Levant.*, pp. 283–96. SMU Press, Dallas.

Solecki, R. S. & R. L. Solecki 1970. 'A new secondary flaking technique at the Nahr Ibrahim Cave site, Lebanon.' *Bulletin du Musée de Beyrouth* 23: 137–142.

Tixier, J., M. Inizan & H. Roche 1980. *Préhistoire de la Pierre Tailée I: Términologie et Téchnologie.* C.R.E.P., Paris.

Yamada, S. 1998. 'Use-wear analysis of selected specimens from Ras el Kelb.' *In* Copeland & Moloney (1998), pp. 176–80.

Lorraine Copeland
Chateau de Marouette
24350 Tocane St. Apre
France
Milcopmar@aol.com

Norah M. Moloney
Institute of Archaeology
University College London
31–34 Gordon Square
London
WC1H 0PY
n.moloney@ucl.ac.uk

3

The Mousterian in Action: Chronology, Mobility, and Middle Palaeolithic Variability

P. B. Pettitt

Keble College, University of Oxford

Abstract

This paper addresses the question of changing Levallois technologies through time and space. It discusses the course of research into, and the nature of, Middle Palaeolithic technology and investigates, in particular, the relationship between Levallois and Quina technologies, raw material and landscape. It conclusion, it proposes a model to account for the interaction of Mousterian lithic technology and Neanderthal mobility.

3.1 Introduction

Contemporary research emphasises the structure of the transport of lithic resources in the Middle Palaeolithic, particularly in the light of 'planning' in the landscape (*e.g.*, Féblot-Augustins 1993, 1997; Geneste 1985, 1989; Kuhn 1992; Roebroeks *et al.* 1988; Wenban-Smith 1998), or of the effects of group subsistence strategies and mobility on technological strategies (Barton 1990; Kuhn 1991, 1992, 1993, 1995; Stiner 1994; Stiner & Kuhn 1992) and in the overall context of hominin network organisation (Gamble 1995, 1998b).

Whilst the raw material movement studies noted above have made important contributions towards understanding hominin behaviour in their own right, a number of claims have been made for 'sophistication' of hominin behaviour in the Middle and Upper Pleistocene as reflected by artefact 'curation' or apparent 'planning' in the landscape (*e.g.*, Pitts & Roberts 1997; Roberts 1996a,b[1]; Roberts *et al.* 1997, 333; Wenban-Smith 1998). These take the important observations of an increase in transported distance over Middle Pleistocene time (Roebroeks *et al.* 1988) to another level of inference, which, as they rest on simple observations, either of artefact movement distance or *inferred* movement distance based on artefact wear, are at best naïve. On a simplistic level proponents of such views rarely define their terms, the effect of which is that words such as 'sophisticated' and 'complex' are meaningless and therefore that resulting interpretations have no intellectual weight. On a more specific level the problem is even more acute. Given that chimpanzees curate items in the landscape, *e.g.*, 0.5km for termite digging sticks, '*hundreds* of metres for nut-cracking stones' (Goodall 1986, 544, her emphasis) and 6km in the case of a fragment of pelvic bone (McGrew 1992), and that organic computer simulations tested against random motion patterns suggest that curation in the order of some tens of kilometres may occur as inadvertent, additive products of short-distance stochastic movement (McGrew 1992), it is both simplistic and inaccurate to use raw material movement *per se* as an indicator of hominin intelligence (Pettitt & Schumann 1993). Similarly, as chimpanzees 'cache' tools at strategic sites for later use (*e.g.*, ant dipping 'wands' at Gombe (Goodall 1986), and nut-cracking stones at Tai (Boesch & Boesch 1984)) neither can such behaviour be seen to suggest necessarily 'complex' behaviour. It would now even appear that regional differences in the modes and organisation of behavioural traits involving material culture — 'cultures' as archaeologists might loosely define them — are not unique to hominins (Whiten *et al.* 1999). As modified items are transported by animals as diverse as birds (nest constituents, food) and rhesus monkeys (stones held in the cheeks), it is surprising that simple arguments as to raw material movement — at least over tens as opposed to hundreds of kilometres — have been used by archaeologists at all in the cognitive realm. Similarly, grandiose claims for the heuristic power of the *chaîne opératoire* have never delivered that which their hyperbole promised (see, for example, Schlanger 1994, who talks of '*unleashing* the *chaîne opératoire* for an archaeology of mind' — my emphasis). The *chaîne opératoire*, having been unleashed, has little more than whimpered.

While I follow the broad paradigm of artefact movement and use in the landscape here, I seek to avoid such simplistic analogies, rather to address a number of questions that I suggest are central to Early (in particular Middle) Palaeolithic research and from which a more meaningful discourse may emerge. I propose a model which invokes a meaningful interaction between Mousterian lithic technology and the mobility of its Neanderthal makers.

The recent history of Middle Palaeolithic studies has tended to emphasise technology, as opposed to typology, and research is demonstrating not only a considerable variability in the Middle Palaeolithic repertoire, but patterns of chronological change that form the core of this paper. In view of this, the one overriding question to ask of the lithic data is why technology varies over time. We are now at the point of appreciating the considerable variability subsumed under the Levallois heading (see papers in Dibble & Bar-Yosef 1995), but few archaeologists have begun to ask why Levallois strategies were actually used at all and why the frequency of their use varied over time. What was the importance of Levallois technology to Neanderthals?

In addition to this overriding question of the nature of technological variability in the Middle Palaeolithic and the importance of Levallois within it, there are a number of characteristics of the Middle Palaeolithic that stand out and demand attention. These may be summarised as follows and are discussed below:

1. Raw material use in the Middle Palaeolithic is almost entirely local, but when material does move over distances greater than 20km it is almost always in the form of Levallois products. Why?

2. Quina technology is always executed on immediately available raw materials. Why?

3. There is a good degree of chronological patterning from Oxygen Isotope Stage (OIS) 6 onwards, which relates to Levallois technology and its relationship to other means of primary flaking. This is the only dynamic element of the Middle Palaeolithic. Why is there chronological variability?

4. Why did the Middle Palaeolithic evolve in the first place, given the apparent success of the Lower Palaeolithic if longevity may be taken to indicate such?

I attempt here to address such questions, and seek to build them into a model of Neanderthal mobility and technology in which, for the first time, technology is meaningfully used as a dynamic means of adaptation. I first discuss the course of research into Middle Palaeolithic technology by way of setting the context. After this I make some generalisations about the nature of Middle Palaeolithic technology, before moving on to a summary of what may be said about chronological patterning. The final section presents the model constructed from these discussions which integrates Neanderthal mobility into technology.

3.2 Middle Palaeolithic technology: some historical aspects and salient points

One of the striking characteristics of the Middle Palaeolithic — one which clearly distinguishes it from the Lower Palaeolithic — is the chronological patterning it displays over at least 150,000 years. Mellars' observations on the chronological patterning of the Ferrassie, Quina, and MTA variants (*e.g.,* Mellars 1969), and Binford's and Rolland's later support for this are well known (Binford 1973; Rolland 1976, 1988). At about the same time as the Bordes–Binford debate was raging, Mellars gathered data which suggested that, contrary to Bordes' own views, the Mousterian variants were synchronous (*e.g.,* Bordes 1953, 465, 1968, 101–5, 1972) and that there was a degree of chronological structure in the Mousterian (Mellars 1965, 1967, 1969, 1970). These observations, which Mellars has continued to refine (*e.g.,* 1986a; 1986b; 1988; 1996, 183–90), were made possible due to a background of active field research initiated by Bordes, because of which evidence of the Mousterian had '... increased strikingly — both in quantity and quality — during the past two decades ... ' (Mellars 1969, 134). Mellars' observations still had some roots in the concerns of the Mousterian debate, since the chronological issues '... evidently have a crucial bearing on several problems relating to the cultural interpretation of the industries which are currently in dispute ... ' (Mellars 1969, 136). He made two main observations on the three most clearly defined Mousterian variants; first, that there was a marked tendency for the MTA to be stratified above the Quina Mousterian (as represented at the time by at least 12 sites, including Combe Grenal, Pech de l'Azé and Roc de Marsal), and often immediately below stratigraphic layers containing Upper Palaeolithic assemblages (no less than nine sites), which suggested to him that, at least in southwest France, MTA assemblages were chronologically later than Quina Mousterian assemblages. Secondly, he observed that at Combe Grenal, at least nine Quina Mousterian assemblages were stratified above five Ferrassie Mousterian assemblages. This stratification of the Quina Mousterian above the Ferrassie was also found at Abri Chadourne, Dordogne, and at Roc-en-Pail, Maine-et-Loire. Moreover, at a number of other sites where at least two levels of one of the two Charentian variants was observable, the technological and typological features of these industries displayed trends consistent with a Ferrassie to Quina transition, *i.e.,* a loss of the Levallois, faceting and laminar indices and an increase in the relative frequencies of (notably transverse) racloirs. To Mellars (1969, 155) this suggested a '... gradual transition from industries of the Ferrassie, to those of the Quina, form', indicative of a '... low but cumulative technological/typological change ... '.

Various criticisms of Mellars' observations were made, notably that it was based on a small number of sites, placed too much emphasis on the apparently discrete clusters of the three relevant variants at Combe Grenal, and that it was impossible to link the various sites together on sedimentological, biostratigraphic, and palynological grounds (Ashton & Cook 1986). However, his observations have held well, with the application of absolute dating methods to Mousterian levels — notably that of Le Moustier (Mellars 1986b) — supporting his views of a late chronostratigraphic position for the southwest French MTA (Mellars 1988). In reply to Mellars' observations Binford (1973) also stressed the tendency for the Typical Mousterian of the early Wurm (*i.e.*, OIS 5) to cluster into discrete groups and emphasised Bordes long-standing division of the MTA on *chronological* grounds. Rolland (1976, 1988) subsequently found further evidence for chronological patterning in the Middle Palaeolithic from the penultimate (Riss) glacial to the last glacial, particularly in terms of the frequency of the Levallois technique, demonstrating that, from a low frequency in the Riss (OIS 6), the Levallois index (IL) rises through the early Wurm and reaches a peak in late OIS 5, whereupon there is a dramatic decline in its use (Mellars' Ferrassie–Quina transition) until the late Mousterian of OIS 3 (Rolland 1988, 166). These chronological trends are discussed below. One can see how, with the impact of absolute dating of Middle Palaeolithic sites still in the future, chronological interpretations of Middle Palaeolithic variability were necessarily dependent upon stratigraphic observation: writing in the mid-1970s, Rolland noted how there was a problem in fixing the chronological extent of the Mousterian, the lower limits of which, although controversial (*open* to Bordes) '... probably antedate Upper Pleistocene times ... ' (Rolland 1976, 1–16).

Chronological patterning of the Mousterian has been established solely on the basis of the deep stratigraphies of enclosed sites, and it is important to note that this takes place against a background both of quartz/quartzite chopper industries which appear to be produced well into the Upper Pleistocene, and a scatter of small and as yet undated assemblages from open sites. The recognition of chronological patterns is therefore to some extent biased, but considering that enclosed sites probably played a pivotal role in Middle Palaeolithic behaviour (Mellars 1996) the data they furnish can be taken to represent major aspects of Middle Palaeolithic technological behaviour.

In a technological sense, part of the patterning recognised by Mellars (the Ferrassie-Quina transition) represents a chronological shift in methods of débitage, *i.e.*, a fall off in the use of Levallois technology probably towards the end of OIS 5 or early in OIS 4 (Mellars 1988). This is one part of a wider, cyclical fluctuation of the Levallois index from the late Middle Pleistocene (*i.e.*, OIS 6) onwards, as recognised by Rolland (1988) as well as in the production of relative frequencies of racloirs and denticulates/notches (Le Tensorer 1978). In a typological sense, Mellars' first observation — that the MTA occurs relatively late in the Mousterian — represents the fluctuation of methods of façonnage, in this case of the reappearance of bifaces. Although the Middle Palaeolithic is essentially an elaboration of methods of débitage it is clear, then, that both principles of débitage and façonnage contribute towards Mousterian variability.

In recent years the laminar nature of many Levallois industries has been increasingly recognised. Such industries may be observed in OIS 7 (*e.g.*, Crayford and Baker's Hole in England, and in certain Yabrudian industries of the Near East — Meignen 1994) and in (the substages of) OIS 5 throughout Europe to the Near East, *e.g.*, Seclin (Révillion & Tuffreau 1994b), Saint-Germaine-des-Vaux/Port Racine (Révillion & Cliquet 1994), Riencourt-Les-Bapaume (Ameloot-van der Heijden 1994), the 'Micoquian' of Vinneuf level N1 (Gouédo 1994), Rocourt, Belgium (Otte 1994), Rheindahlen and Tönchesberg in Germany (Conard 1990), Gorham's Cave, Gibraltar (Waechter 1951, 1964, and pers. obs.) and, possibly, Bau de l'Aubesier, Ardèche (Buisson-Catil 1994). In southwest France, the Ferrassie Mousterian variant fits into this general technological and chronological grouping of assemblages, as in the Ardèche, *e.g.*, as at Bérigoule and Baume-des-Peyrards (Buisson-Catil 1994). Révillion (1995) has noted that many of such laminar assemblages dating to before the last glaciation tend to be produced using non-Levallois *chaînes opératoires*, which are mainly developed in the early stages of the Last Glaciation, and has suggested that Middle Palaeolithic laminar assemblages can be divided into two groups, a *phase ancienne* consisting of mainly non-Levallois industries of OIS 8–6, and a *phase récente* dating to the early Last Glacial, in the chronological context of technological rupture and continuity.

It has long been recognised that the Quina Mousterian seems to be confined to the harsh conditions of OIS 4 (*e.g.*, Binford 1989; Mellars 1969), while racloir-dominated Charentian Mousterian-like assemblages are also found in the similarly cold conditions of OIS 6. By contrast, Laminar Levallois assemblages are found again in OIS 3, *e.g.*, at the Abri du Maras, Ardèche (Moncel 1994) as well as assemblages in which principles of façonnage are pronounced, notably in the MTA(A) (Mellars 1969) and the leaf-point industries of Central Europe (Kozlowski 1990).

It follows that there appears to be a broad degree of chronological fluctuation over at least 150,000 years in western Europe, in both techniques of débitage and of façonnage. Whether this has a broad correlation with climate, *e.g.*, Levallois with warmer interglacial or interstadial conditions, non-Levallois débitage with colder, stadial conditions, remains to be demonstrated, but it follows that there is some degree of diachronic dynamism in the Middle Palaeolithic in general. This is not innovative in the sense of the creation of new artefact forms as in the Upper Palaeolithic, but rather reflects the cyclical fluctuations of elements of a general technological repertoire (White & Pettitt 1995).

A degree of chronological patterning of technological behaviour and subsistence strategies has been noted by Stiner (1994) and Kuhn (1995) in the Pontinian of Latium, Italy. This is essentially a Mousterian technocomplex in that it displays many of the technological *chaînes opératoires* noted above and contains typically Mousterian implement forms, but it is specifically adapted to small beach-pebble raw material and all products therefore tend to be diminutive in size. Both Stiner and Kuhn pick up a shift around 55,000 years ago, before which cores were worked in a centripetal (discoidal) fashion and scavenging formed the basis for most ungulate procurement strategies, but after which a shift to more economical, parallel (uni- and bipolar) methods of primary flaking accompanied a shift to ambush hunting. A relatively high transport of 'exotic' lithic material and retouched flakes before this date was interpreted as a system of high-mobility scavenging of diffuse resources with provisioning of *individuals.* By contrast, the more economical use of (more local) raw materials was interpreted as ambush hunting in the context of more prolonged and substantial occupations in which places are provisioned. Although the Electron Spin Resonance (ESR) dating of the Pontinian assemblages used by Stiner and Kuhn and the integrity of their resulting chronological pattern is open to question (Aldhouse-Green 1996) their research is another potential indicator of diachronic patterning of the Middle Palaeolithic.

3.3 Middle Palaeolithic technological variability

In technological terms, the Middle Palaeolithic *sensu lato* is defined on the basis of an extensive array of *chaînes opératoires* organised either around the principles of façonnage, whereby a mass of material (nodule or flake) is reduced in a progressive manner towards an envisaged final form such as a biface, or débitage, whereby a mass of material is divided up by one of a number of *chaînes opératoires* into usable flakes of different shapes and sizes (Boëda *et al.* 1990). It is notably the elaboration of systems of débitage that defines the Middle Palaeolithic in technological terms, in which sense it is no surprise that this occurs as a developmental continuum from the Lower Palaeolithic (White & Pettitt 1995). Middle Palaeolithic systems of façonnage and débitage were organised either in relation to a secant plane, in which flakes were removed in relation to a conceived fixed plane which usually took the form of a continuous striking platform, notably in Levallois technology (Boëda 1988, 1993a), discoidal/centripetal flaking (Boëda 1993b) and in the production of bifaces inherited from the Lower Palaeolithic (White & Pettitt 1995), or they were more randomly organised in a number ways utilising multiple striking platforms (Boëda *et al.* 1990) in ways also derived from the Lower Palaeolithic which may be classified as migrating plane technology (White & Pettitt 1995, 29). At a coarse level, the relative frequency of Levallois systems of débitage as opposed to non-Levallois systems is reflected by the Levallois Index (IL; Bordes 1953) which appears to fluctuate over the course of OIS 6–3 (Mellars 1969, 1988; Rolland 1988).

Levallois technology is the classic example of secant plane organisation, and probably developed out of the conceptually similar production of bifaces (Tuffreau 1994). If correct, then this can be defined as the conceptual development of variable systems of débitage out of a system of façonnage aimed at the production of bifacial tools: this can be seen as the only major technological innovation of Middle Pleistocene Europe (White & Pettitt 1995, 33). Whilst Bordes (1980) recognised the variable systems of débitage encompassed by the Levallois term, European Levallois *chaînes opératoires* have recently come under much scrutiny, notably by Boëda on northern French late Middle Pleistocene sites (Boëda 1988, 1993a,b), who has noted a considerable diversity of approaches subsumed under this heading. This has been supported by observations outside of Europe, *e.g.,* in North Africa (van Peer 1992) and the Near East to Mongolia and Japan, and can therefore be taken to indicate that it is a general later Middle Pleistocene technological phenomenon amidst late archaic hominin populations (see papers in Dibble & Bar-Yosef 1995). The result of this recent attention has been both to push the recognition of Levallois *chaînes opératoires* back in time (possibly as far back as OIS 12 — Tuffreau 1994 — certainly to OIS 8) and to broaden the definition to the point that the term 'Levallois' may have lost its classificatory meaning except when applied to the more 'classic' techniques. In a broad sense, Levallois technology can be defined as the production of flakes, points and/or blades of usually predetermined form through a series of *chaînes opératoires* organised around a secant plane. As Boëda (1988, 1993b) has noted, such schemes may be organised in one or two directions *along* the surface of removal (unipolar or bipolar) or may be centripetal, with flakes removed from *around* the perimeter of the striking platform. They may also be worked in a single process of removal geared towards the removal of a solitary flake or a restricted set of flakes (linear) or towards a repeated series of flakes in a recurrent fashion, often with intervening phases of repreparation of the striking surface. It has become increasingly clear that some Levallois *sensu lato* industries, notably belonging to OIS 5, have a distinctly laminar aspect (see papers in Révillion & Tuffreau 1994a, and see below). Levallois technology may be used in all of Bordes' Mousterian variants, but in the Charentian Mousterian it is either absent or very infrequent as it is in the Quina variant and high in the Ferrassie variant (*e.g.*, Bordes 1953). Although the greater part of Mousterian technology is executed on material available at distances of under 5km from the site from which it is recovered, it is interesting to note that when material does move greater distances in the landscape — up to 100km in France, possibly 300km in Central Europe — it is mainly Levallois products that do so (Féblot-Augustins 1993; Geneste 1985). This observation in particular will be developed below.

Recurrent centripetal Levallois primary flaking may be confused with the discoidal core technique, which has for some time been recognised as a classic Mousterian primary flaking technique (*e.g.*, Bordes 1950, 1968, 243; Boëda 1993a; Mellars 1996, 69–73). The distinction between the two is that with the discoidal technique, which is itself a systematic approach to raw material, there is an absence of either core rejuvenation or the removal of a 'privileged' flake (Boëda 1993a) As well as the characteristic cores, which are usually worked in multi-directional fashion on both faces and often pyramidal in cross-section with a relatively high edge-angle between the dorsal and ventral surfaces, this technique yields two characteristic products, the Pseudo-Levallois Point and the *éclat débordant*, the form of which is determined by the dorsal scar pattern produced by centripetal flaking (Cook & Jacobi 1998; Locht & Swinnen 1994). As with Levallois technology, the discoidal technique is found in all of the Mousterian variants.

In recent years, Alain Turq has identified a number of technological strategies in use in the southwest French Mousterian, which may be defined as *Quina technology* for the reason that they are found most commonly in the Quina Mousterian (Turq 1989, 1992). This technology is almost exclusively executed on locally (less than 5km radius) available raw material. The sites which form the basis for Turq's study are located in the Haute-Agenais (Turq 1984) and Dordogne, including Combe Grenal and Roc de Marsal (Turq 1985, 1988). In all of Turq's *chaînes opératoires*, a relatively small 'sausage shaped' nodule is reduced from one end in a number of ways. Turq emphasises that the Quina techniques of primary flaking are a systematic and economic approach to small, elongated nodules of raw material, and may have been deliberately geared towards the production of distinctive Quina tools where maximum thickness was located directly opposite the retouched edge of the tool (Turq 1992, 75). Whether this intentional aspect is true or not, products of Quina technology tend to be highly cortical, relatively thick, and triangular in cross-section, characteristics that may affect the type and location of subsequent retouch (Pettitt 1999).

An aspect of Middle Palaeolithic technology that has been neglected in the literature is the simple, *ad hoc* flaking of locally available quartz and quartzite cobbles into 'chopper/chopping tool' industries. Such assemblages have mainly been found on open air sites such as Les Planes, Lot-et-Garonne (Collina-Girard & Turq 1991), and the specialised bovid-hunting/processing sites of La Borde, Lot (Jaubert *et al.* 1990) and Mauran, Haute-Garonne (Farizy *et al.* 1994), although a quartzite element can be found in many enclosed site Mousterian assemblages. The only systematic study of such industries remains that of Turq (1978) for the Dordogne and Lot. Although it is unclear whether these industries represent the deliberate manufacture of 'core-tools' such as choppers (that is, whether they are primarily systems of façonnage) or whether they are instead systems of débitage geared towards the simple removal of small numbers of usable flakes (Ashton *et al.* 1992) they stand in clear contrast to the technological systems generally taken to be characteristic of the Mousterian. In view of this, it is important to consider how biased our view of the Mousterian is in favour of the classic variants of the enclosed cave and rockshelter sites, and that these 'patches' occur in the context of a 'scatter' of small, open air assemblages generally too small to be classified in terms of Bordes' variants, such as those from the Dronne and Isle valleys studied by Geneste (1985).

3.4 Chronological range and patterning of the Middle Palaeolithic

Traditionally, the term 'Mousterian' has been used to refer strictly to Middle Palaeolithic assemblages dating to the Upper Pleistocene, that is, to the Last Glacial period (Mellars 1967). This definition is followed here, with assemblages dating to the Middle Pleistocene being referred to under the general term Middle Palaeolithic. Since Bordes (1977, 37) noted that the Mousterian has always had an open chronology, the duration of the Middle Palaeolithic[2] has been extended considerably in both directions, notably back into the Middle Pleistocene. At the Orgnac 3 aven, Ardèche (Moncel & Combier 1992), a sequence of flake industries dating to between 350,000 and 300,000 BP on the basis of sedimentology, biostratigraphy, uranium-series and ESR dates, displays the growth in use of the Levallois technique to produce usable flakes, about 25% of which are used to produce flake tools of Middle Palaeolithic type, which clearly dominate over pebble-tools and bifaces. If one defines the Middle Palaeolithic in one sense as containing more than a handful of Levallois products, then the strong Levallois product and flake-tool component in the upper levels indicates the existence of Middle Palaeolithic industries here by 300,000 BP. At the Maastricht-Belvédère loess and gravel pit, Netherlands, a flake industry with Levallois element from site K has been dated to 250,000 BP by TL and

ESR, in which single and convergent racloirs dominate (de Loecker 1992). Other Middle Palaeolithic assemblages have been found in the same unit 4 fluviatile deposits at Maastricht such as the major site C which has been dated by TL to 225–315,000 BP (Huxtable & Aitken 1992). These assemblages indicate human occupation of the south Netherlands during a warm temperate phase of the Saalian interglacial complex (van Kolfschoten & Roebroeks 1985), evidence for which has been found elsewhere, for example at Mesvin, Belgium by 250,000 BP (Cahen 1984), and at Pontnewydd, Wales, where a Middle Palaeolithic assemblage produced on pebbles of volcanic rock (Green 1984, 1988) and dominated by Levallois products, occasional flake-blades, single and transverse racloirs has been dated to older than 191,000 BP (this is a weighted mean assuming all of the assemblages at Pontnewydd reflect one occupation; numerous dates are much earlier, including a TL date of 200,000 BP on a burnt flint core). The chronology at Pontneweydd has been developed using U-series, TL, ESR and ^{14}C, and, taken as a whole, indicates Middle Palaeolithic occupation of an open landscape during a cool period, probably over several millennia between 200,000 and 220,000 BP, possibly at the close of OIS 7 (Aldhouse-Green 1995; Green 1984). In this time period one can also place the 'laminar Levallois' assemblages of Crayford (Cook 1986) and Baker's Hole (Robinson 1986; Wenban-Smith 1992). Various Middle Palaeolithic assemblages dating to OIS 6 have been recovered from the loess of northern France and Belgium, notably at Beauvais, Oise, where flake supports were produced using the discoidal technique (Locht & Swinnen 1994) and Pucheuil, Seine-Maritime, level 6a, where supports were produced using a unipolar recurrent Levallois *chaîne opératoire* (Delagnes 1993), which are similar to assemblages dating to the early part of the Last Glacial (OIS 5), *e.g.*, Hermies, Pas-de-Calais (Masson & Vallin 1993), Houppeville (Vallin 1992) and Vollezele-Congoberg, Belgium (Vynckier *et al.* 1986).

In southwest France, an assemblage displaying a unipolar Levallois *chaîne opératoire* has been found in bed 51 at the Abri Suard, Charente, which in Bordean terms is a Typical Mousterian variant rich in racloirs (Delagnes 1990). This has been dated to late OIS 6 on sedimentological and biostratigraphic grounds. Similar industries continue alongside quartz pebble industries in the upper levels of the site and in the neighbouring Abri Bourgeois-Delaunay and Grotte Duport into the beginning of the last glacial period (Delagnes 1992). Whilst Middle Palaeolithic assemblages are known from OIS 7 in the north of France (*e.g.*, Biache-Saint-Vaast; Tuffreau & Sommé 1988) it is, in fact, not until OIS 6 that one finds non-Acheulean, flake/flake-tool assemblages in the south of France. These late Middle Pleistocene, Middle Palaeolithic assemblages are usually rich in racloirs, and may be defined as a Typical Mousterian of Levallois facies, *e.g.*, Abri Suard, and Grotte Vauffrey, Dordogne, beds VIII, VII, VI, V, where the Middle Palaeolithic industries continued to be deposited possibly in the Last Interglacial (bed IV) and into the early Last Glacial (beds III and IIs; Rigaud 1988). Others can be defined as a Charentian Mousterian, *e.g.*, Baume-Bonne, Provence, bed D (Delagnes 1992), Les Tares, Dordogne (Rigaud & Texier 1981) and Ranc Pointu, Ardèche (Combier 1967), which are similar both to the Rissian industry of La Micoque layer 3 and to Upper Pleistocene Quina Mousterian industries, or Bérigoule, northwest Provence, level II which is similar technologically to the Upper Pleistocene Ferrassie Mousterian assemblages of bed M2 at La Ferrassie and bed D at Abri Chadourne (Buisson-Catil 1994). Although such industries have been classified variously as 'Premousterian', 'Protocharentian', 'Micoquian' or 'Tayacian' (*e.g.*, La Micoque beds 3 and 4), they are technologically and typologically very similar to the Upper Pleistocene Quina industries in that they are essentially characterised by a high frequency of racloirs to the detriment of denticulates and notches, relatively high frequencies of stepped 'Quina' retouch, a very low frequency or absence of bifaces, and a variety of *chaînes opératoires*. They can be taken to represent the 'Mousterianisation' of southern French lithic technology (Delagnes 1992), and form part of a cyclical fluctuation of Quina Mousterian industries in southern France (Le Tensorer 1978). This occurs in the context of variable (but usually low) frequencies of bifaces in a continuum of later Acheulean assemblages through to the early Mousterian assemblages of OIS 5 (Chauchat 1992). The inclusion of such variable but low frequencies of bifaces in the development of Mousterian industries has been seen as an evolution of flake technology with a 'Micoquian influence' (Moncel 1995).

3.5 Technology and mobility

By the time that groups of Neanderthals occupied southwest France and left occupational debris in the caves and rockshelters of its river valleys, racloirs had been around for a long time. Pre-modern human groups had, for example, produced and discarded almost identical racloirs to the Mousterian forms some half a million years ago at High Lodge, Suffolk (Ashton 1992). Whilst the generalised tool forms that are so characteristic of pre-modern human adaptations fail to change in a dynamic sense — which is perhaps not surprising given that they are, as Foley (1987) states, relatively fixed phylogenetic traits — the *technology* of the Early Palaeolithic, whilst being generally non-innovative, changes in a much clearer, cyclical pattern.

It is conceivable that *primary flaking* was the major way in which Neanderthals responded to fluctuation in raw material supply or nature (Pettitt 1999). By contrast, the production of tool forms can be seen as a relatively fixed process perhaps stimulated by functional demand. If the demands are of function, *i.e.,* butchery and maintenance of (one assumes) a wood-based technology, it is clear that certain, at least generalised, tool forms would always be required. A cutting edge (bifaces, racloirs, unretouched flakes), points for use as knives and perhaps tips for thrusting spears, high-angled notched forms for spokeshaving, denticulated edges for heavy-duty cutting and a variety of heavier chopper forms for tasks requiring the heaviest investment of kinetic energy, would always be necessary. Many of these, of course, must be made on supports produced by primary flaking. In view of this, and as tools per se are often fixed in form once made, *tool production itself is incapable of responding to variability in the nature of raw material.*

By contrast, the obvious way in which to approach a block of material with an economy in mind is to divide it up in as economic means as possible, *i.e.,* to maximise desired output. This can occur in a number of ways. Certain techniques of primary flaking are more suited to certain types of material, *e.g.,* Quina technologies to the small, elongated river cobbles so common in the Dordogne and Vézère river valleys, in no way different to certain bifaces that may be suited to certain types of material (White 1998a). Levallois techniques may keep relatively large products thin, and in this sense the fact that it was Levallois products that Neanderthals carried around the landscape is interesting (see above). It was presumably better to carry a pouch full of relatively thin (and therefore light) Levallois flakes that were large enough to fulfil a variety of tasks or take a variety of modifications than to carry half a dozen flakes from a discoidal core (*éclats débordants* — translated here as encircling flakes) or a Quina racloir. This raises the question as to how far Mousterian technology varies with, or in relation to, Neanderthal mobility strategies

Some generalisations can be made about technologies from OIS 6–3. The relative importance of Levallois technologies from *c.* 300,000 BP down to the later part of OIS 5 is striking. Although as noted above the term Levallois encompasses many different *chaînes opératoires*, many of which bear only some resemblance to the 'classic' Levallois technique (Révillion 1995), thus the term 'Levallois *sensu lato*' is employed here as a means to distinguish these Levallois and recurrent laminar technologies from discoidal and 'Quina' type technologies.

3.5.1 OIS 6

Although use of the discoidal technique as a 'generalised' primary flaking technique is documented at this time, *e.g.,* at Beauvais, Oise (Locht & Swinnen 1994), there seems to be a clearly dichotomous nature of technology in this broad period. On the one hand many assemblages are dominated by Levallois *sensu lato* techonology often very similar to those of OIS 5. These include assemblages from Pucheuil level 6a, Seine Maritime (Delagnes 1993), bed 51 of Abri Suard, Charente (Delagnes 1990) and Grotte Vaffrey beds VIII, VII, VI and V (Rigaud 1988).

By contrast, a number of OIS 6 assemblages display clearly developed characteristics of 'Quina' technology, notably Baume-Bonne bed D, Provence (Delagnes 1992), Les Tares, Dordogne (Rigaud & Texier 1981), Ranc Pointu, Ardèche (Combier 1967), La Micoque layer 3, Dordogne and Bérigoule level II, northwest Provence (Buisson-Catil 1994) which Delagnes (1992) takes to represent the 'Mousterianisation' of southern French technology.

In addition to this dichotomous lithic technology, a small 'background' of assemblages apparently centred on façonnage (*i.e.,* biface assemblages usually referred to as 'Late Acheulean') which persist until early Stage 5 (Moncel 1995). Thus, by OIS 6, the major Mousterian technological axis — Levallois and Quina technology is apparent, in addition to an indication of the apparent fluctuation in use of systems of façonnage persisting from the Lower Palaeolithic.

3.5.2 OIS 5

Although a number of taphonomic factors may affect our picture of OIS substage 5e (Ashton *pers. comm.*) more chronological and climatic control is evident in OIS 5 *sensu lato*, with the result that one is able to observe with much improved clarity some degree of chronological patterning that one cannot observe for OIS 6 (it may be expected, hypothetically, that there is a chronological patterning of Levallois *sensu lato* and Quina technologies within OIS 6 as one observes in OIS 5). In early OIS 5 the dominance of Levallois *sensu lato* industries is striking, *e.g.,* as at Hermies, Pas de Calais (Masson & Vallin 1993), Houppeville, northern France (Vallin 1992), Abri Caminade (beds M2i–M3s), Roc en Pail, La Ferrassie (beds C, D1 and D2), Combe Grenal (levels 35, 33–27), and beds D and C at Abri Chadourne, Dordogne. The latter are most similar to the OIS 6 industries of La Micoque layer 3 and Bérigoule as noted above. In addition to the technological similarities, an actual continuity of OIS 6 Levallois *sensu lato* industries at Grotte Vauffrey is observable through the Last Interglacial (5e: bed IV) to later parts of OIS 5 (beds III, IIs; Rigaud 1988).

The technological shift from Levallois-dominated to non-Levallois Quina technologies in later OIS 5 or at the beginning of OIS 4 is well-known (the 'Ferrassie-Quina Transition' of Mellars 1969), and has been identified at a number of sites including Abri Caminade, Combe Grenal, Abri Chadourne, Roc en Pail (see Mellars 1996). In addition, a 'Quinarisation' of technology can be observed in some assemblages, notably Roc de Marsal, in which there is a fall-off in Levallois frequency through the sequence (Le Tensorer 1978; Turq 1985) accompanied by a shift from a use of predominantly local raw material to exclusively local material (pers. obs.) in sedimentological and faunal contexts indicative of a late OIS 5 or (more probably) OIS 4 position, also as at Petit-Puymoyen. The dominance of systems of *débitage* organised along non-Levallois principles in OIS 4 is striking.

3.5.3 OIS 3

By contrast, the dominance of systems of *façonnage* in OIS 3 — whether centred upon the production of bifaces (the MTA) or (later, it would seem) leaf points — is striking. In a technological sense the archaeology of OIS 3 represents in many respects a return to principles of *façonnage* not employed since the later Acheulean assemblages of OIS 6 or early OIS 5. It should be noted, however, that some elaboration in systems of *débitage* is evident, notably in the restructuring of primary flaking in the Italian Pontinian around 55,000 BP (Kuhn 1995). This shift apparently accompanies changes in resource procurement and concomitant mobility strategies (Stiner 1994).

Overall, the lithic technology of OIS 6–4 relates to the overwhelming dominance of *débitage* as an organising principle, and is structured by a dichotomous use of Levallois and Quina technologies. Occasional use of the ubiquitous discoidal core technology (which, I suggest, may itself be viewed as variant of Quina technology) and even *façonnage* (the odd biface) are evident. Before OIS 5 chronological resolution is too coarse to examine whether the apparently cyclical fluctuation between the two technologies in OIS 5–4 can be pushed back further in time (I suggest that it can in a more general technological sense, *e.g.*, by technological eccentricities apparently confined to certain periods such as the twisted ovate bifaces of OIS 11/10; White 1998b). I predict that, as our chronological resolution improves, technological fluctuation beyond biface production will be identified back well into the Lower Palaeolithic, at least in rudimentary form such as that advocated by Wenban-Smith (1998). By contrast to the technology of OIS 6–4, OIS 3 lithic technology is organised on radically different principles, *i.e.*, clearly employed systems of *façonnage*, notably the MTA (*e.g.*, see Mellars 1996). One might include leaf-points here also. In the following section, an interpretation is forwarded which links this apparent patterning with hypothesised variations in Neanderthal mobility strategies.

3.6 Mobility and technology in Neanderthal societies: an interpretative model

As noted at the start of this paper, I am concerned with questions as to why it tends to be Levallois technology that does move around in the landscape, why, conversely, Quina technology tends to be *ad hoc* and local, why technological variability exists from at least OIS 6 onwards and, ultimately why the 'Middle' Palaeolithic 'arose' in the first place. I move now to an interpretative model accounting for these questions.

I suggest that, from OIS 6 (or perhaps 7 in northern Europe) hominins began for the first time to exapt[3] elements of technology (Levallois *sensu lato*) for purposeful incorporation into varying mobility strategies, geared around the principle of flexibility in mobility systems that were relatively wide-ranging due to dispersed resources in which, therefore, relatively unfamiliar environments were routinely encountered where raw material procurement could not be taken as a given. By contrast, such flexibility need not be important in situations of lowered mobility of OIS 4 (dictated by a hostile environment) and OIS 3 (permitted by a rich resource base) in which unfamiliarity with the environment was not a factor and the mainly *ad hoc* behaviour characteristic of the Lower Palaeolithic would suffice, or in which at best the curation of a single, inflexible tool (the biface) over short-durations would be reliable enough. Table 3.1 presents the contrasts that I suggest can be discerned between the technological and mobility strategies of OIS 5, 4 and 3, as determined by environment and resources.

How does this relate to the questions posed at the beginning of the paper? The model proposes that it is the conjunction of variable climate, faunal composition and distribution and environment that will determine range size and behaviour (Gamble & Steele 1999) and that this will, in turn, cause certain elements of the Middle Palaeolithic technological repertoire to be exapted for use in such milieux. Thus, limited mobility may either be *determined* by the harsh environments of OIS 4, or *permitted* by a relatively rich and nucleated faunal biomass, *e.g.*, with the rich 'Coygan' faunas of OIS 3 (Currant & Jacobi 1997). By contrast, the relatively closed environments that pertained for large parts of OIS 5 (Roebroeks *et al.* 1992) and concomitantly dispersed faunal biomass determined a wider ranging behaviour in which technological elements were exapted to provide the flexibility demanded by such environments, *e.g.*, sufficient inbuilt flexibility as to allow responses to any demand as it may arise in relatively unpredictable environments. I suggest that Levallois *sensu lato* strategies are eminently suited to this: they tend, through core preparation, to produce relatively large flakes/blades which are nevertheless relatively thin and therefore light: carrying around a small amount of such pieces permits great flexibility (they can be used unretouched, be retouched into any of the generalised tool forms employed by Neanderthals, pointed to rehaft as knives, *etc.*). This makes much more sense that carrying around one small biface or a couple of large, weighty Quina racloirs, the nature of whose support precludes further resharpening or modification.

The environments of OIS 4 were unequivocally hostile according to Greenland ice (GRIP, GISP2: Anklin *et al.* 1993; Taylor *et al.* 1993) and Antarctic ice (Vos-

Stage	Technical input	Raw material procurement, curation	Flexibility	Environment and resources	Mobility
5	*débitage* systems (Levallois)	relatively wide; some products move up to 100km	very; exaption of Levallois products	relatively closed, thus resources are more dispersed	relatively mobile by necessity
4	*débitage* systems (non-Levallois)	local; products rarely move at all	low; flexibility pertains only in direct response to raw material	open, harsh; restricted fauna	relatively low and restricted
3	*façonnage* systems	local; products move on within restricted areas (< 30km)	inflexible; system geared towards short-term curation of a tool (biface)	relative open; rich ('Coygan') fauna	rich fauna allows limited mobility

Table 3.1: Technological-mobility strategies from Oxygen Isotope Stages 5–3.

tok) cores (Petit *et al.* 1999), and supported a fauna that was much reduced in diversity — *e.g.*, as represented by the dominance of reindeer in the Quina Mousterian levels at Combe Grenal. Such hostile environments may have artificially restricted range sizes to the extent in which the relatively large ranging patterns of OIS 5 became impossible. In such a context — still of high mobility but within very small, circumscribed territories — the flexibility of Levallois strategies became redundant and technology was instead focused upon the most parsimonious means of reducing the raw materials that were available in the immediate landscape: thus a process of 'Quinarisation' may have begun as a response to a gradual lowering of range mobility distances. This can be seen, for example, in the Quina Mousterian sequence at Combe Grenal (levels 26–17). Bordes (1972) was the first to notice an 'evolution' through these levels basically reflecting a gradual decline of the Levallois technique (as reflected by the IL or Levallois Index). Turq (1992) elaborated upon Bordes' observations, noting a concomitant lowering of the faceting index and general laminarity of the assemblages. This 'Quinarisation' is presented in Figure 3.1 (left graph) and compared to the representation of immediately available raw material (grey flint) and flint available at a distance of less than 30km (chalcedonic flint) and *c.* 50km (Bergerac flint) (my own data). It can be seen that the two are correlated: there is a gradual trend of increasing use of locally available raw material at the expense of the more distant sources. In combining the two datasets it is apparent that as more local material is used and behaviour becomes arguably more 'situational' and localised, less use is made of the Levallois technique. While scrapers are still the predominantly manufactured tool forms, Quina strategies are increasingly employed to produce the flakes upon which to make them.

From *c.* 60,000 BP, the relatively rich 'Coygan' faunas of OIS 3 (Currant & Jacobi 1997) may have permitted relatively small range sizes in which, once again, lithic flexibility as a response to uncertainty was not a prerequisite as environments that were encountered on a regular basis were familiar to Neanderthal groups and may perhaps for the first time be referred to as *territories.* In contrast to the technological strategies of OIS 4, however, Neanderthal strategies of OIS 3 were organised around principles of *façonnage*, as opposed to *débitage*, reflecting the re-emergence of the dominant use of a single, reliable tool (the biface, or later the leaf-point) in contexts of high mobility within small, circumscribed ranges in which, once again, uncertainty in resource procurement was not a crucial factor, although as Swainston (*pers. comm.*) has emphasised, the technology of leaf-point production may itself be related to technological flexibility.

Thus from at least OIS 7 in northern Europe and 6 in southern France, the distribution of faunal resources structured the organisation and extent of Neanderthal mobility in the landscape. No doubt this was so for earlier hominin groups; the difference from around this time is that hominins exapt technology in a meaningful way as part of their response to the landscape. To me, this is the reason for the origins of the Middle Palaeolithic: an important social element (Gamble 1996, 1998a) is not inconsistent with such a model, and the potential use of Levallois technological gestures in holding together socially relatively mobile intimate and effective networks of OIS 5 is particularly intriguing. The concomitant uncertainties of access to lithic resources determined the primary flaking strategies which Neanderthals employed in order to interact with the landscape. Put another way: if Mousterian technological variability comes down to one thing, then it is the *frequency of exaption of Levallois technologies.*

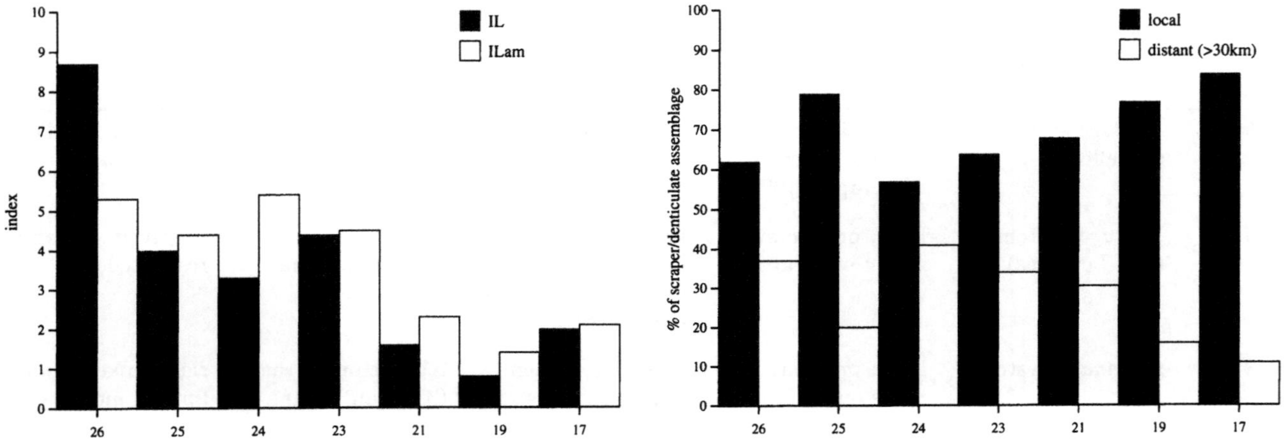

Figure 3.1: Levallois (IL) and laminarity (ILam) indices through the Quina sequence at Combe Grenal (left-hand graph; after Turq) and frequencies of local and 'exotic' raw material use for racloirs in the same sequence (right-hand graph — my data).

3.7 Conclusion: the Mousterian in action

The Middle Palaeolithic record was by all accounts created by variably mobile groups of Neanderthals operating in localised territorial regions that varied in extent as a direct function of climate and environment. Making stone tools and, in certain cases and to a small extent at least, carrying these around the landscape in both structured and unstructured ways, was a major part of their behavioural adaptation, one which is divorced from our own by deep time in the same way as their biological adaptation is by evolutionary divergence. It seems obvious that the Neanderthals had at their disposal a number of *clear* and *deliberate* approaches to raw materials structured as purposeful yet internally variable *chaînes opératoires*, which could be drawn upon if certain circumstances prevailed, such as a rarity of suitable materials or restricted access to these. Whilst not organising themselves around principles of resharpening of flakes and implements to varying degrees, their systematic approach to lithic technology, I suggest, centred on two pivotal concerns: first, primary flaking, by which means Neanderthals responded to raw material in the most appropriate ways and in a parsimonious manner if necessary, and, secondly, by the creation of certain generalised implement forms, by which Neanderthals carried out the tasks which demanded technological aids in the first place. This latter pivotal concern was unchanging.

Understanding the way in which these two structuring principles were played out in the landscape is the key to understanding Neanderthal behaviour. To do this we need to view the technological systems of Neanderthals in the context of the fluctuating landscape, especially in terms of the lithic forms which were carried around as (presumably) personal gear which form the point of meaningful contact between the environment as determinant of action on the one hand and the conscious Neanderthal response to situational pressures on the other. Models of lithic assemblage variability invoking immediate factors such as the intensity of occupation of enclosed sites are not enough by far: Neanderthals could and did move, as did the resources on which they were dependent. It follows that we must, if at all possible, ascertain the manner in which they moved, and how lithic technology was concomitantly shaped.

I suggest that a number of clear areas for future research emerge from the above:

1. The chronology of OIS 7 and 6. These Stages are crucial in the emergence of the chronological patterning of technology characteristic of the Middle Palaeolithic of the Upper Pleistocene.
2. Detailed characterisation of the nature of Levallois products, both of laminar and more 'classic' facies. Can they really be differentiated from non-Levallois products in the ways suggested above on the basis of intuition?
3. More detailed elucidation of the form in which lithic items move around the landscape, in conjunction, where possible, with identification of distance of transport and the faunal base.
4. Analysis of the topographic context of industries: how does topography affect mobility and therefore, potentially technology (*cf.* Barton 1990).
5. Comparison of 'Lower' and 'Middle' Palaeolithic assemblages in the light of mobility, fauna and the form of transported items, where possible, in order to clarify the process of origin of the Middle Palaeolithic. Further elaboration in the context of the mosaic appearance of Neanderthal biological traits and set against the increasing amplitude of climatic fluctuations from *c.* 600,000 BP.

Acknowledgements

I am very grateful to Nick Ashton and Mark White who had much to say on a previous draft and to Norah Mcloney and Mike Shott for inviting me to contribute to this volume. Part of this work arose in the context of a doctoral thesis presented to the University of Cambridge, which was supported by a British Academy major studentship. I am grateful for the constant encouragement and advice of Paul Mellars during this time, and to my thesis examiners Clive Gamble and John Gowlett for their comments on the relevant areas from which this paper arose.

Notes

1. Discussing soft hammers, Roberts (1996a, 9) notes: 'The amount of wear on these hammers shows they were used for long periods, and therefore that they must have been kept and carefully looked after over this time. I find it hard to envisage ... these activities taking place without a form of communication that was more advanced than that of our primate relatives. Hunting and butchering large mammals in open environments would have, in hominids, *required a level of planning and co-operation that could only be serviced by speech*' (my emphasis). Later on in the paper Roberts refers to the hominid behaviour at Boxgrove as 'partly modern'. All of Roberts' points — that wear on soft hammers indicates long-term curation and careful storage (as opposed to heavy use in one or a few episodes), levels of planning, co-operation and speech, are debatable. 'Partly modern' is a meaningless concept: by Roberts' argument chimpanzees are 'partly modern' in that they cache tools at nut-cracking sites. We would not, though, expect to have a conversation with them about this.

2. I define assemblages as Middle Palaeolithic if they are dominated by flakes generally produced by discoidal, Levallois *sensu lato* and occasionally opposed platform technology and only rarely by alternate, alternating and random techniques producing multiplatform 'globular' cores (as Ashton, *pers. comm.* has noted, a broadening of core technology from the Lower Palaeolithic); flake tools dominated by scrapers, occasionally by denticulates, often associated with points and occasionally with 'Upper Palaeolithic' tool forms; with a very low importance or total absence of bifaces, which, if present tend to be small relative to Middle Pleistocene biface forms and usually of sub-triangular or triangular form.

3. I follow Gamble (1993, 7) in his biological definition of exaption as the ' ... diversification of new species through the effect of phenotypic and behavioural potential being coopted in colonizing situations'. I would extend this to the artefactual (*i.e.*, lithic) realm and to situations other than colonising in which less-familiar landscapes were being encountered.

References

ALDHOUSE-GREEN, S. 1995. 'Pontnewydd cave, Wales, a later Middle Pleistocene hominid and archaeological site: a review of the stratigraphy, dating, taphonomy and interpretation.' *In* Bermúdez de Castro *et al.* (1995), pp. 37–55.

ALDHOUSE-GREEN, S. 1996. 'Review of Kuhn, S., *Mousterian Lithic technology.*' *Annals of Human Biology* 23(4): 340–2.

AMELOOT-VAN DER HEIJDEN, N. 1994. 'L'ensemble lithique du niveau CA du gisement de Riencourt-Les-Bapaume (Pas-de-Calais).' *In* Révillion & Tuffreau (1994a), pp. 63–75.

ANKLIN, M. *et al.* 1993. 'Climate instability during the last interglacial period recorded in the GRIP ice core.' *Nature* 364(6434): 203–7.

ASHTON, N., F. HEALY & P. PETTITT (eds.) 1998. *Stone Age Archaeology: Essays in honour of John Wymer.* Lithic Studies Society Occasional Paper No. 6. Oxbow, Oxford.

ASHTON, N. M. 1992. 'The High Lodge flint industries.' *In* N. M. Ashton, J. Cook, S. G. Lewis & J. Rose (eds.), *High Lodge: Excavations by G. de G. Sieveking, 1962–68 and J. Cook, 1988*, pp. 124–63. British Museum Press, London.

ASHTON, N. M. & J. COOK 1986. 'Dating and correlating the French Mousterian.' *Nature* 324(13): 113.

ASHTON, N. M., J. MCNABB & S. PARFITT 1992. 'Choppers and the Clactonian: a reinvestigation.' *Proceedings of the Prehistoric Society* 58: 21–8.

BARTON, C. M. 1990. 'Stone tools and Palaeolithic settlement in the Iberian peninsula.' *Proceedings of the Prehistoric Society* 56: 15–32.

BERMÚDEZ DE CASTRO, J. M., J. L. ARSUAGA & E. CARBONELL (eds.) 1995. *Evolución Humana en Europa y los Yacimentos de la Sierra de Atapuercas.* Junta de Castilla y León, Valladolid.

BINFORD, L. R. 1973. 'Interassemblage variability: the Mousterian and the functional argument.' *In* C. Renfrew (ed.), *The Explanation of Culture Change: Models in Prehistory*, pp. 227–54. Duckworth, London.

BINFORD, L. R. 1989. 'Styles of style.' *Journal of Anthropological Archaeology* 8: 51–67.

BOËDA, E. 1988. 'Le concept Levallois et évaluation de son champ d'application.' *In* Otte (1988), pp. 13–26.

BOËDA, E. 1993a. *Le Concept Levallois: Variabilité des Méthodes.* Centre de Recherches Archéologiques Monograph 9. Centre National de la Recherche Scientifique, Paris.

BOËDA, E. 1993b. 'Le débitage discoïde et le débitage Levallois récurrente centripète.' *Bulletin de la Société Préhistorique Française* 90(6): 392–404.

BOËDA, E., J. M. GENESTE & L. MEIGNEN 1990. 'Identification de Chaînes Opératoires lithiques du Paléolithique Ancien et Moyen.' *Paléo* 2: 43–80.

BOESCH, C. & H. BOESCH 1984. 'Mental map in wild chimpanzees: an analysis of hammer transports for nut-cracking.' *Primates* 25: 160–70.

BORDES, F. 1950. 'Principes d'une méthode d'étude des techniques de débitage et de la typologie du Paléolithique Ancien et Moyen.' *L'Anthropologie* 54: 19–34.

BORDES, F. 1953. 'Essai de classification des industries 'mousteriennes'.' *Bulletin de la Société Préhistorique Française* 50: 457–66.

BORDES, F. 1968. *The Old Stone Age.* World University Library, London.

BORDES, F. 1972. *A Tale of Two Caves.* Harper and Row, London.

BORDES, F. 1977. 'Time and space limits of the Mousterian.' *In* R. Wright (ed.), *Stone Tools as Cultural Markers*, pp. 37–9. Humanities Press, New Jersey.

BORDES, F. 1980. 'Le débitage Levallois et ses variants.' *Bulletin de la Société Préhistorique Française* 77(2): 45–9.

BUISSON-CATIL, J. (ed.) 1994. *Le Paléolithique Moyen en Vaucluse: à la rencontre des chasseurs néandertaliens de Provence nord-occidentale.* Notices d'Archéologie Vauclusienne 3. Service d'Archéologique en Vaucluse, Avignon.

CAHEN, D. 1984. 'Paléolithique Inferieur et Moyen en Belgique.' *In* D. Cahen & P. Haesaerts (eds.), *Peuples Chasseurs de la Belgique Préhistorique dans leur Cadre Naturel*, pp. 133–56. Institut des Sciences Naturelles, Brussels.

CHAUCHAT, C. 1992. 'Réflexions sur les outillages du Paléolithiques Moyen.' *Paléo* 4: 61–7.

COLLCUTT, S. N. (ed.) 1986. *The Palaeolithic of Britain and its Nearest Neighbours: Recent Trends.* University of Sheffield, Sheffield.

COLLINA-GIRARD, J. & A. TURQ 1991. 'Le Paléolithique Moyen sur galets de la station Des Planes, Commune de Montayral (Lot-et-Garonne).' *Paléo* 3: 49–74.

COMBIER, J. 1967. *Le Paléolithique de l'Ardèche, dans son cadre paléoclimatique.* Delmas, Bordeaux.

CONARD, N. J. 1990. 'Laminar lithic assemblages from the Last Interglacial Complex in northwestern Europe.' *Journal of Anthropological Research* 46: 243–62.

COOK, J. 1986. 'A blade industry from Stoneham's Pit, Crayford.' *In* Collcutt (1986), pp. 16–19.

COOK, J. & R. JACOBI 1998. 'Discoidal core technology in the Palaeolithic at Oldbury, Kent.' *In* Ashton *et al.* (1998), pp. 124–36.

CURRANT, A. & R. JACOBI 1997. 'Vertebrate faunas of the British Late Pleistocene and the chronology of human settlement.' *Quaternary Newsletter* 82: 1–8.

DE LOECKER, D. 1992. 'Site K: A Middle Palaeolithic site at Maastricht-Belvédère (Limburg, The Netherlands).' *Archäologisches Korrespondenzblatt* 22: 449–60.

DELAGNES, A. 1990. 'Analyse technologique de la méthode de débitage de l'Abri Suard (La Chaise-de-Vouthon, Charente).' *Paléo* 2: 81–8.

DELAGNES, A. 1992. *L'Organisation de la Production Lithique au Paléolithique Moyen: Approche technologique a Partir de l'Etude des Industries de La Chaise-de-Vouthon (Charente).* Ph.D. thesis, University of Paris X.

DELAGNES, A. 1993. 'Un mode de production inedit au Paléolithique Moyen dans l'industrie du niveau 6e du Pucheuil (Seine-Maritime).' *Paléo* 5: 111–20.

DIBBLE, H. L. & O. BAR-YOSEF (eds.) 1995. *The Definition and Interpretation of Levallois Technology.* Monographs in World Archaeology 23. Prehistory Press, Madison.

FARIZY, C., J. DAVID & J. JAUBERT 1994. *Hommes et Bisons du Paléolithique Moyen à Mauran (Haute-Garonne).* Centre National de la Recherche Scientifique, Paris. *Gallia-Préhistoire* Supplement.

FÉBLOT-AUGUSTINS, J. 1993. 'Mobility strategies in the late Middle Palaeolithic of Central Europe and Western Europe: elements of stability and variability.' *Journal of Anthropological Archaeology* 12: 211–65.

FÉBLOT-AUGUSTINS, J. 1997. *La Circulation des Matières Premières au Paléolithique.* Etudes et Recherches Archéologiques de l'Université de Liège, Liège.

FOLEY, R. 1987. 'Hominid species and stone-tool assemblages: how are they related?' *Antiquity* 61: 380–92.

GAMBLE, C. 1993. *Timewalkers: The Prehistory of Global Colonisation.* Alan Sutton, Bath.

GAMBLE, C. 1995. 'Raw materials, technology and variability in Middle Pleistocene Europe.' *In* Bermúdez de Castro *et al.* (1995), pp. 387–402.

GAMBLE, C. 1996. 'Hominid behaviour in the Middle Pleistocene: an English perspective.' *In* C. Gamble & A. Lawson (eds.), *The English Palaeolithic Reviewed*, pp. 63–71. Trust for Wessex Archaeology, Salisbury.

GAMBLE, C. 1998a. 'Handaxes and palaeolithic individuals.' *In* Ashton *et al.* (1998), pp. 104–9.

GAMBLE, C. 1998b. 'Palaeolithic society and the release from proximity: a network approach to social relations.' *World Archaeology* 29(3): 426–49.

GAMBLE, C. & J. STEELE 1999. 'Hominid ranging patterns and dietary strategies.' *In* H. Ullrich (ed.), *Hominid Evolution. Lifestyles and Survival Strategies*, pp. 396–409. Edition Archaea, Gottingen.

GENESTE, J. M. 1985. *Analyse lithique d'industries Moustériennes du Périgord; une approche téchnologique de comportement des groupes humaines au Paléolithique Moyen.* Ph.D. thesis, Université de Bordeaux I.

GENESTE, J. M. 1989. 'Economie des resources lithiques dans le Moustérien du Sud-Ouest de la France.' *In* M. Otte (ed.), *L'Homme de Néanderthal, 6. La Subsistence*, Etudes et Recherches Archéologiques de l'Université de Liège, pp. 75–97. Université de Liège, Lièege.

GOODALL, J. 1986. *The Chimpanzees of Gombe: Patterns of Behavior.* Belknapp, Harvard.

GOUÉDO, J. M. 1994. 'Remontage d'un nucleus à lames du gisement micoquien de vinneuf (yonne).' *In* Révillion & Tuffreau (1994a), pp. 77–102.

GREEN, H. S. 1984. *Pontnewydd Cave: A Lower Palaeolithic Hominid Site in Wales. The First Report.* National Museum of Wales, Cardiff.

GREEN, H. S. 1988. 'Pontnewydd Cave: the selection of raw materials for artefact manufacture and the question of natural damage.' *In* R. J. MacRae & N. Moloney (eds.), *Non-Flint Stone Tools and the Palaeolithic Occupation of Britain*, pp. 223–42. British Archaeological Reports British Series 189, Oxford.

HUXTABLE, J. & M. J. AITKEN 1992. 'Thermoluminescence dating results for the Palaeolithic site Maastricht-Belvédère.' *Analecta Praehistorica Leidensia* 18: 41–4.

JAUBERT, J., M. LORBLANCHET, H. LAVILLE, R. SLOTT-MOLLER, A. TURQ & J. P. BRUGAL 1990. *Les Chasseurs d'Aurochs de la Borde: un site de Paléolithique Moyen (Livernon, Lot).* Documents d'Archéologie Française 27. Maison des Sciences de l'Homme, Paris.

KOZLOWSKI, J. K. 1990. *Feuilles de Pierre: Les Industries à Pointes Foliacées du Paléolithique Supérieur Européen.* Etudes et Recherches Archéologiques de l'Université de Liège. Université de Liège, Liège.

KUHN, S. L. 1991. ''Unpacking' reduction: lithic raw material economy in the Mousterian of west-central Italy.' *Journal of Anthropological Archaeology* 10: 76–106.

KUHN, S. L. 1992. 'On planning and curated technologies in the Middle Palaeolithic.' *Journal of Anthropological Research* 48(3): 185–214.

KUHN, S. L. 1993. 'Mousterian technology as adaptive response: a case study.' *In* G. Larsen-Peterkin, H. M. Bricker & P. A. Mellars (eds.), *Hunting and Animal Exploitation in the Later Palaeolithic and Mesolithic of Eurasia*, Archaeological Papers of the American Anthropological Association 4, pp. 25–31. American Anthropological Association, Arlington.

KUHN, S. L. 1995. *Mousterian Lithic Technology: An Ecological Perspective.* Princeton University Press, Princeton.

LE TENSORER, J. M. 1978. 'Le Moustérien type Quina et son évolution dans le sud de la France.' *Bulletin de la Société Préhistorique Française* 75(5): 141–9.

LOCHT, J. L. & C. SWINNEN 1994. 'Le débitage du gisement de Beauvaise (Oise): aspects de la *chaîne opératoire* au travers de quelques remontages.' *Paléo* 6: 89–104.

MASSON, B. & L. VALLIN 1993. 'Un atelier de débitage Levallois intact au sein des loess Weicheliens du nord de la France à Hermies (Pas-de-Calais).' *Bulletin de la Société Préhistorique Française* 90(4): 265–8.

MCGREW, W. C. 1992. *Chimpanzee Material Culture.* Cambridge University Press, Cambridge.

MEIGNEN, L. 1994. 'Paléolithique Moyen au Proche-Orient: le phénomène laminaire.' *In* Révillion & Tuffreau (1994a), pp. 125–59.

MELLARS, P. A. 1965. 'The sequence and development of Mousterian traditions in south-western France.' *Nature* 205: 626–7.

MELLARS, P. A. 1967. *The Mousterian succession in South-West France.* Ph.D. thesis, University of Cambridge.

MELLARS, P. A. 1969. 'The chronology of Mousterian industries in the Périgord region of Southwest France.' *Proceedings of the Prehistoric Society* 6: 134–71.

MELLARS, P. A. 1970. 'Some comments on the notion of 'functional variability' in stone-tool assemblages.' *World Archaeology* 2(1): 74–89.

MELLARS, P. A. 1986a. 'A new chronology for the French Mousterian period.' *Nature* 322: 410–11.

MELLARS, P. A. 1986b. 'Reply to Ashton and Cook.' *Nature* 324: 113–14.

MELLARS, P. A. 1988. 'The chronology of the South-West French Mousterian: a review of the current debate.' *In* Otte (1988), pp. 97–120.

MELLARS, P. A. 1996. *The Neanderthal Legacy: An Archaeological Perspective from Western Europe.* Princeton University Press, Princeton.

MONCEL, M. H. 1994. 'L'industrie lithique des trois niveaux supérieurs de l'Abri du maras (ardèche).' *In* Révillion & Tuffreau (1994a), pp. 117–23.

MONCEL, M. H. 1995. 'Biface et outil-biface du Paléolithique Moyen Ancien: réflexion à partir de sites d'Ardèche, Orgnac 3 et Payre.' *Paléo* 7: 157–69.

MONCEL, M. H. & J. COMBIER 1992. 'L'outillage sur éclat dans l'industrie lithique du site Pléistocéne Moyen d'Orgnac 3 (Ardèche, France).' *L'Anthropologie* 96(1): 5–48.

OTTE, M. (ed.) 1988. *L'Homme de Néanderthal, 4. La téchnique.* Etudes et Recherches Archéologiques de l'Université de Liège 31. Université de Liège, Liège.

OTTE, M. 1994. 'Rocourt (Liège, Belgique): industrie laminaire ancienne.' *In* Révillion & Tuffreau (1994a), pp. 179–86.

PETIT, J. R., J. JOUZEL, D. RAYNAUD, N. I. BARKOV, J. M. BARNOLA, I. BASILE, M. BENDER, J. CHAPPELLAZ, M. DAVIS, G. DELAYGUE, M. DELMOTTE, V. M. KOTYLAKOV, M. LEGRAND, V. Y. LIPENKOV, C. LORIUS, L. PÉPIN, C. RITZ, E. SALTZMAN & M. STIEVENARD 1999. 'Climate and atmospheric history of the past 420,000 years from the Vostok ice core, Antarctica.' *Nature* 399: 429–36.

PETTITT, P. B. 1999. *Tool Reduction Models, Primary Flaking, and Lithic Assemblage Variability in the Middle Palaeolithic of Southwest France.* Ph.D. thesis, University of Cambridge.

PETTITT, P. B. & B. A. SCHUMANN 1993. 'Falling into history: hominid conceptions of time at the Middle to Upper Palaeolithic transition.' *Archaeological Review from Cambridge* 12(2): 25–50.

PITTS, M. & M. ROBERTS 1997. *Fairweather Eden.* Century, London.

RÉVILLION, S. 1995. 'Technologie du débitage laminaire au Paléolithique Moyen en Europe sepentrionale: état de la question.' *Bulletin de la Société Préhistorique Française* 92(4): 425–41.

RÉVILLION, S. & D. CLIQUET 1994. 'Technologie du débitage laminaire du gisement Paléolithique Moyen de Saint-Germaine-des-Vaux / Port Racine (secteur 1) dans le contexte des industries du Paléolithique Moyen du Massif Armoricain.' *In* Révillion & Tuffreau (1994a), pp. 45–62.

RÉVILLION, S. & A. TUFFREAU (eds.) 1994a. *Les Industries Laminaires au Paléolithique Moyen.* Dossiers de Documentation Archéologiques 18. Centre National de la Recherche Scientifique, Paris.

RÉVILLION, S. & A. TUFFREAU 1994b. 'Valeur et signification du débitage laminaire du gisement Paléolithique Moyen de Seclin (Nord).' *In* Révillion & Tuffreau (1994a), pp. 19–43.

RIGAUD, J. P. 1988. *La Grotte Vaufrey: Paléoenvironnement, Chronologie, Activités Humaines.* Mémoires de la Société Préhistorique Française XIX. Société Préhistorique Française, Paris.

RIGAUD, J. P. & J. P. TEXIER 1981. 'A propos des particularités techniques et typologiques du gisement des Tares, commune de Sourzac (Dordogne).' *Bulletin de la Société Préhistorique Française* 78(4): 109–17.

ROBERTS, M. B. 1996a. 'And then came clothing and speech.' *British Archaeology* 19: 8–9.

ROBERTS, M. B. 1996b. ''Man the hunter' returns at Boxgrove.' *British Archaeology* 18: 8–9.

ROBERTS, M. B., S. A. PARFITT, M. I. POPE & F. F. WENBAN-SMITH 1997. 'Boxgrove, West Sussex: rescue excavations of a Lower Palaeolithic landsurface (Boxgrove Project B, 1989–91).' *Proceedings of the Prehistoric Society* 63: 303–58.

ROBINSON, P. 1986. 'An introduction to the Levallois industry at Baker's Hole (Kent), with a description of two flake cleavers.' *In* Collcutt (1986), pp. 20–22.

ROEBROEKS, W., N. CONARD & T. VAN KOLFSCHOTEN 1992. 'Dense forests, cold steppes, and the Palaeolithic settlement of Europe.' *Current Anthropology* 33(5): 551–86.

ROEBROEKS, W., J. KOLEN & E. RENSINK 1988. 'Planning depth, anticipation and the organization of Middle Palaeolithic technology: the 'archaic natives' meet Eve's descendants.' *Helinium* 28(1): 17–34.

ROLLAND, N. 1976. *The Antecedents and Emergence of the Middle Palaeolithic Industrial Complex in Western Europe.* Ph.D. thesis, University of Cambridge.

ROLLAND, N. 1988. 'Observations on some Middle Palaeolithic time-series in southern France.' *In* H. L. Dibble & A. Montet-White (eds.), *Upper Pleistocene Prehistory of Western Eurasia*, University Museum Monograph 54. University Symposium Series Vol. I, pp. 161–80. The University Museum, Philadelphia.

SCHLANGER, N. 1994. 'Mindful technology: unleashing the *Chaîne Opératoire* for an archaeology of mind.' *In* C. Renfrew & E. B. W. Zubrow (eds.), *The Ancient Mind: Elements of Cognitive Archaeology*, pp. 143–51. Cambridge University Press, Cambridge.

STINER, M. C. 1994. *Honor Among Thieves: A Zooarchaeological Study of Neanderthal Ecology.* Princeton University Press., Princeton.

STINER, M. C. & S. L. KUHN 1992. 'Subsistence, technology and adaptive variation in Middle Palaeolithic Italy.' *American Anthropologist* 94: 306–39.

TAYLOR, K. C., G. W. LAMOREY, G. A. DOYLE, R. B. ALLEY, P. M. GROOTES, P. A. MAYEWSKI, J. W. WHITE & L. K. BARLOW 1993. 'The 'flickering switch' of late Pleistocene climate change.' *Nature* 361: 432–6.

TUFFREAU, A. 1994. *Continental comparisons.* Unpublished paper presented at the international symposium on *The English Palaeolithic Reviewed* held at the Society of Antiquaries of London, 28th October 1994.

TUFFREAU, A. & J. SOMMÉ 1988. 'Le gisement Paléolithique Moyen de Biache Saint-Vaast (Pas-de-Calais): stratigraphie, environnement, études archéologiques.' *Mémoires de la Société Préhistorique Française* 21.

TURQ, A. 1978. 'Note préliminaire sur l'outillage en quartzite entre Dordogne et Lot.' *Bulletin de la Société Préhistorique Française* 75(5): 136–40.

TURQ, A. 1984. 'Observations sur le Paléolithique Moyen du Haute-Agenais.' *Bulletin de la Société Préhistorique Française* 81(2): 50–2.

TURQ, A. 1985. 'Le Moustérien de type Quina du Roc de Marsal (Dordogne).' *Bulletin de la Société Préhistorique Française* 82(2): 45–51.

TURQ, A. 1988. 'Le Moustérien de type Quina de Roc de Marsal à Campagne (Dordogne): contexte stratigraphique, analyse lithologique et technologique.' *Documents d'Archéologie Périgourdine* 3: 5–30.

TURQ, A. 1989. 'Approche technologique et économique du faciès Moustérien de type Quina: étude préliminaire.' *Bulletin de la Société Préhistorique Française* 86(8): 244–56.

TURQ, A. 1992. 'Raw material and technological studies of the Quina Mousterian in Périgord.' *In* H. L. Dibble & P. Mellars (eds.), *The Middle Paleolithic: Adaptations, Behaviour and Variability*, pp. 75–85. University Museum of Philadelphia Museum Monograph 78, Philadelphia.

VALLIN, L. 1992. 'Le gisement Moustérien d'Houppeville–Les Hautes Terres Sud (Seine-Maritime): Étude d'un assemblage lithique en milieu loessique.' *Revue Archéologique de l'Ouest* 9: 5–37.

VAN KOLFSCHOTEN, T. & W. ROEBROEKS (eds.) 1985. *Maastricht-Belvédère: stratigraphy, palaeoenvironment and archaeology of the Middle and Late Pleistocene deposits.* Analecta Praehistorica Leidensia 18. Leiden University Press, Leiden.

VAN PEER, P. 1992. *The Levallois reduction strategy.* Monographs in World Archaeology 13. Prehistory Press, Madison, Wisconsin.

VYNCKIER, P., P. M. VERMEERSCH & L. BEECKMANS 1986. 'A Mousterian site at Vollezele-Congoberg (Galmaarden, Belgium).' *Notae Praehistoricae* 6: 5–14.

WAECHTER, J. D. 1951. 'Excavations at Gorham's Cave, Gibraltar.' *Proceedings of the Prehistoric Society* 17: 83–92.

WAECHTER, J. D. 1964. 'The excavations at Gorham's Cave, Gibraltar, 1951–1954.' *Bulletin of the Institute of Archaeology of London* 4: 189–221.

WENBAN-SMITH, F. F. 1992. 'Early Palaeolithic cultural facies and the Levalloisian at Baker's Hole.' *Papers from the Institute of Archaeology* 3: 1–10.

WENBAN-SMITH, F. F. 1998. 'Clactonian and Acheulian industries in Britain: their chronology and significance reconsidered.' *In* Ashton *et al.* (1998), pp. 90–7.

WHITE, M. J. 1998a. 'On the significance of Acheulean biface variability in southern Britain.' *Proceedings of the Prehistoric Society* 64: 15–44.

WHITE, M. J. 1998b. 'Twisted ovate bifaces in the British Lower Palaeolithic: some observations and implications.' *In* Ashton *et al.* (1998), pp. 98–104.

WHITE, M. J. & P. B. PETTITT 1995. 'Technology of Early Palaeolithic western Europe: innovation, variability and a unified framework.' *Lithics* 16: 27–40.

WHITEN, A., J. GOODALL, W. C. MCGREW, T. NISHIDA, V. REYNOLDS, Y. SUGIYAMA, C. E. G. TUTIN, R. W. WRANGHAM & C. BOESCH 1999. 'Cultures in chimpanzees.' *Nature* 399: 682–5.

P. B. Pettitt
Keble College
University of Oxford
Oxford OX1 3PG
Great Britain
paul.pettitt@bristol.ac.uk

4

A Late Upper Palaeolithic Flint Workshop from Rogalinek near Poznań, West Poland

Piotr Dmochowski
Instytut Prahistorii, Poznań

Abstract

Research into Prehistoric stone age technology has benefitted from the use and analysis of refitting studies as well as experimental flint knapping. Such analyses can be applied not only to modern flint assemblages but also to old materials which have never been subjected to similar investigations. A careful policy of saving archaeological material is conducive to analysing or re-analysing materials from earlier excavations. Today, the knowledge and experience gained from many generations of prehistorians has enabled us to reveal some of the mysteries held within old collections. The present paper presents the results of a recent re-investigation of a flint collection found almost a century ago which at that time was of no particular interest.

4.1 Introduction and circumstances of discovery

The site of Rogalinek is located in midwestern Poland, 18km south of Poznań (Fig. 4.1). An old drawing of the site shows that it covered a small hill by the Warta River and is probably part of a Late Glacial Warta River terrace. Today, this area is covered with a forest of 40-year old pine trees. Two recent field surveys undertaken on the site produced no results.

According to notes in the Poznań Archaeological Museum, the materials of this study were discovered by the famous Polish archaeologist Józef Kostrzewski, around 1918 and which for the last 80 years have been stored in the museum warehouse. At the time of their original discovery they were dated to the Neolithic. Such a date was understandable at the beginning of the 20th century when a chronological framework for the Stone Age was being constructed. Today, however, their Neolithic age cannot be accepted.

4.2 Description of materials

The material consists of 823 fragments of raw material, cores and debitage products such as blades, flakes, numerous chips and very few retouched pieces. The number of burnt and patinated flint pieces is also very low. Due to the incomplete nature of the material and the method of initial recovery, I consider it useless to present here a detailed description of each category. Furthermore, the fragmentary nature of the material and method of core reduction employed makes classification difficult so that my own preliminary descriptions often had to be changed during the refitting exercise.

The raw material consists of erratic flint and chert nodules of different shapes and sizes ranging from 10–15cm. Traces of long distance glacial transportation are evident in the form of frost cracking in the material. Similar types of flint can still be obtained close to the site on the surface of Terrace VII. Nodules could have been collected from that area and brought to the site although the presence of raw material directly at the site is also likely.

4.3 Refitting

Although the surface collection was incomplete and refits usually difficult to accomplish, the refitting process proved unexpectedly successful. Three hundred and six pieces were refitted into 64 groups of from 2-34 pieces in the largest block and forming 37% of the total number of pieces. The results of the refitting and a consideration of raw material selection indicate that the assemblage consists of no fewer than 22 different blocks of flint or chert. There are 16 cores in the assemblage, some of which were so badly fragmented that they were only recognisable after refitting. The four most interesting blocks will be described in detail to illustrate the specific method of core reduction on the site.

4.3.1 Block 1

Thirty four fragments of flakes, blades and one truncation were fitted to this core (Fig. 4.2). The original block of dark blue flint was flat with a smooth, slightly

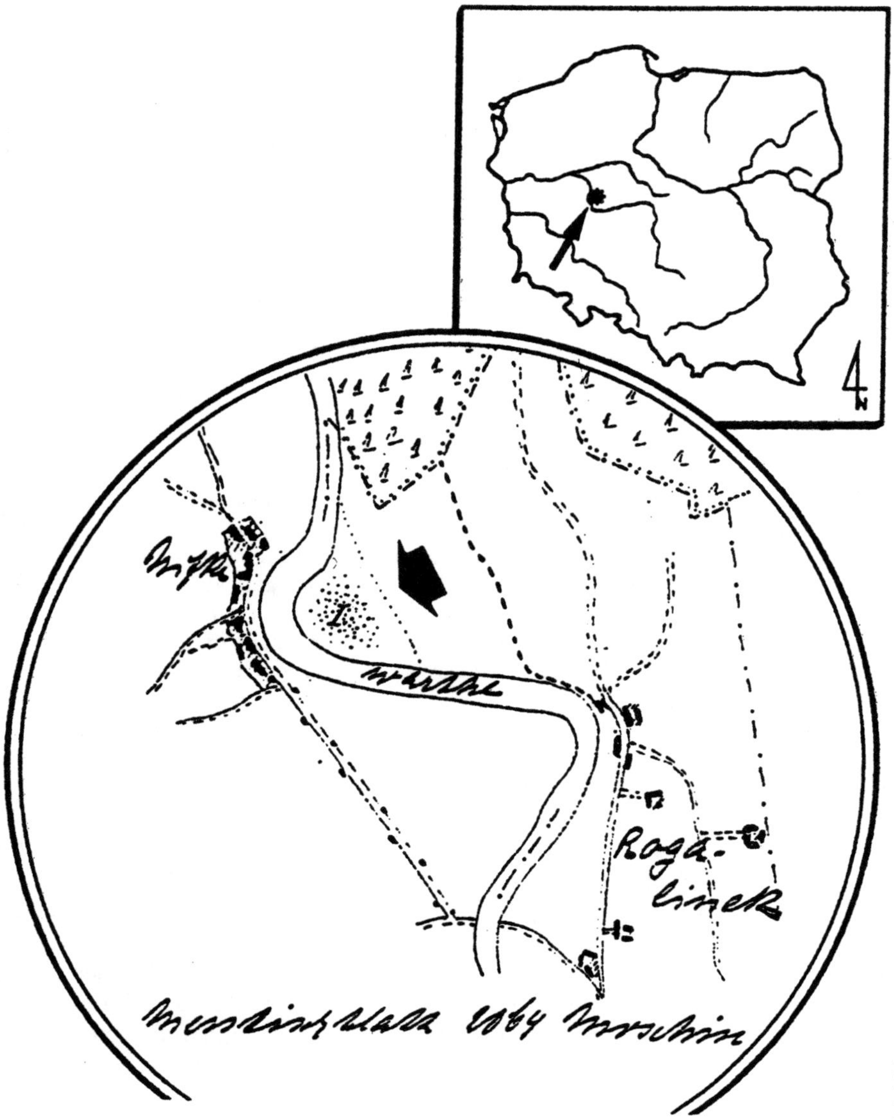

Figure 4.1: The location of Rogalinek I. Part of the original 1918 sketch.

weathered cortex. The first removal was made along a natural ridge of the nodule (natural crest) and one thick plunging blade was detached and preserved in a number of pieces.This blade has a well-prepared, faceted butt. After detaching about four to six further blades, the core was rotated 180° and one blade was detached from the opposite end after which it was rotated once again to its previous position. The subsequent removals revealed ancient break surfaces which reduced the size of nodule. The core was rotated once more and a new striking platform was set up on the flat back of the nodule. Flaking then continued with about ten blades and flakes being detached, during which time the platform was often faceted. A small bladelet from this series was truncated (Figs. 4.2c & 4.8a). During this stage several blades terminated in hinge fractures destroying the flaking face so that the core had to be rotated again. Two final blades were detached before the exhausted core was abandoned.

4.3.2 Block 2

The refitting of 7 blades and blade-like flakes formed a relatively large, single platform core with traces of intensive faceting of the striking platform (Fig. 4.3). The blanks are irregular and curve to one side. While some blades from the middle stages of the debitage process are missing, it is estimated that this core produced around 12–15 blanks.

4.3.3 Block 3

The core is made from a grey blue flint nodule (Fig. 4.4). Fourteen fragments have been refitted including a series of small chips resulting from platform faceting. Debitage began with the production of two large plunging blades. There are no signs of cresting, but evidence of platform faceting can be seen on the butts of both blades. The blades are broken in the middle which appears to be the reason for their discard. As in Block 2 the best blades are missing. The

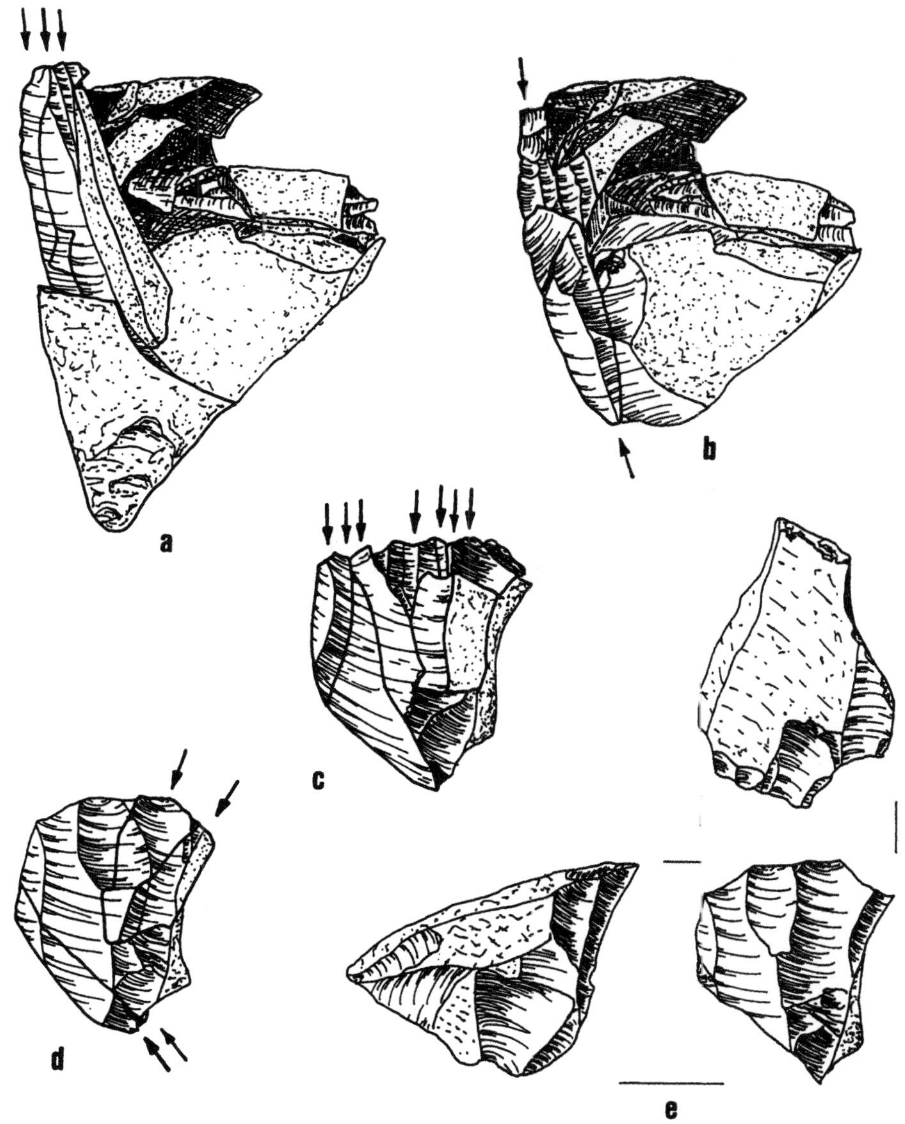

Figure 4.2: Block 1: sequence of four successive reduction stages: **a**: fully refitted core; **b**: preparation of the flaking face from the opposite end; **c**: core after the first rotation; **d**: final removals; **e**: exhausted core (scale 75%).

knapping of this core ceased after two flat flakes were detached from the middle of flaking face.

4.3.4 Block 4

Twenty-four fragments were refitted onto this block of dark blue flint which still has some areas of thick cortex (Fig. 4.5). The exhausted core looks like a typical flake core but after refitting it became clearly obvious, from the position of the flaking face on the longer surface of the nodule, that the knapper wanted to obtain blades. Butts of the initial blade-like flakes are faceted but debitage ended more quickly than previously planned because it was impossible to produce real blades. The core was reoriented to form a new platform on the old flaking face. The final series of flakes consisted of ten pieces including six which could be refitted onto the block. The butts of those flakes show no traces of preparation.

4.4 Method of knapping

Refitting shows changes in knapping strategies which occur between knapper and the nodule of raw material. Many similarities between individual blocks suggest that they are all the result of the same method of flint knapping. The most important factor in this method is that of recurrent faceting of the core platform; when the chance of obtaining a blade is present, faceting is always visible on the platform (Fig. 4.6a). The desirable angle between the striking platform and the flaking face is around 75° but when it reaches 90° hinge fractures often occur which cause the flaking face to shatter. As platforms were not rejuvenated, the only way to continue knapping was to change core

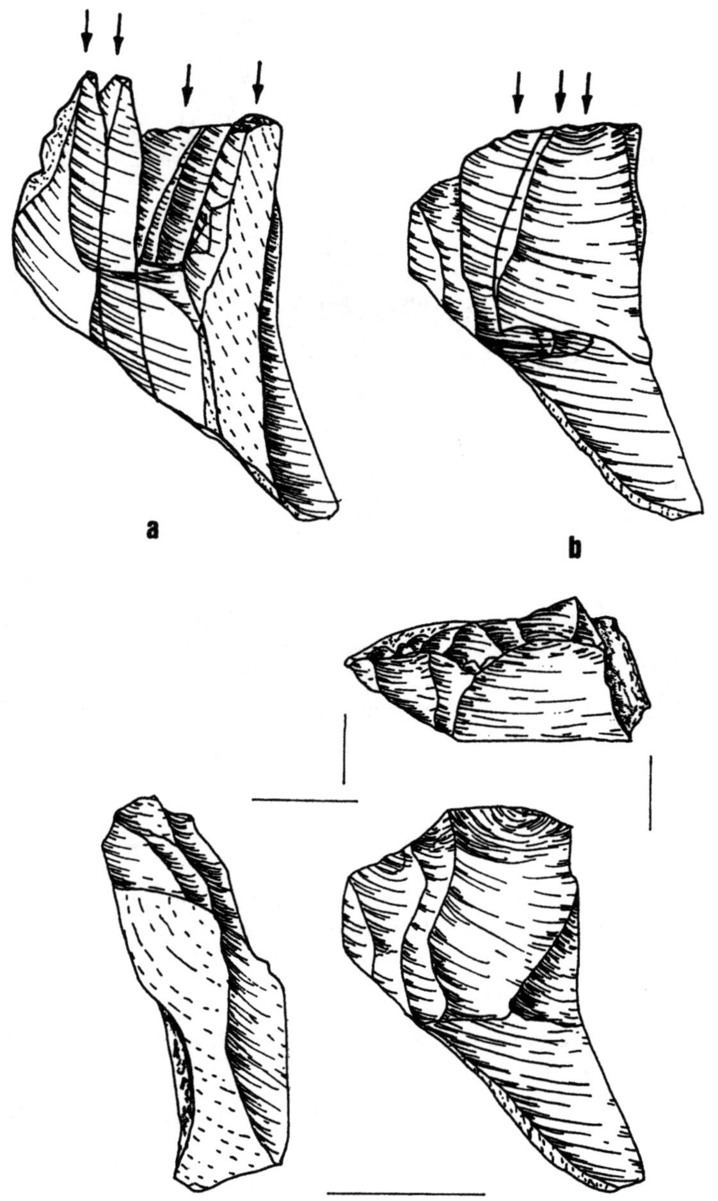

Figure 4.3: Block 2: **a**: refitted core; **b**: the final blade-like flakes; **c**: discarded core with faceted striking platform (scale 75%).

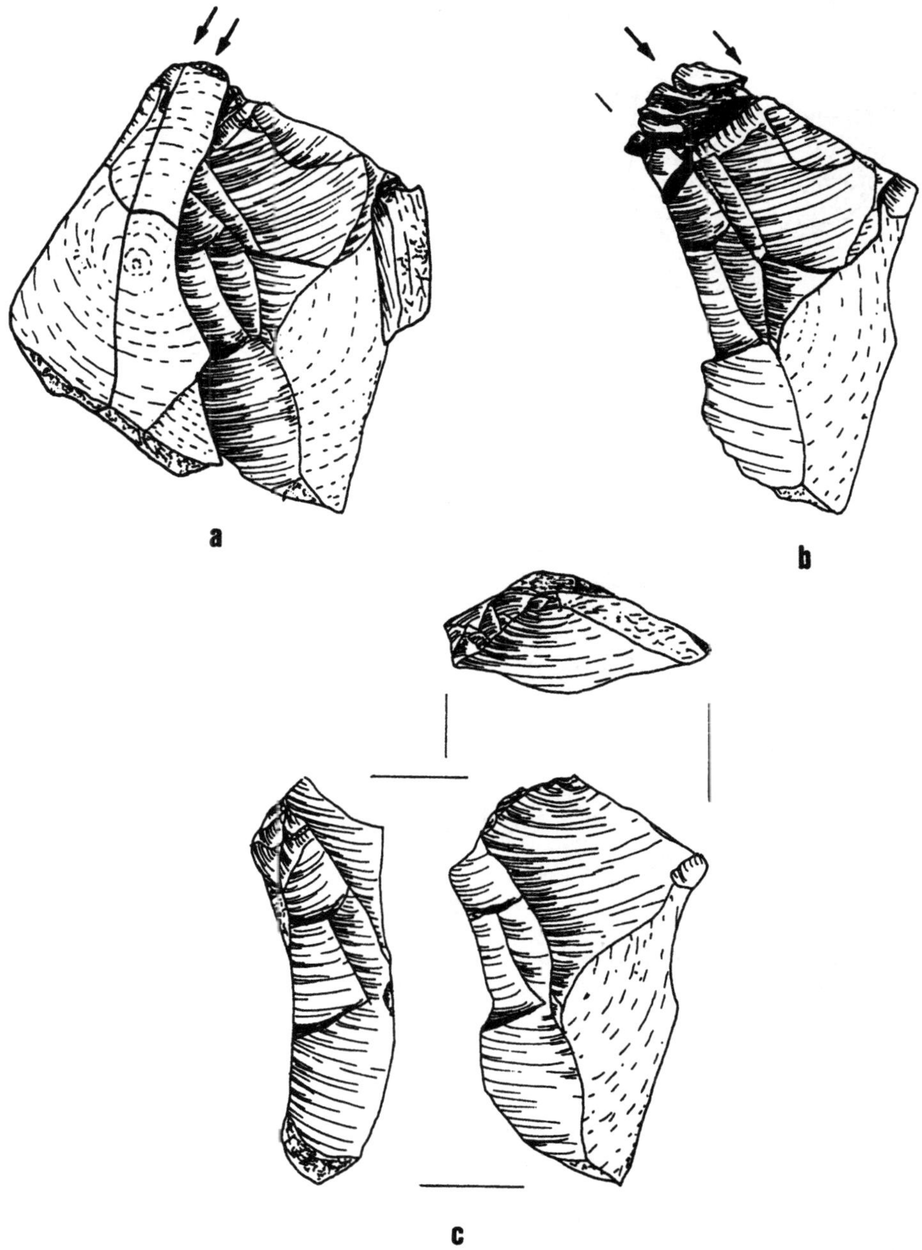

Figure 4.4: Block 3: **a**: refitted core with two plunging blades; **b**: the final flakes and series of refitted flakes from platform faceting; **c**: exhausted core with faceted platform and with hinge shattered flaking face (scale 75%).

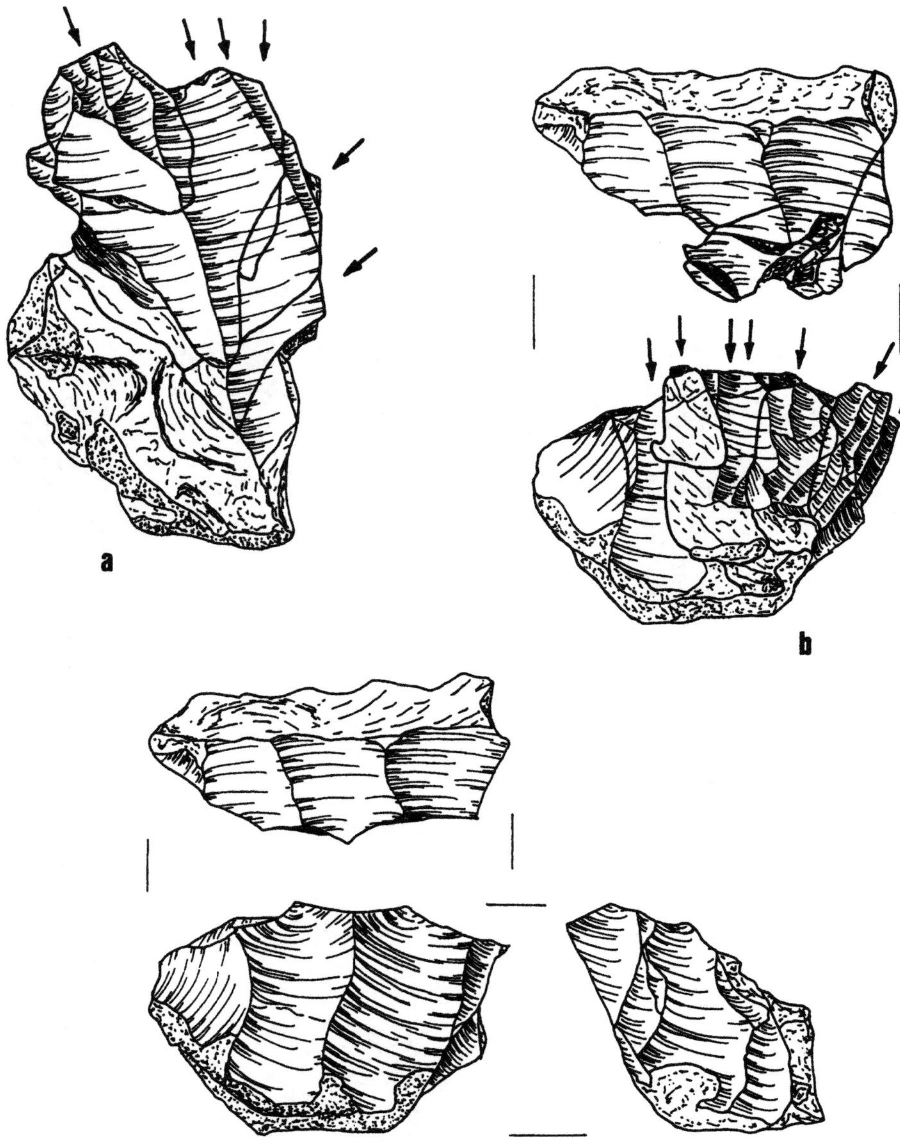

Figure 4.5: Block 4: **a**: fully refitted core showing elongated flaking face; **b**: core after rotation, the final stage of reduction; **c**: exhausted core (scale 75%).

orientation.

Attributes of remnant platforms and butts clearly indicate hard hammer technique. It would seem that the stone hammer technique was one of careful force; as a result, besides butts with visible impact points (Fig. 4.6f) or with pronounced bulbs (Fig. 4.6e), and butts with éraillure scars and collapsed platforms (Fig. 4.6h), there are numerous butts with a very delicate lip rather closer to the bulb (Fig. 4.6g) than occurs with soft hammer techniques. This has been achieved experimentally by a solid impact causing associated 'tear out' (Madsen 1996).

In conclusion, we can assume that this method of knapping is not particularly precise as it exploits the natural shape of nodules without recourse to any cresting. When core preparation is limited to striking platforms only, the resulting debitage is crude in appearance and non-standard. Therefore, the differentiation between flakes and real blades is very difficult to determine.

4.5 Site interpretation and dating

The site of Rogalinek I has features of a workshop with a lot of waste, very few proper debitage products, a few burnt elements and few tools. It could have been the site of a short stay aimed at the preparation of blanks or tools for future use. We can assume that some of the tools could have been replaced, some could have been destroyed during manufacture and some could have been lost or rejected as poor specimens. Today, due to the incomplete nature of

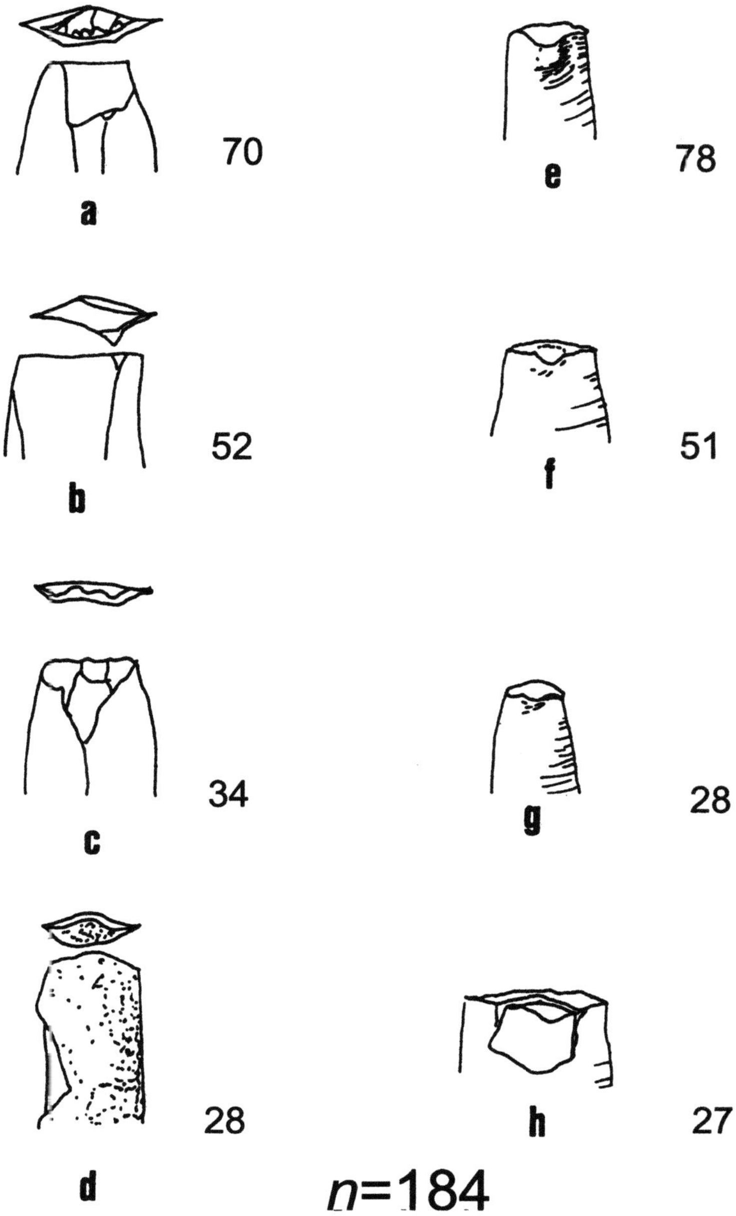

Figure 4.6: Frequency of butt attributes in the assemblage: **a**: faceted butt; **b**: plain unprepared butt; **c**: linear or shattered butt; **d**: unprepared old surface; **e**: well developed bulb; **f**: impact point; **g**: bulb and lateral lip; **h**: scaly bulbar fracture.

the material, it is impossible to study many detailed questions, among them style, quantity of processed and selected material and distribution. However, at the present stage of research, the potential of this site appears to be very promising.

The materials from Rogalinek I have not yet been dated securely but their homogeneous content and differences to many well known archaeological assemblages offer some ideas about their origin. We can say that they are neither Mesolithic nor Neolithic; they show no relation to Mesolithic technology, and although Neolithic cores and blades are often faceted they are made with a punch which allows higher precision and standardisation of debitage products.

The Late Upper Palaeolithic Hamburgian and Mazovian assemblages can be also excluded as their flint technology differs from that described above. Cresting, grinding, platform rejuvenation and the common occurrence of opposed platform cores are typical of these technocomplexes (Fiedorczuk 1992; Kobusiewicz 1973; Schild 1980).

A final attempt to identify the material from Rogalinek is to compare it with the Arched Backed Piece complex or Federmesser assemblages. The presence of one backed blade and one truncation in our material might indicate this as a possible line of investigation. The Federmesser sites in Poland have not been completely researched and they were published long ago. However, there are some features of debitage which are common to Federmesser lithic technology and the material described in this paper. Debitage products are non-standard and seem to be characterised by flakes

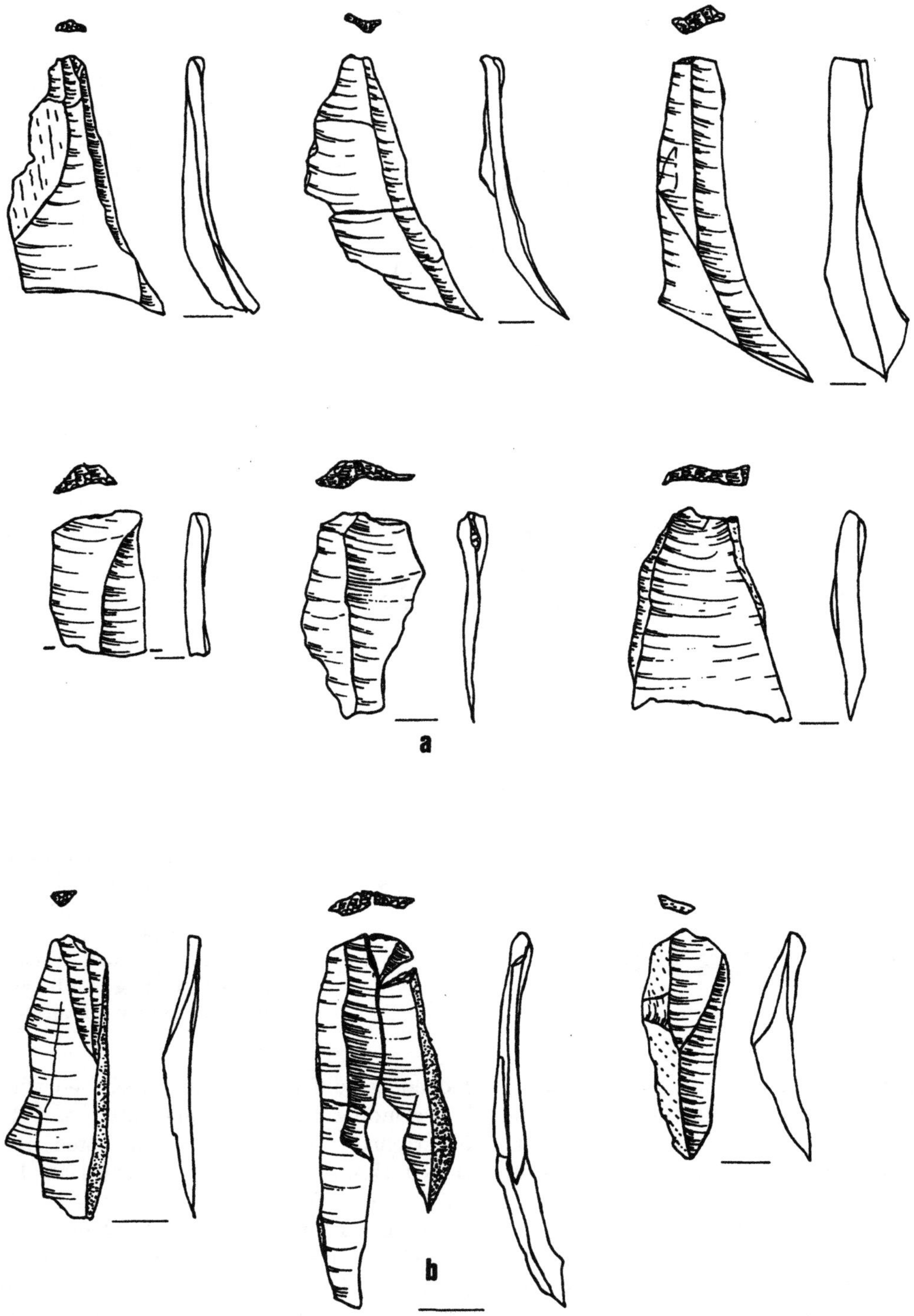

Figure 4.7: The most ideal blades as seen from refitting. **a**: blades and flakes from Block 2, all have faceted butts; **b**: blades from Block 1, not all are faceted (scale 75%).

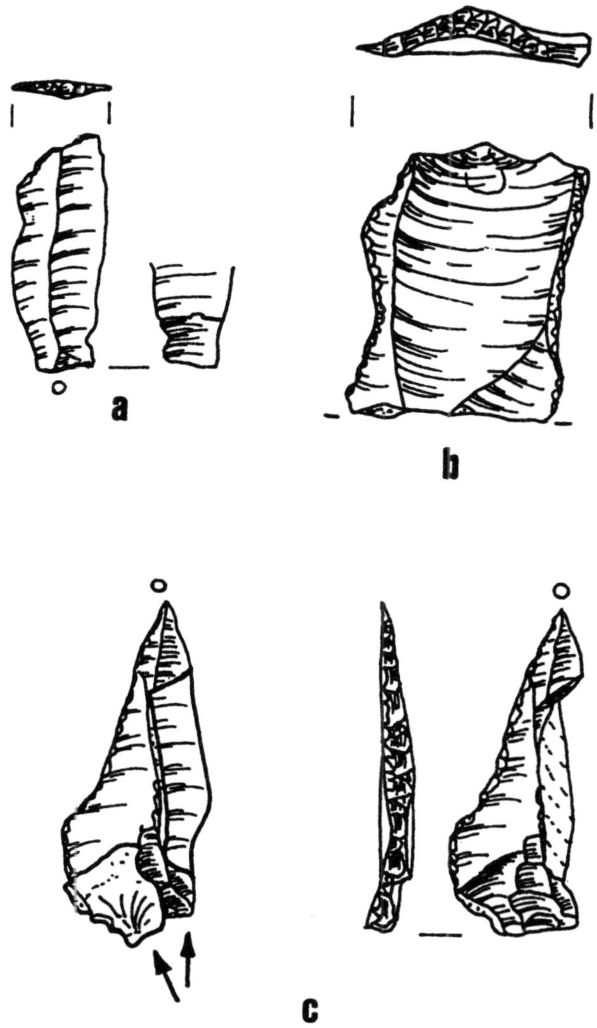

Figure 4.8: Retouched tools: **a**: truncation from Block 1; **b**: retouched flake with faceted butt; **c**: refitted backed blade (scale 1:1).

(Fiedorczuk 1992). Platform faceting is sometimes apparent but without refitting it is difficult to understand the process fully (Schild 1975).

Naturally, some differences can be observed which derive from other raw material, a different time period or non-homogeneity of assemblages. These aspects require additional studies based on original material, not on publications. It is interesting that more similarities to our materials can be found in North West Europe at such Federmesser sites as Egtwed (Fischer 1988), Ahrenshsft (Hartz 1987) or Hengistbury Head (Barton 1992). I am sure that this is not a coincidence; firstly it shows that the scope of Late Upper Palaeolithic technologies is very wide and secondly that in Poland we still have much work to do in the area of stone age technology in contrast to the many years spent on the development of typological studies.

4.6 Conclusion

An archival flint collection from Rogalinek which was rediscovered after almost a century, can be seen from a new vantage point as a result of refitting studies and the knowledge gained from experimental flint knapping. This material allows us to understand flint knapping methods of one of the oldest human settlements in the Polish Lowland. The refitting analysis indicates a method of blade production which can be described by the following features:

- hard hammer technique;
- recurrent facetting of the core platform;
- frequent change of the core orientation;
- total lack of other core preparation methods such as trimming, grinding or cresting.

The research on the site Rogalinek I has just started. It is quite probable that this will be the beginning of a new investigation into the Federmesser in Poland. It is hoped that the preliminary hypotheses presented in this paper are correct and will be supported by further studies resulting form excavations of the site.

Acknowledgements

Translated by Elżbieta Michałkiewicz

References

BARTON, R. N. E. 1992. *Hengistbury Head, Dorset. Vol. 2: The Late Upper Palaeolithic and Mesolithic Sites.* OUCA Monograph 34. Oxford University Committee for Archaeology, Oxford.

FIEDORCZUK, J. 1992. 'Późnopaleolityczne zespoły krzemienne ze stanowiska rydno IV 57 w świetle metody składanek.' *Przegląd Archeologiczny* 39: 13–65.

FISCHER, A. 1988. 'A Late Palaeolithic flint workshop at Egtved, East Jutland.' *Journal of Danish Archaeology* 7: 7–23.

HARTZ, S. 1987. 'Neue spätpaläolitische Fundplätze bei Ahrenshöft, Kreis Nordfriesland.' *Offa* 44: 5–52.

KOBUSIEWICZ, M. 1973. 'Problems concerning Hamburgian culture in Central Europe.' *Przegląd Archeologiczny* 21: 65–92.

MADSEN, B. 1996. 'Late Palaeolithic cultures of south Scandinavia: Tools, traditions and technology.' *In* L. Larsson (ed.), *The Earliest Settlement of Scandinavia*, pp. 61–73. Almquist and Wiksell International, Stockholm.

SCHILD, R. 1975. 'Późny paleolit.' *In* W. Chmielewskiego & W. Hensla (eds.), *Prahistoria Ziem Polskich. Tom I: Paleolit i Mezolit*, pp. 159–338. Zakład Narodowy Imienia Ossolińskich, Wrocław, Warszawa, Kraków, Gdańsk.

SCHILD, R. 1980. 'Introduction to dynamic technological analysis of chipped stone assemblages.' *In* R. Schild (ed.), *Unconventional Archaeology. New Approaches and Goals in Polish Archaeology*, pp. 57–85. Zakład Narodowy Imienia Ossolińskich, Wrocław, Warszawa, Kraków and Gdańsk.

Piotr Dmochowski
Instytut Prahistorii
UAM ul. Sw. Marcin 78
61-809, Poznań
Poland
dmocho@amu.edu.pl

5
The Çatalhöyük Obsidian Hoards: a Contextual Analysis of Technology

James Conolly

Institute of Archaeology, University College London

Abstract

During the course of excavations at Çatalhöyük a number of hoards of obsidian artefacts have been found buried under house floors. The primary purpose of this paper is to document and examine their technological and contextual characteristics. It is then argued that the hoards are both technologically distinct from non-hoarded lithics, and that the act of their hoarding has some ritual significance because of their preferential burial near hearths and in the southern part of houses.

5.1 Introduction

Unlike the deposition of objects into graves, hoarding can be a private and secular act (Rissman 1988, 209) and, as such, is often associated with the hiding of personal or family wealth. Yet Bradley notes a variety of types of hoarding behaviour in prehistory, such as votive deposits, secular hoards of craftsmen's material, industrial hoards, and personal storage hoards, to name a few (Bradley 1990, 10–12). Moreover, he points out that the particulars of the social context in which hoards are deposited vary (1990, 192), so there is little value in gross cross-cultural comparisons between, for example, Roman coin hoards in Britain with Palaeo-Indian fluted-point hoards from North America or Chalcolithic metal hoards from Israel.

This paper is concerned with the phenomenon of obsidian hoards at the Neolithic site of Çatalhöyük, Turkey (*c.* 8,500–7,500 bp). Comparative data for this period are relatively hard to find: few Near Eastern Neolithic habitation sites possess what could be termed a lithic 'hoard' of any sort. Evidence for ritual deposits is more forthcoming: on the southern edge of the Judean Desert, for example, the Neolithic cave of Nahal Hemar contained a variety of inorganic and organic objects, including finished stone tools such as 'Nahal Hemar knives' (Bar-Yosef & Alon 1988). The excavators argue that between 6,900–6,300 bp the cave was used as a storage facility for objects of ritual significance, of which these knives are considered to be a part (Bar-Yosef & Alon 1988, 27). Perhaps some comparison could be drawn between this and the Neolithic stone axe hoards from Western Europe (*e.g.*, Le Rouzic 1927; Nielsen 1977; Rech 1979) where the phenomenon has been interpreted as a means by which Neolithic groups could distinguish themselves through the destruction of wealth (Bradley 1990, 75). Elsewhere in the Near East, there is the example of 'cached' (but not buried) lithic artefacts at PPNB Beidha, where 115 pieces of flint were found in the remains of a wooden box on a house floor (Mortensen 1988). In this case, the objects appear to have been preserved in situ as a result of a house fire and a collapsed roof. Here the depositional context is less interesting than the fact that the box contained raw material as well a finished pieces, suggesting that the handling and working of lithic raw material was performed on site. In fact, Mortensen goes on to suggest that — together with the transformation of other raw materials such as animals and plants — lithic material was being 'domesticated' (Mortensen 1988, 201; *cf.* Hodder 1992, 31–43) through the process of raw material being brought into the settlement, and transformed into cultural objects. I shall return to this idea later in the paper.

5.2 The Çatalhöyük obsidian hoards

5.2.1 1961–1965

In the original Çatalhöyük site reports there are several accounts of knapped stone artefacts being found in caches or hoards in some of the buildings. For instance, in Bialor (1962, 74) there is a reference to five tanged arrowheads found together in a pit in Level VIII of the settlement, and in Level III 'a cache of eleven blades and one projectile point was found in the storeroom of House 2' (1962, 95) (Fig. 5.1). In House 1 of Level II 'a cache of twelve laurel-leaf blades and two projectile points was found on the lower floor

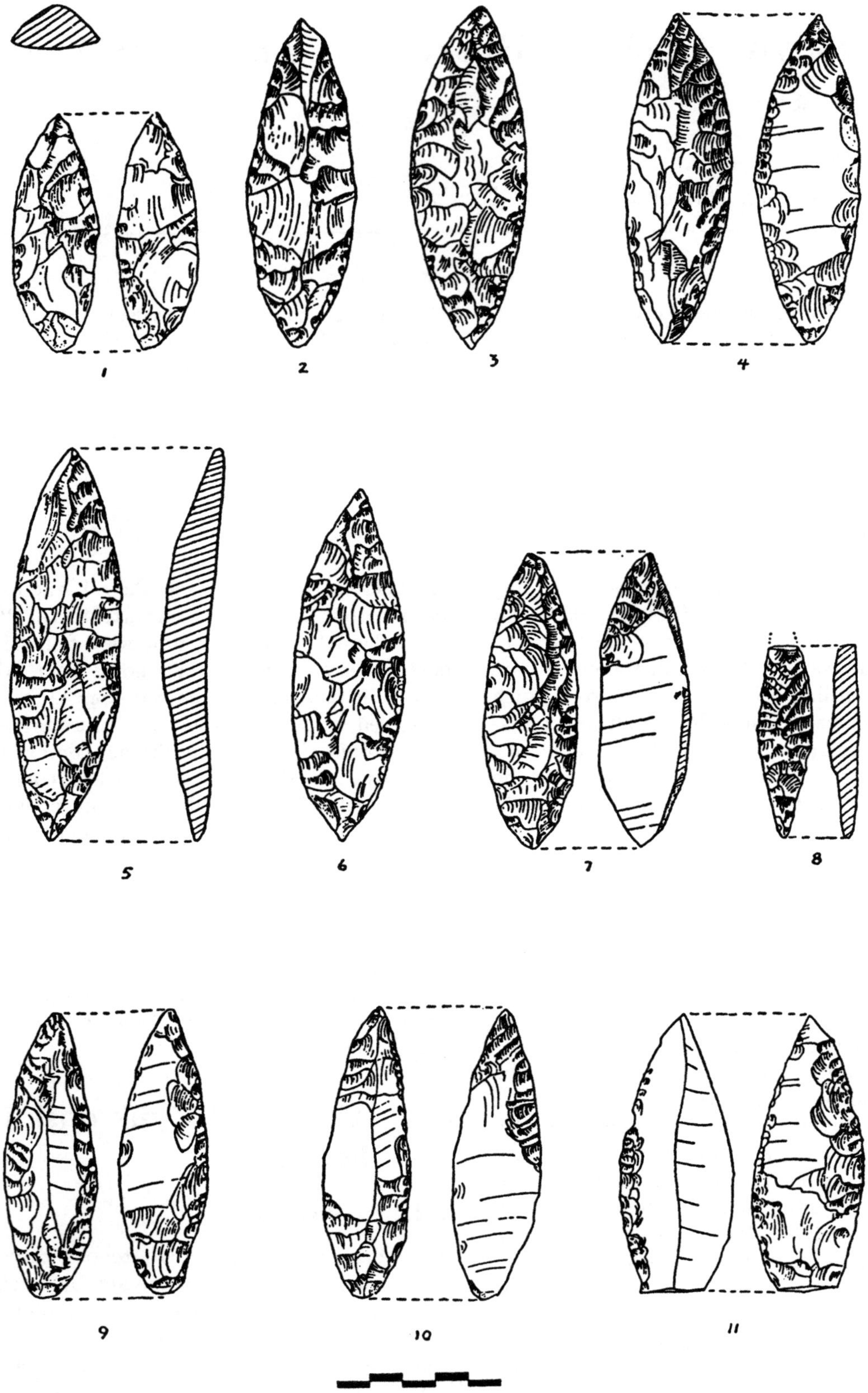

Figure 5.1: Hoard from House 2, Level III (from Bialor 1962, 94).

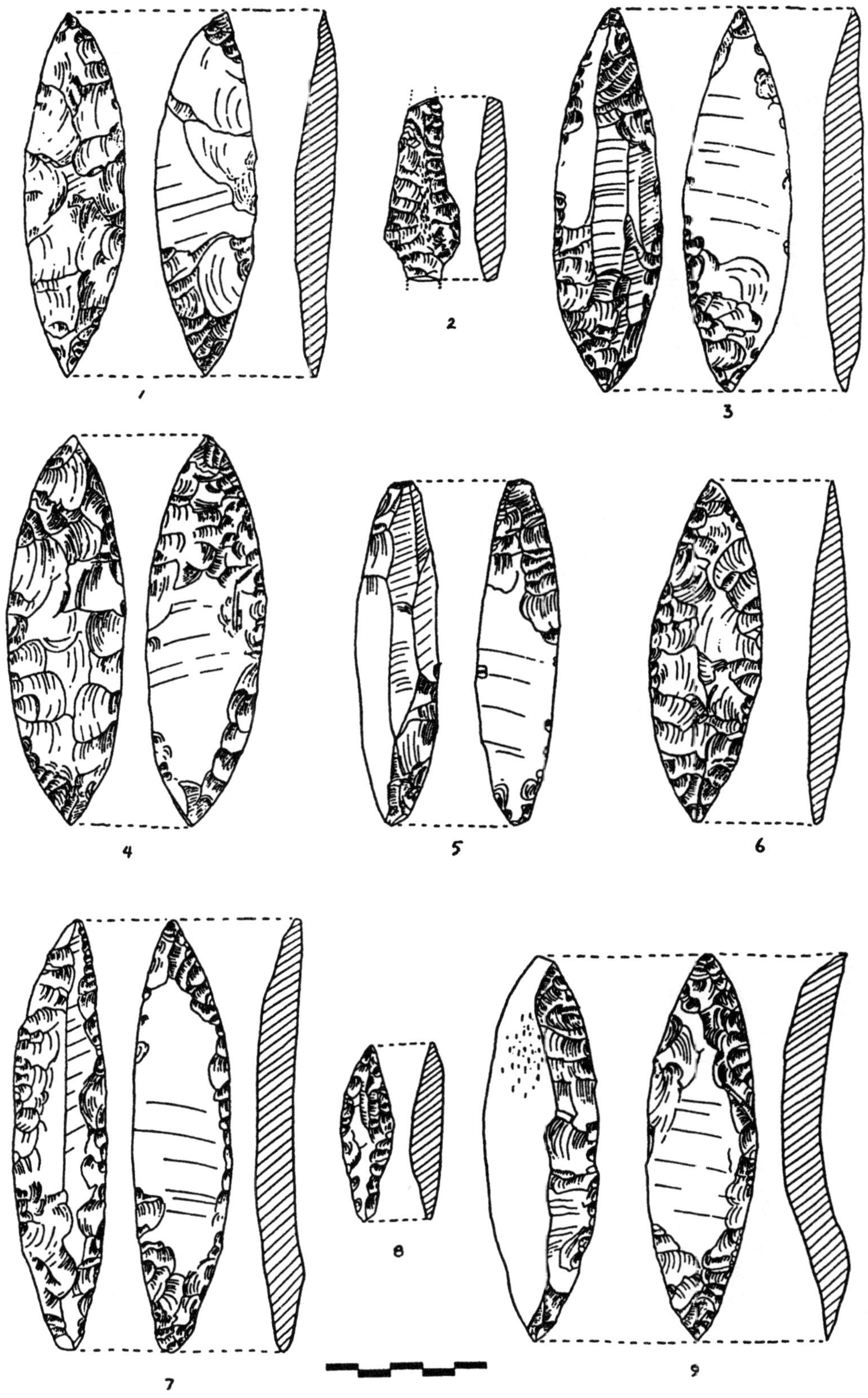

Figure 5.2: Hoard from House 1, Level II (from Bialor 1962, 101–2).

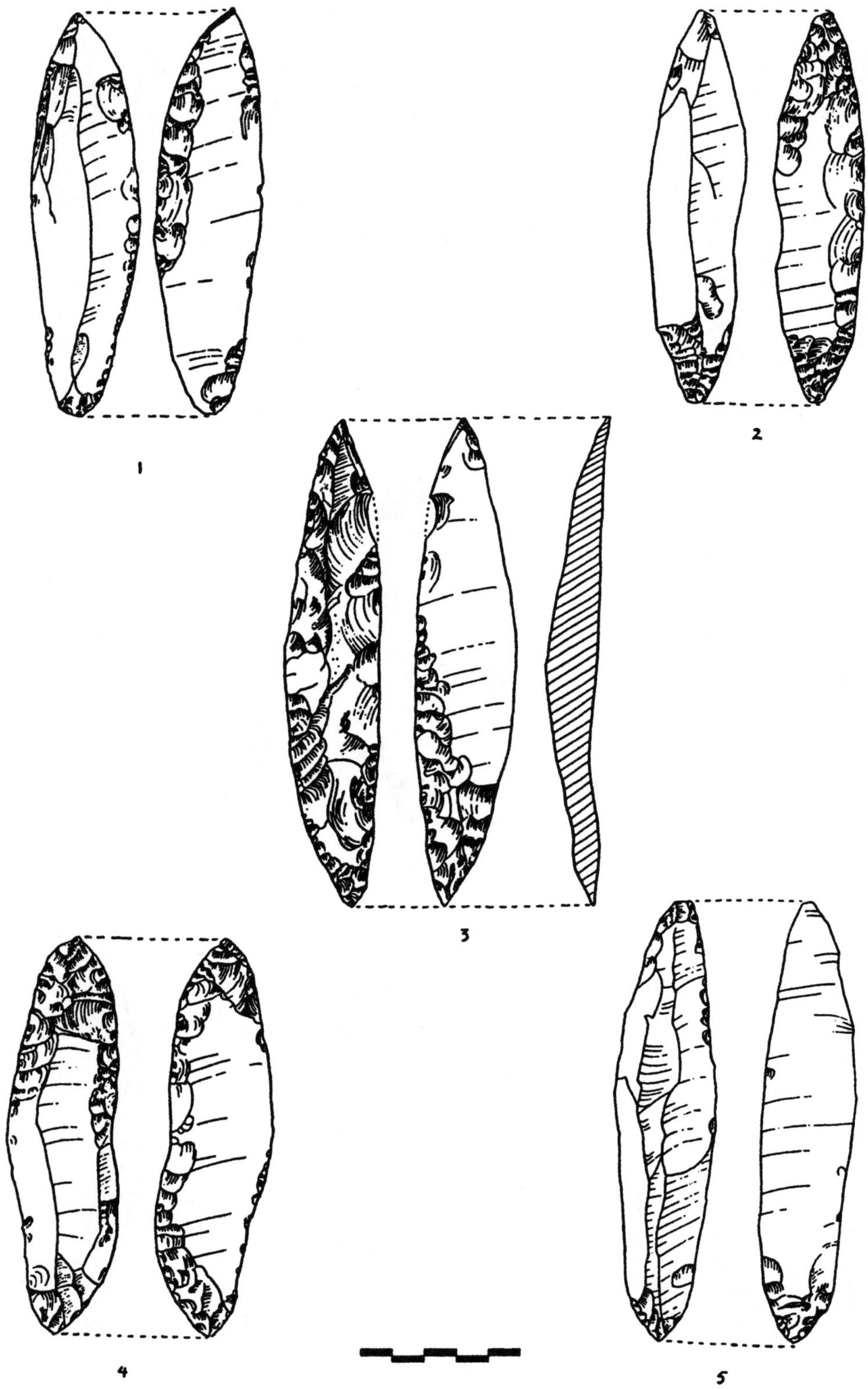

Figure 5.3: Hoard from House 1, Level II (continued from Fig. 5.2).

of the storeroom ... ' (1962, 98) (Figs. 5.2–5.3), and these 'formed a neat pile and may therefore have been contained by a string or cloth bag which has since disappeared and left no trace' (1962, 98–100). In the following year's report, Mellaart draws attention to the 'numerous hoards of weapons, buried below the floor, probably in bags' (Mellaart 1963, 101). Then, in his report of the 1963 season, he gives six separate examples of obsidian hoards found beneath the floors (Mellaart 1964, 103).[1] In most cases these hoards were said to consist primarily of some of the larger implements found at Çatalhöyük, such as large bifaces, blades and projectile points. There are other instances, however, where deposits of smaller tools were found. For example, a 'microlithic' hoard of from Level II 'near a hearth on the floor of a small room' consisted of forty-five or 50 pieces (Fig. 5.4). Another deposit of fifty small tools, or reused broken weapons found beneath a floor in Level VII, was interpreted as 'probably a votive deposit or an offering' (Mellaart 1964, 105–107) (Fig. 5.5).

It is difficult to gather further details about these hoards beyond what has been already published; some are stored and displayed at the Archaeological Museum in Konya, others are in the Museum of Anatolian Civilisations in Ankara, and all are difficult to examine closely. Contextual information is also particularly limited, and we are essentially dependent on the drawings and descriptions provided by Bialor and Mellaart. What is clear, however, is that not all buildings at Çatalhöyük have obsidian hoards. To this end it is also worth drawing attention to some other differences between Çatalhöyük buildings, such as the differentiation Mellaart consistently made between what he termed 'shrines' and 'houses'. The distinction is founded largely on the basis that some buildings have greater amounts of wall paintings and sculptures, as well as large numbers of elaborate artefacts such as obsidian mirrors, projectile points, and (or) figurines of 'goddesses'. Last (1998), however, has argued that the art was an integral part of the domestic routine within living spaces, rather than being constrained to special areas, such as shrines. This is consistent with current arguments maintaining that the separation of ritual and domestic in early prehistory is anachronistic, as ritual appears to have been a component of even the most mundane domestic activities. Nevertheless, the fact that some buildings at Çatalhöyük do have relatively more art and elaborate portable material culture suggests that there are ritual differences between contemporaneous buildings (see Conolly 1996; Hodder 1996b). The buildings that Mellaart identified as having hoards lie at the more complex range of the spectrum, and it may be that the presence of a hoard was one contributory reason for Mellaart defining a building as a shrine. I have argued elsewhere for a concentration of certain productive activities within certain buildings in the latter part of the settlement's history, particularly obsidian pressure blade production (Conolly 1999b), which may also account for some of the differences in building contents.

5.2.2 1993–1997

In 1993, the first season of renewed fieldwork some thirty years after Mellaart had finished his short tenure at Çatalhöyük, a cache of five very large flakes was found under what appeared to be a floor surface in the corner of a room (Fig. 5.6). Since then several more hoards of obsidian have been found, which presents an opportunity to investigate the context of the phenomenon in much greater detail than the 1960s material has permitted, simply because we now have particularly rich information about depositional contexts. It is these recently recovered deposits that I will concentrate on in my attempt to explain the phenomenon. As with many of the Mellaart examples, most of the new hoards were buried in the southern halves of buildings, and in areas that typically contained 'domestic' features like grinding areas, hearths and ovens.

The new hoards consist of the following (in no particular order):

Hoard 1461 The hoard from Building 1 (unit 1461, $n = 12$; 217g) is the most distinctive of the nine new hoards under discussion; it is a remarkable collection of blades struck from a large opposed-platform core. Six of the blades are from one end of the core, and six from the other; the latter group are what have been referred to as 'upsilon' blades (Ataman 1988), the former group possess extremely sharp pointed edges that require only modification of the proximal end to form one characteristic Çatalhöyük projectile type (Fig. 5.7). The hoard was located in the southern end of the main room underneath a grinding area (F27) with an upturned quern that appears to have been positioned deliberately as part of the abandonment process of the house (Matthews 1996) (Fig. 5.8).

Hoard 1836 A blade and flake hoard from Space 113 (unit 1834/1836, $n = 32$; 96g) shows a similar type of technology insofar as the blades have been produced from an opposed-platform core, although they are not of the same standard as those of the Building I example (not illustrated). Unlike Building I, however, there are many flakes ($n = 21$), some of which are very small. Many of the objects in this hoard have been retouched. As with the example from Building I, this hoard was similarly located in the south-eastern corner of the original building.

Hoard 2039 A large flake hoard was also found in Space 113 (units 2038/2039, $n = 50$;[2] 2309g) in which some of the pieces had been quite substantially modified, as with some of the hoards discovered in the 1960s (Figs. 5.9–5.13). In this

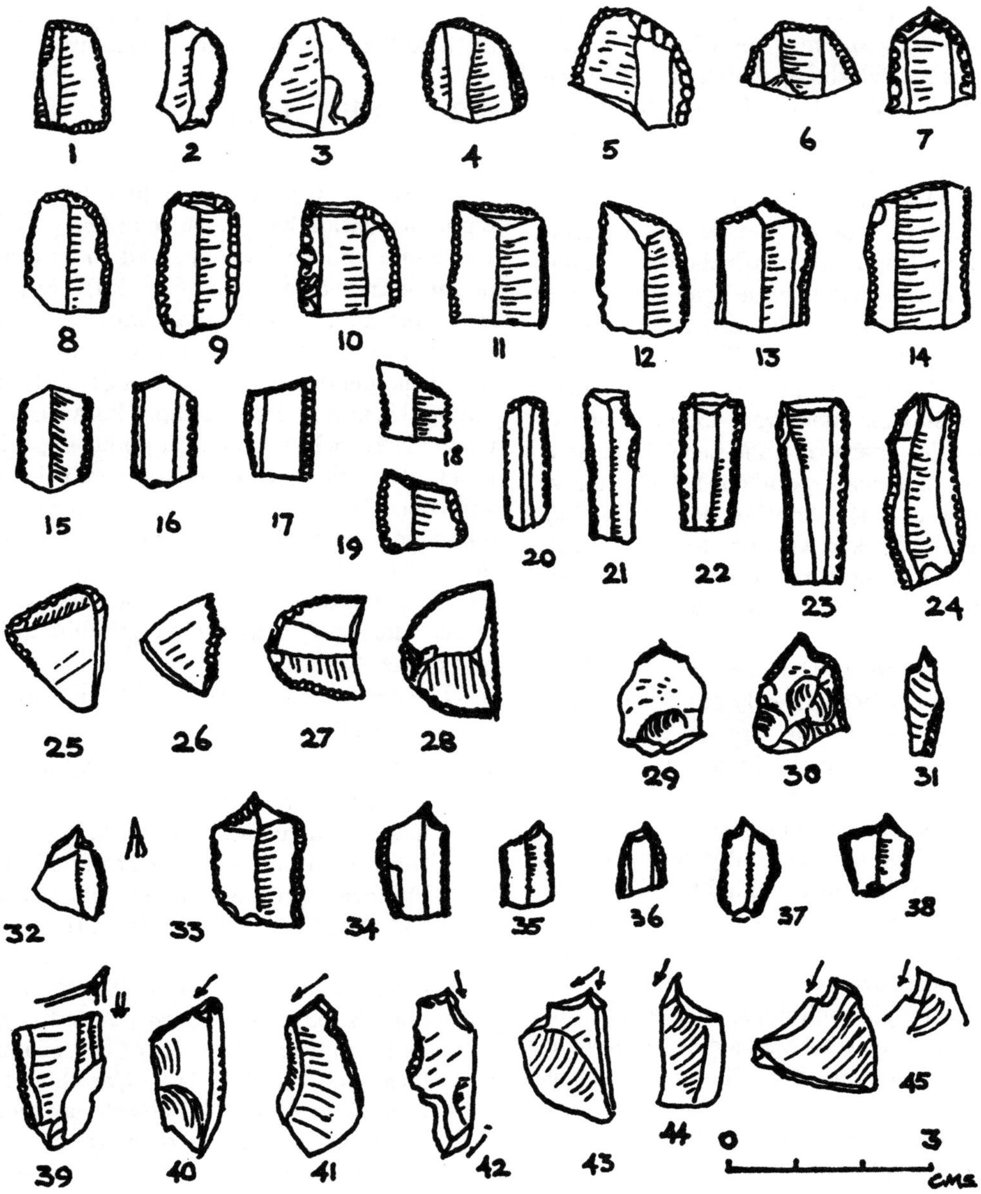

Figure 5.4: 'Microlithic' hoard from House 1, Level II (from Mellaart 1964, 106).

case, there were two depositional events; one in which 35 pieces were placed in the ground, and a second cutting into the first, in which a further 15 pieces were added (Farid 1997). One piece (Fig. 5.11, 24) has an interesting technological history; it would appear that an individual attempted to make a projectile-like object by bifacially retouching the distal end of a large blade, but encountered a problem with an inclusion and so abandoned the process. Certain pieces display ventral or bifacial retouch of a type that suggests they were used for scraping or chopping prior to being deposited. Some of the flakes, however, show little or no subsequent modification, and a few retain their cortical surfaces. This hoard was buried early in the sequence of Space 113, adjacent to an earlier still blocked doorway (Fig. 5.14).

Hoard 2812 A small hoard (unit 2812, $n = 7$; weight n/a) from an adjacent area (Space 109) was also found underneath a floor. It is a collection of large, bifacially retouched flakes very much of the same character as that from space 113 (Fig. 5.15). It is in the south-eastern part of the building, within a cut that was dug immediately in front of a large oven (Fig. 5.16).

Hoard 4280 In the south-eastern part of what was to become 'Building 63'[3] another hoard (unit 4280, $n = 18$; 1135g) was found associated with a hearth-area, in a deposit that cut through the ashy rake-out from an earlier cleaning event

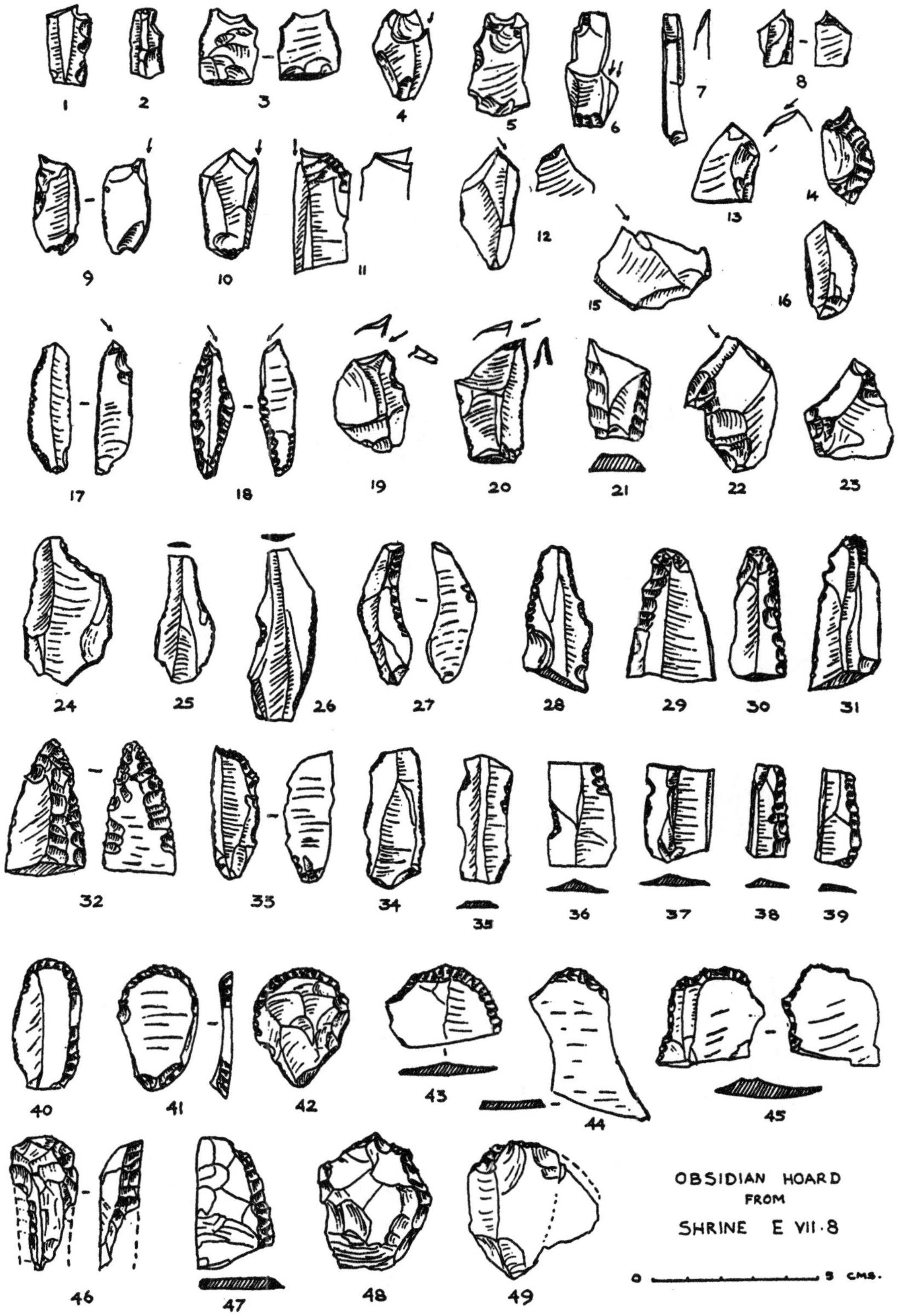

Figure 5.5: 'Obsidian hoard from Shrine E VII.8' (from Mellaart 1964, 110).

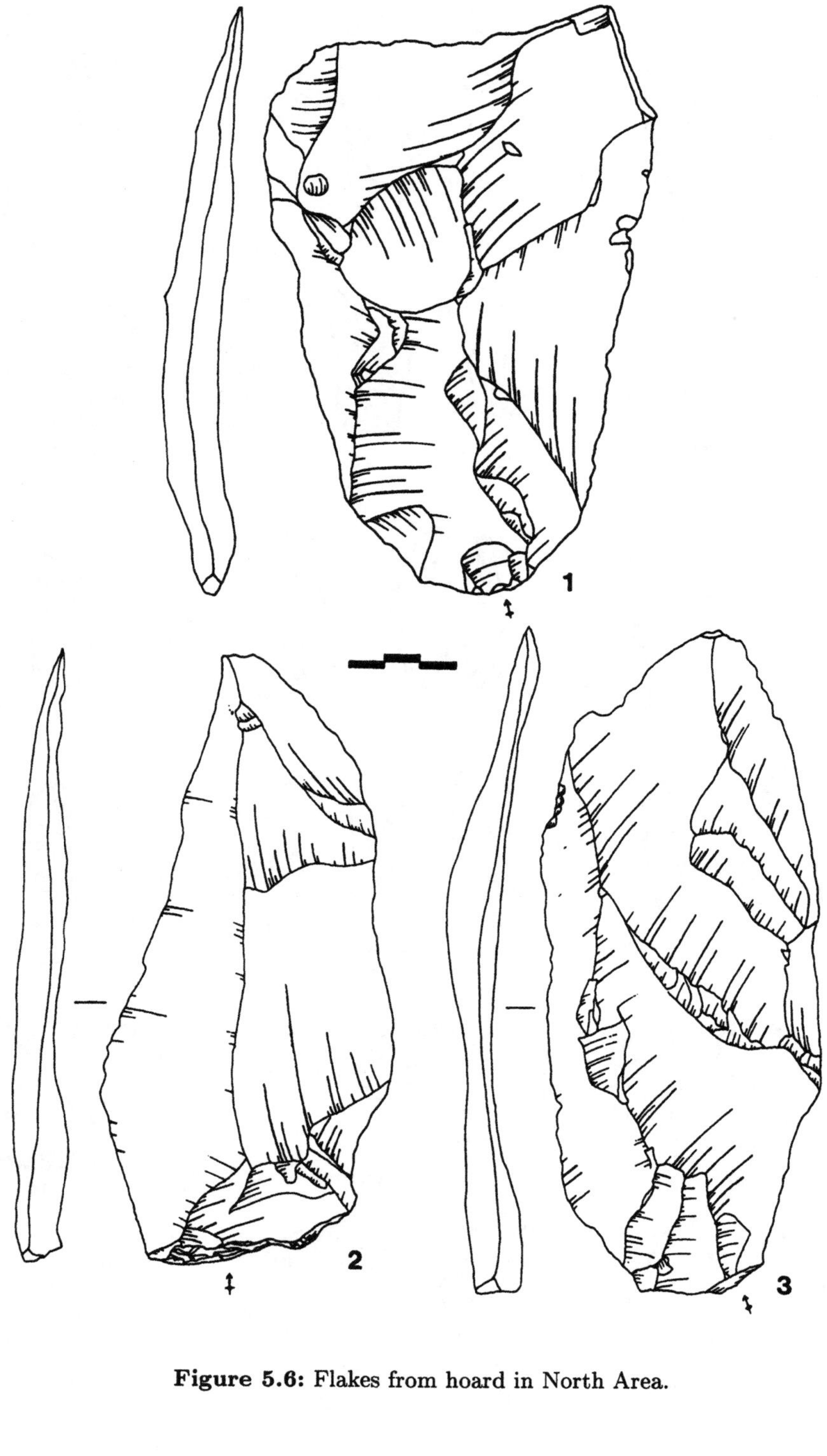

Figure 5.6: Flakes from hoard in North Area.

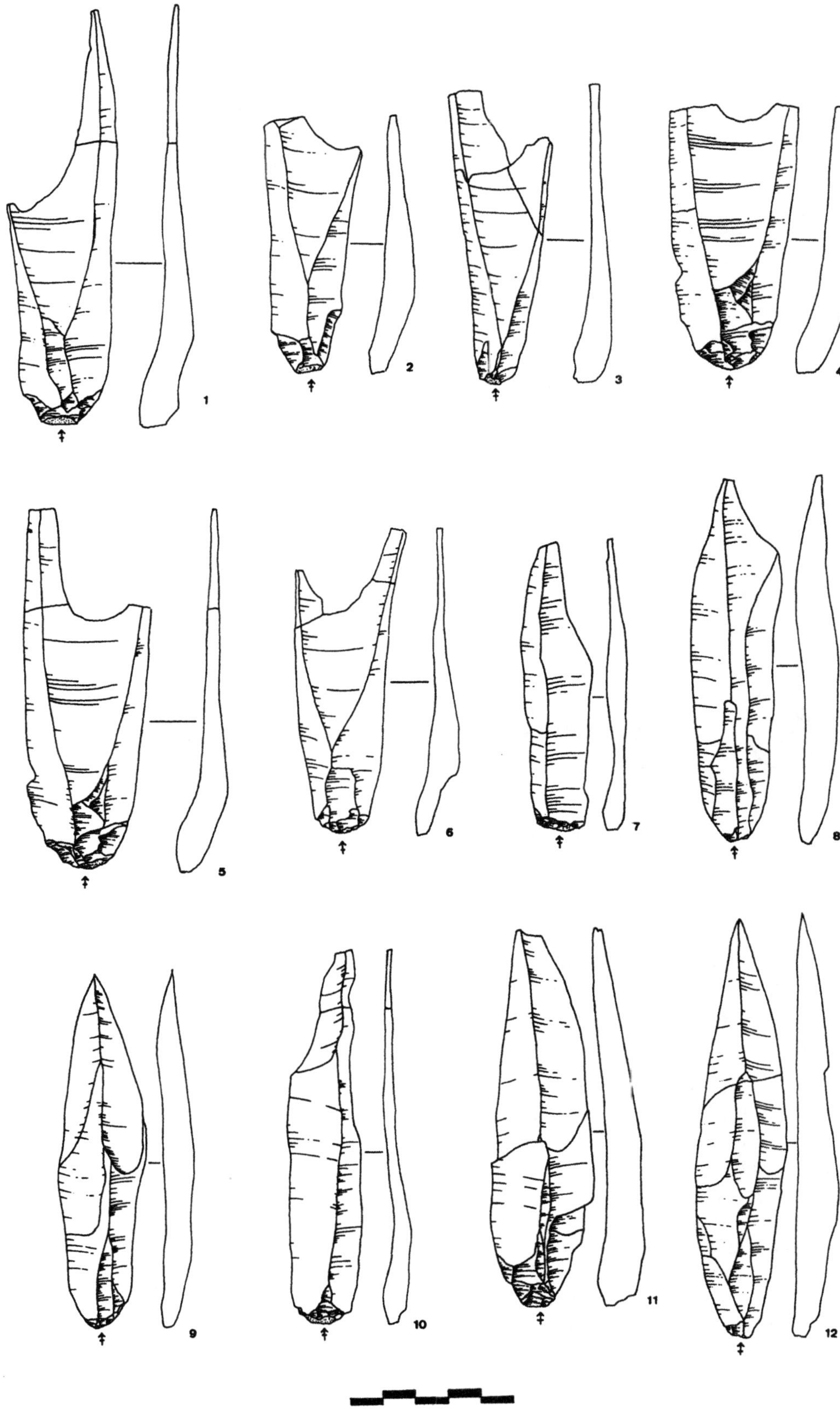

Figure 5.7: Hoard 1461.

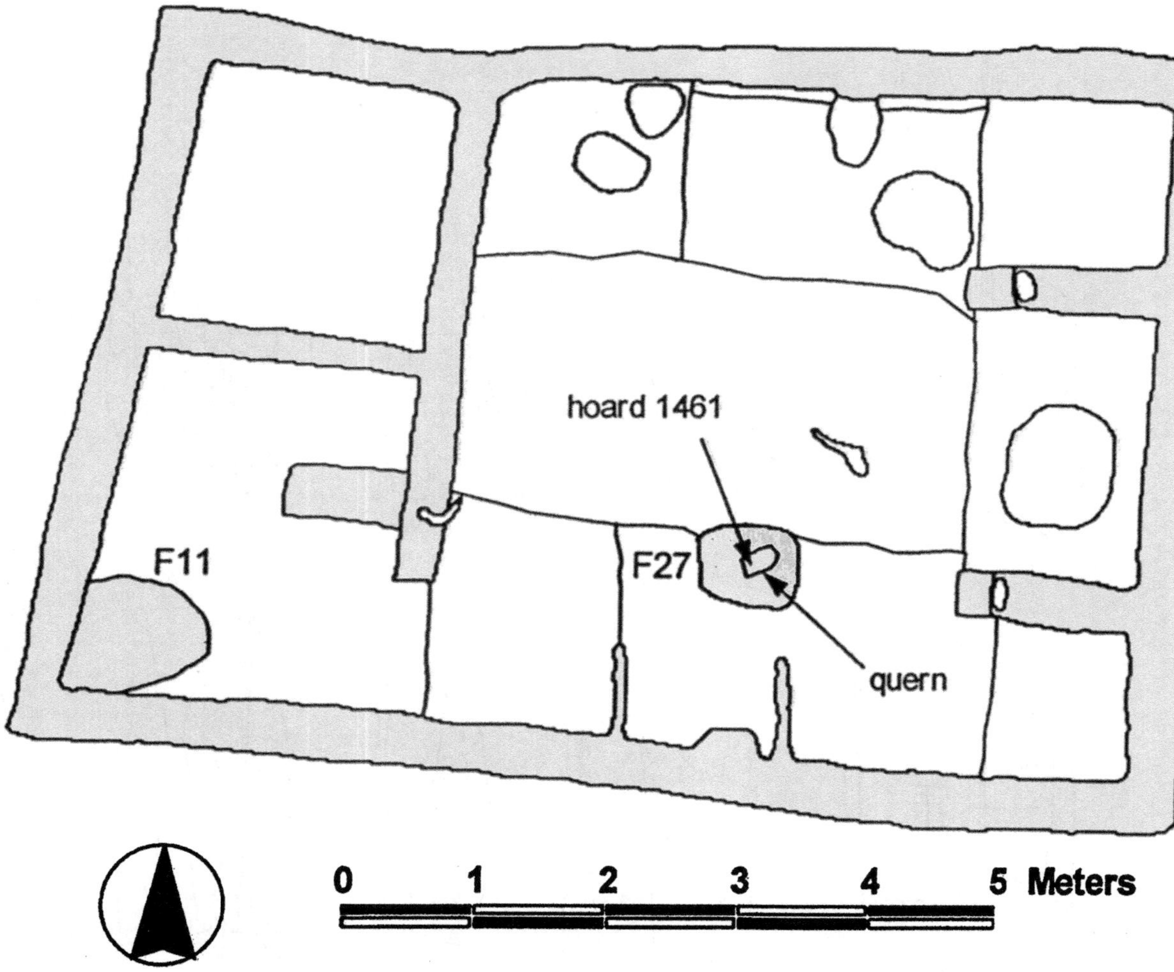

Figure 5.8: Location of Hoard 1461.

(Fig. 5.17). The amount of dorsal retouch on these pieces varies between 'covering' and 'marginal', and four have remnant cortical surfaces. At least six of the objects have retouch of a character that suggests they could have been used as tools.

Hoard 4276 Also in the south-eastern part of what was to become Building 6, and near to the unit 4280 hoard, three more pieces of obsidian (two pieces refit, and were probably broken post-depositionally) were found at the base of what may have been either an ashy rake-out from ovens or material deposited as part of a levelling event prior to the construction of the floors (unit 4276, $n = 2$; 189g) (Fig. 5.18). The excavators noted that this was possibly in some kind of container which had long since disintegrated. Both objects have been bifacially retouched and are of a shape that suggests they may be 'preforms' for large projectile points.

Hoard 4138 An unusual hoard containing flakes and blades was found beneath Space 117 of Building 2. The pieces were tightly packed and lying in a slight hollow, with no clear sign of a cut (unit 4138, $n = 16$; 170g) (Fig. 5.19) and within a series of packed and compacted ashy surfaces that may represent worn floor surfaces and oven rake-outs. The hoard was deposited within the early occupation of the building and is similar in nature to the one from Building I (unit 1460, above). This area of Building 2 was particularly complex, with nine separate ovens or 'fire installations' built during (at least) seven occupation phases. Among the sixteen pieces in the hoard, there are six definite blades (one of which is flint), all of which show retouch along their lateral edges. A complete (and still productive) bipolar blade core and one single-platform blade core fragment were also deposited in the hoard. One of the flakes (unretouched) has a small area of remnant cortical surface.

Hoard 4209 A larger hoard was discovered very near to the hoard from unit 4138 (above) within an irregular cut in the infilling and levelling prior to the construction of Building 2 (unit 4209, n=42;

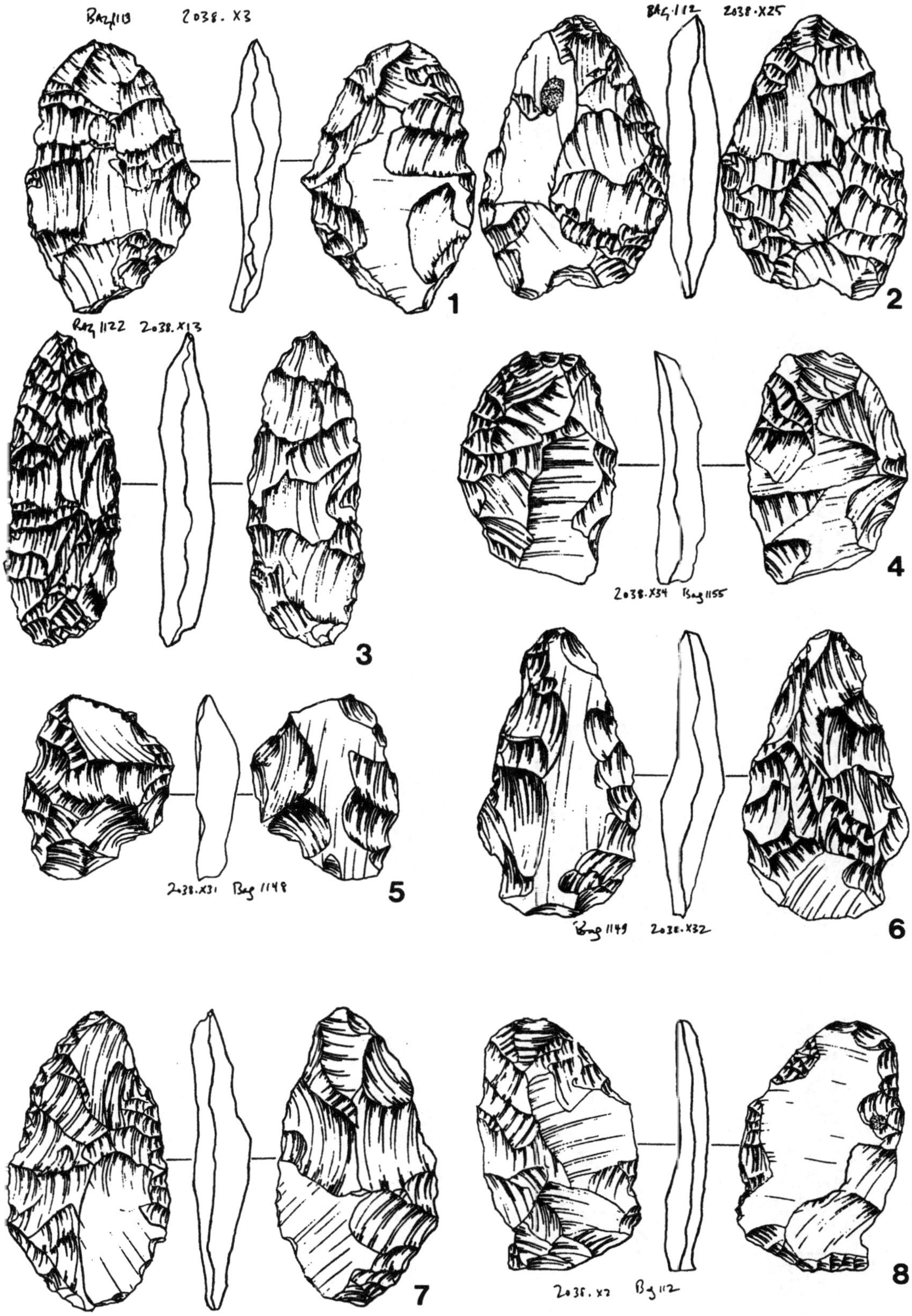

Figure 5.9: Hoard 2039.

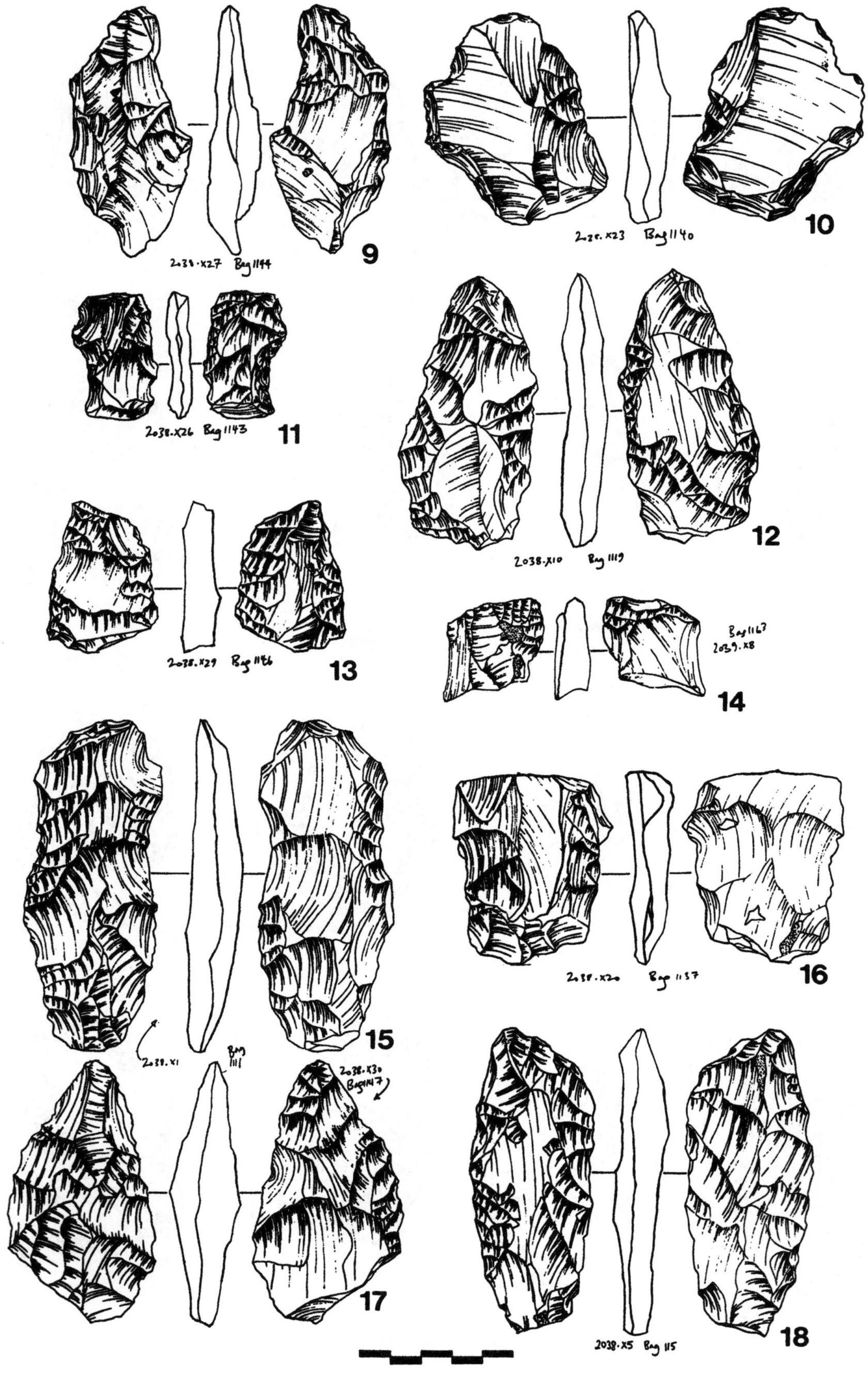

Figure 5.10: Hoard 2039 (cont.).

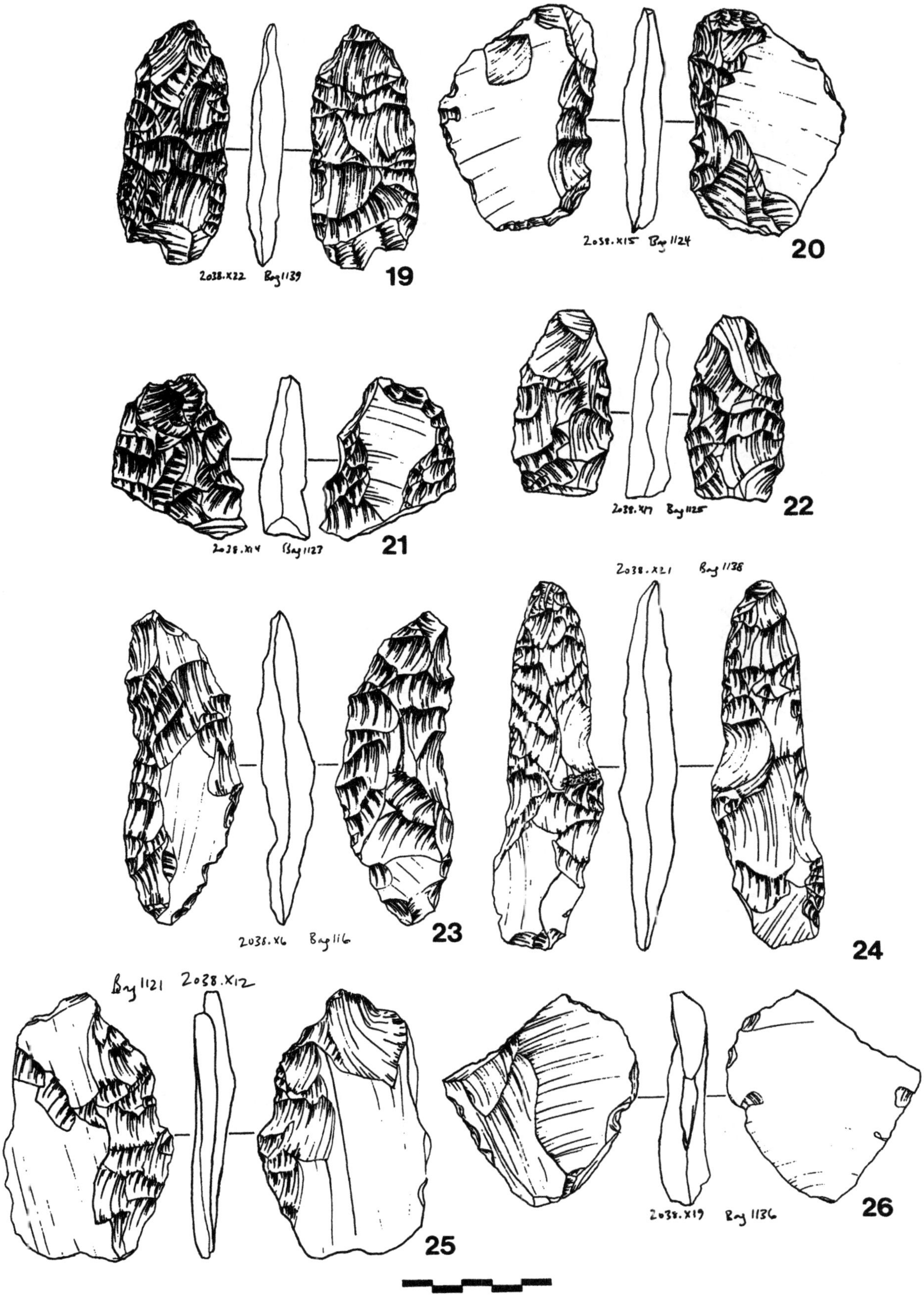

Figure 5.11: Hoard 2039 (cont.).

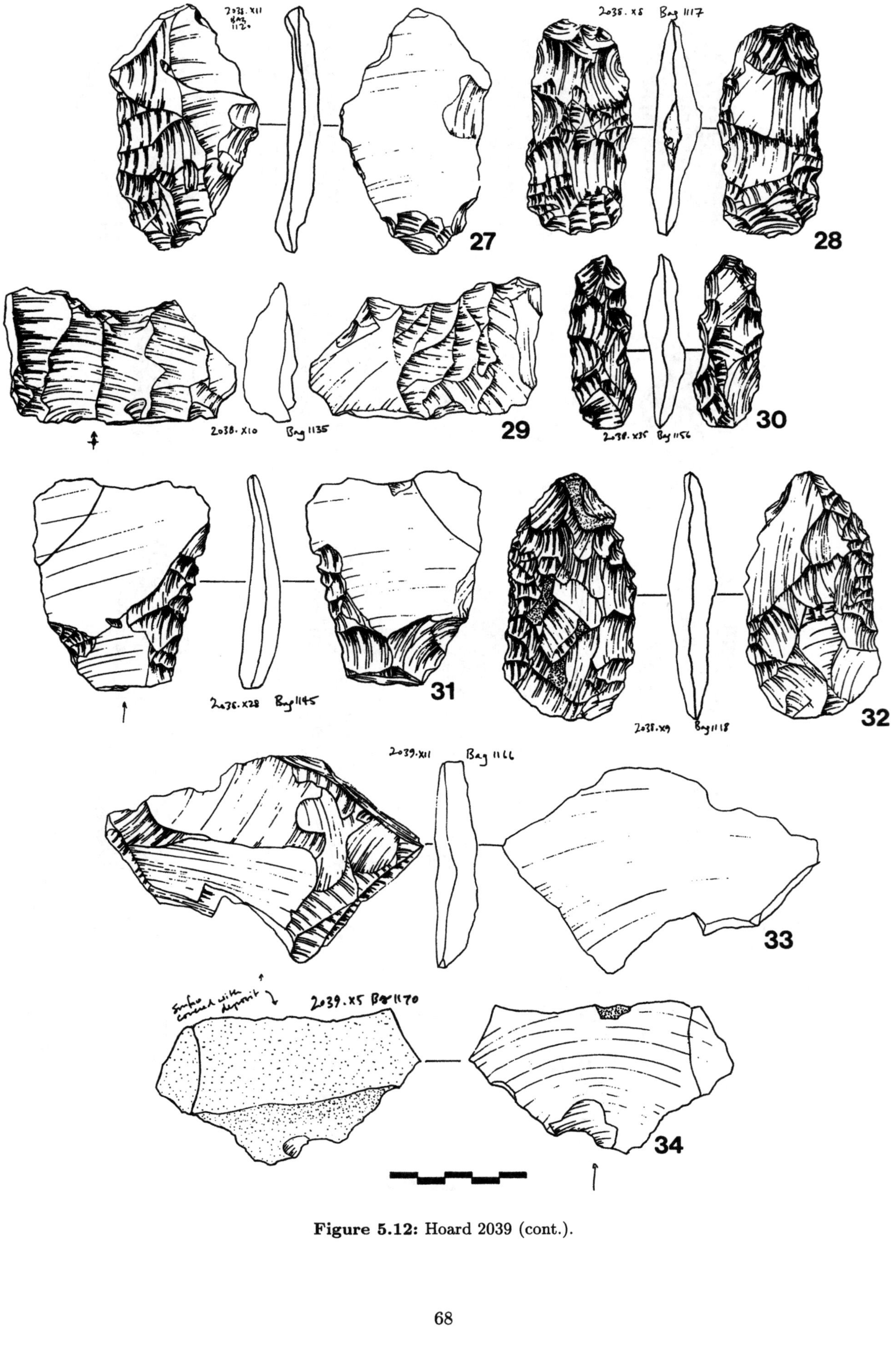

Figure 5.12: Hoard 2039 (cont.).

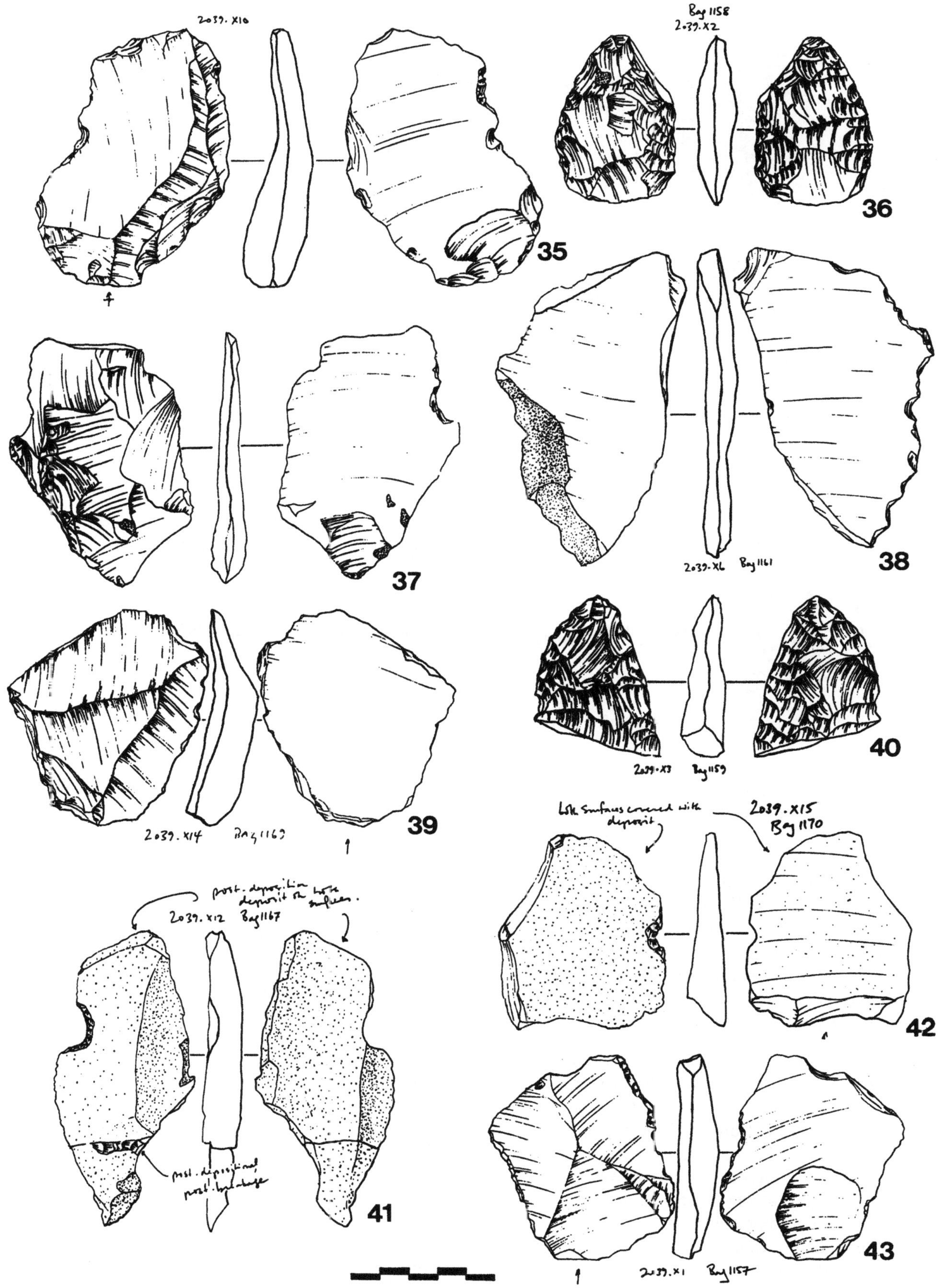

Figure 5.13: Hoard 2039 (cont.).

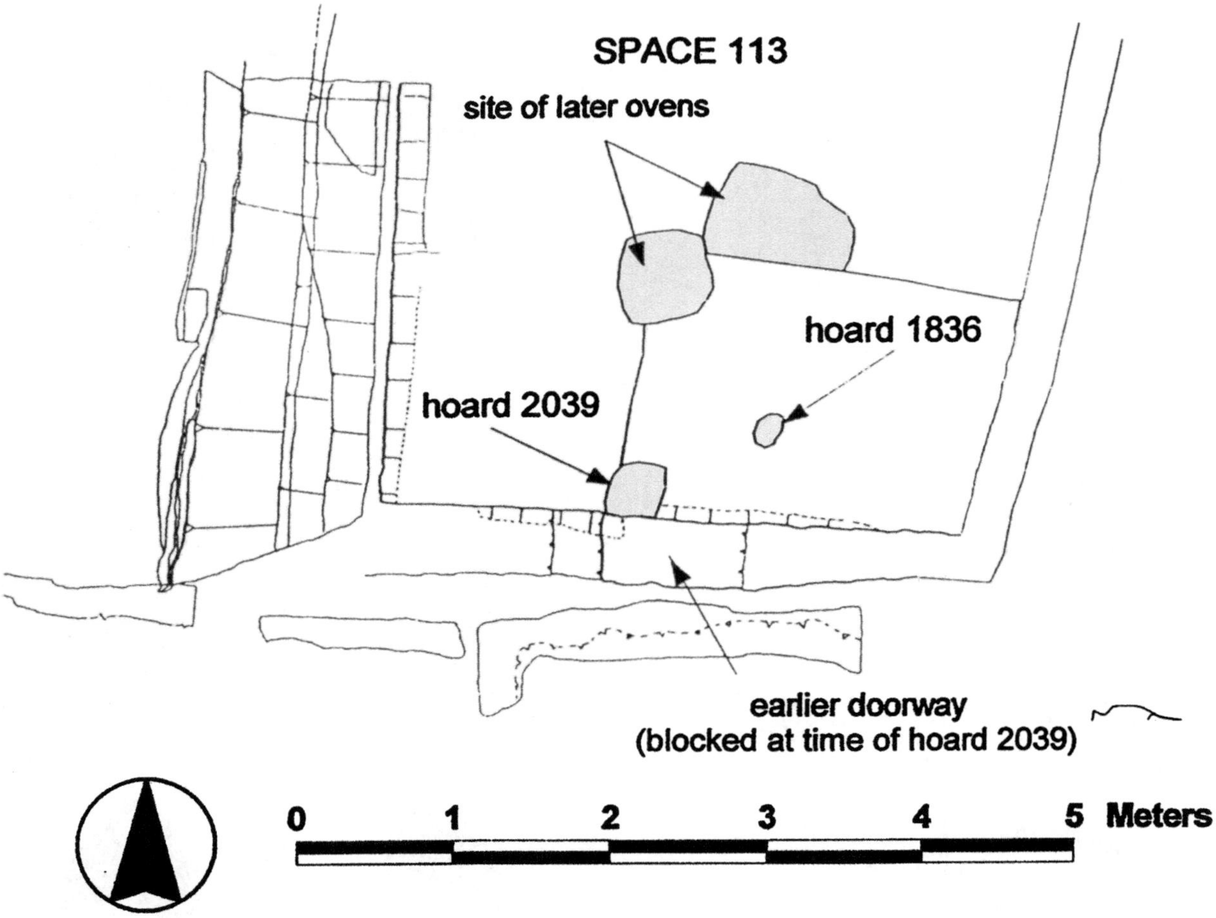

Figure 5.14: Location of Hoards 1836 and 2039.

2686g) (Fig. 5.20). It was positioned below what was to become the south-western quadrant of Building 2, directly below the hearths that were eventually to be build in this part of the house. Out of the total of forty-two pieces, eighteen have no clear evidence of retouch, but eleven have significant amounts (near-covering to covering) of bifacial scalar retouch on their ventral surfaces, with the remainder having marginal retouch on one or both faces. Thirteen pieces have some degree of cortex on their dorsal face.

Hoard 4210 Another smaller hoard 25cm south-west of the one previously described was found associated with the same infilling event prior to the construction of Building 2 (unit 4210, n=8; 346g) (Fig. 5.21). The two hoards may thus be contemporary. The excavator notes the possibility that they may have been deposited within an organic container, although no cut was found. The spatial association below the hearths of the later building should again be noted. This group of pieces includes a fine-grained flint blade fragment and two other obsidian blades (one of the latter shows some lateral retouch). A large *pièce-esquillée* (interpreted as a chisel-like implement by most analysts), a retouched flake, and three bifaces (the larger of which has a very sharp retouched point suitable for use as a piercer) complete the hoard.

5.3 Discussion

A number of strategies of flint and obsidian production have been identified at Çatalhöyük. Elsewhere I have grouped these into strategies of flake production, percussive blade production, pressure blade production, as well as strategies that apply to bipolar blade production for projectile point manufacture, flint dagger working, and obsidian mirror making (Conolly 1999a). Generally speaking, obsidian at Çatalhöyük does not appear to have been a rare resource; it is not worked to exhaustion and there is little evidence of any economising behaviour. The overwhelming absence of cortical surfaces suggests that much of the obsidian was fairly well prepared before it was brought into Çatalhöyük. However, when viewed alongside the other material, hoarded obsidian is obviously at the early stages of the reduction sequence. For ex-

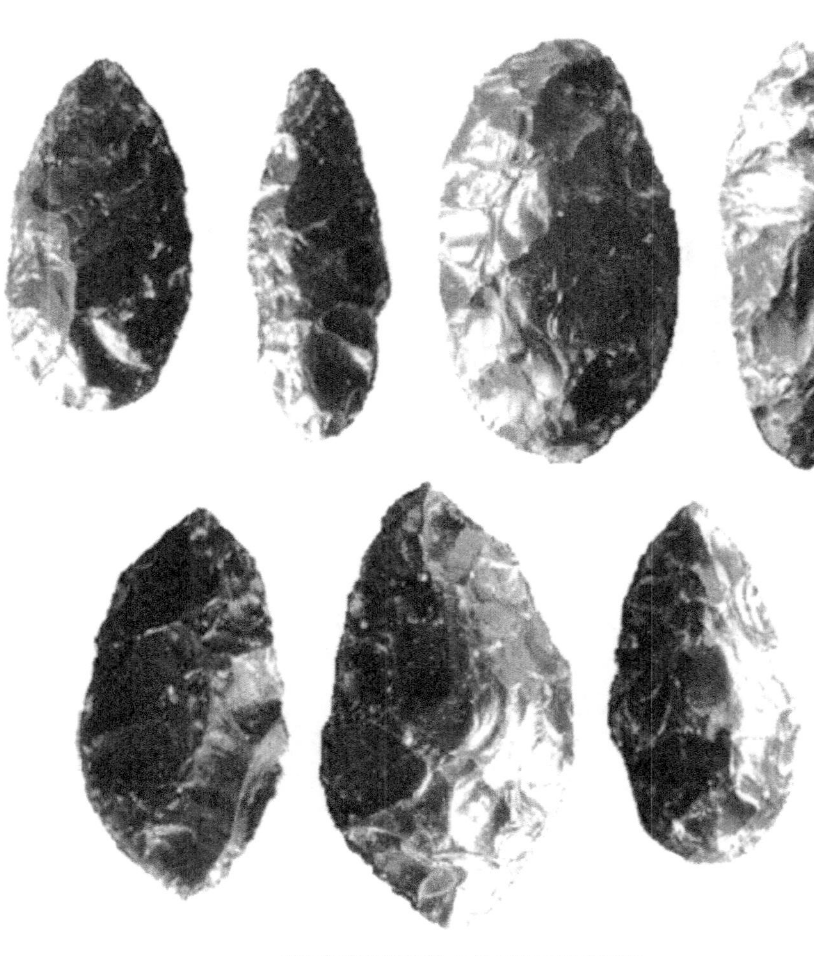

Figure 5.15: Hoard 2812.

ample, although still relatively rare, there are many more hoarded flakes with cortical surfaces than found in non-hoarded contexts. The large size of hoarded pieces compared to the small size of the non-hoarded assemblage also suggests that they have not been reduced to the same extent. Evidence for on-site opposed-platform blade production is completely lacking: there have been no large opposed platform cores or associated debitage found on site. Evidence for on-site biface manufacture is also missing: cortical flakes are all but absent outside of the hoards. In terms of the larger 'bifaces', while we have found quantities of bifacial 'thinning' flakes (Conolly 1996, 195), these completely lack cortical surfaces and are thus attributed to the latter stages of biface reduction. I think that most of the hoards contain material that has come directly from the obsidian sources without much, if any, on-site modification and thus are the 'rawest' category of obsidian introduced into the settlement.

Furthermore, even without taking into account the distinctive depositional contexts from which they derive, all of the hoards are, in differing ways, technologically peculiar when viewed against the ubiquitous flint and obsidian artefacts found in non-hoard contexts at Çatalhöyük. For example, the flakes share a common reduction strategy insofar as they are all large (and occasionally cortical) flakes that have been struck from large obsidian cores. Some of the flakes were then subsequently modified by having smaller thinning flakes removed from one or more margins of the blank. The best explanation that I can offer for why some blanks were modified in this way and others not, is that the manufacturers were attempting to standardise the size of the final product as much as possible. Even the most substantially modified flakes are what would be commonly called rough-outs, which otherwise absent from non-hoarded contexts. The hoarded blades similarly share a common technological process distinct from non-hoarded blades as they have all been produced from opposed platform cores. The reduction sequence appears to be similar in process to that of the *Naviform* process; the shape of the proximal end of the blades, and the degree of platform preparation is very similar to *Naviform* blades of the Levantine PPNB (*e.g.,* Wilke & Quintero 1994). Without going into unnecessary detail, the sequence that appears to have been followed involves the production of a crested ovoid core, the removal of one of the crests, followed by an alternating sequence of blade removals from opposed ends of the core. This process produces a distinctive series of 'positive' blades that possess conver-

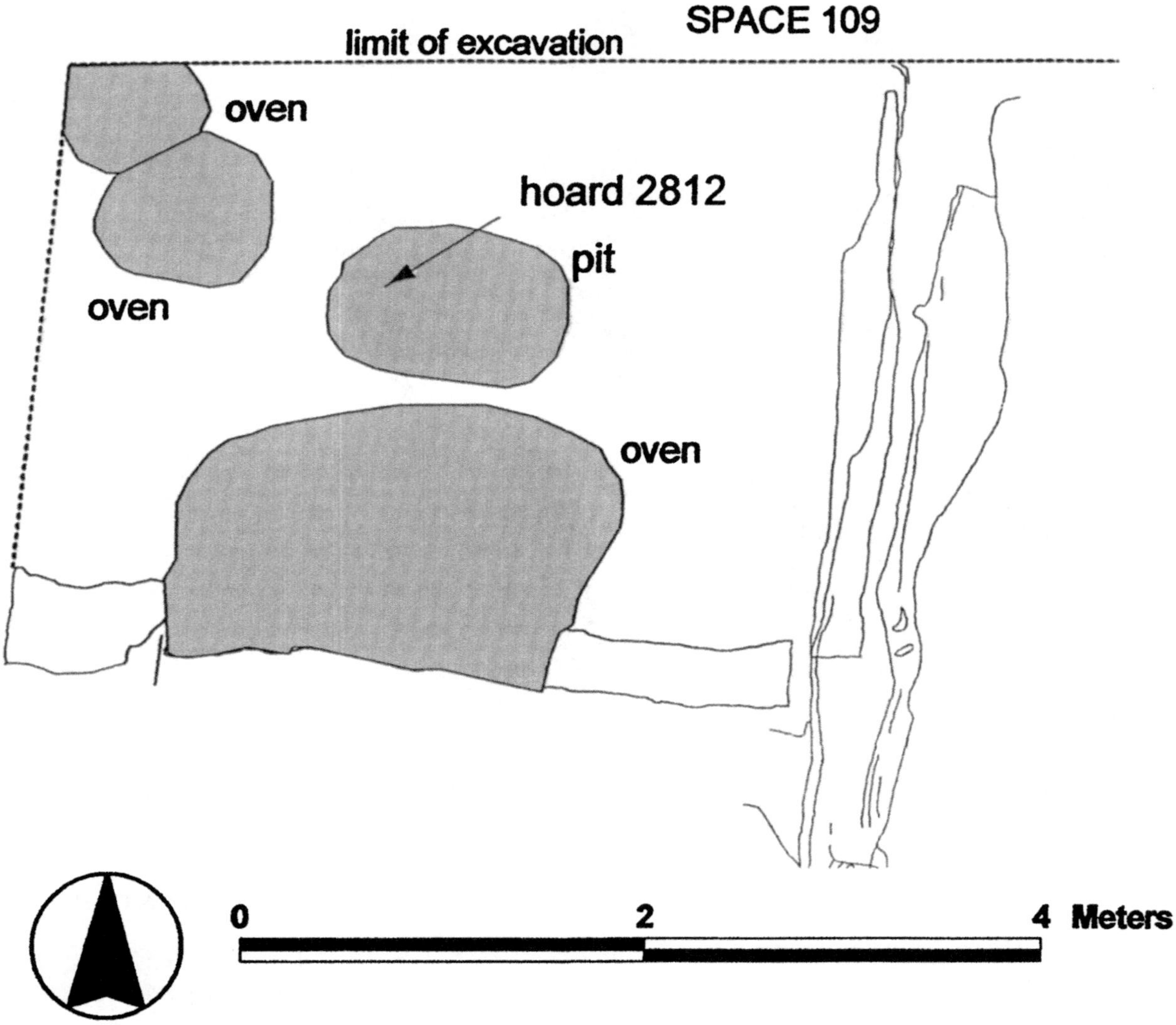

Figure 5.16: Location of Hoard 2812.

gent distal ends, and a series of 'negatives' from the other end of the core which possess characteristically 'upsilon' distal ends.

What appears to be happening for some elements of the obsidian economy, therefore, is that opposed-platform blades, large flake 'blanks', and flake bifaces, are being imported ready-made into Çatalhöyük. This is corroborated by recent findings from quarry sites at the Central Anatolian obsidian sources in the region Göllü Dagi. In particular, cores and bifaces of a form that appear to be related to the Çatalhöyük hoards have been found the site of Kaletepe (Fig. 5.22) (Balkan-Atli & Der Aprahamian 1998). This strongly suggests that bifaces and opposed platform blades were produced at the obsidian sources and then carried to Çatalhöyük. The reason for some of the substantial bifacial modification at source may be to standardise the size of the obsidian prior to transport. It similarly may be the case that it is easier to transport ready knapped blades, than large cores or blocks of material. These are then cached under floors and abandoned for one reason or another, or are initially buried, then retrieved at a later date, or immediately enter life above the floors. If they are used for tool manufacture, then they are suitable for a range of different things (from large projectiles and scrapers, to the smallest flake tools) required by the inhabitants. As noted earlier, however, the hoarded blanks do not articulate with all of the strategies of production identified at Çatalhöyük; flint and obsidian small flake-cores and single platform blade (percussive and pressure) cores are found *in situ*, although not in great quantity.

It may be, therefore, that the act of hoarding large pieces of obsidian represents a process of safeguarding family wealth. Yet it would appear that many of these hoards were never meant to be recovered; the evidence from Hoards 4209 and 4210 suggests that some hoards

Figure 5.17: Hoard 4280.

Figure 5.18: Hoard 4276.

Figure 5.19: Hoard 4138.

were buried before the construction of a house, effectively eliminating any clear spatial clues to their precise location for easy recovery, in contrast to hoards buried underneath the floor of an occupied dwelling where location can be related to extant features. Furthermore, the common depositional context of all the hoards is suggestive of meaning beyond the functional 'buried personal wealth' argument. Their spatial positioning and association in the south-east near ovens and grinding areas may be important clues to their symbolic meaning.

There is not enough detail from the Mellaart reports to construct precisely where in each building the hoards were buried, and the recent discoveries are too few to claim that there is a definite pattern for buried obsidian. Nevertheless the evidence presented here does point to a recurring spatial pattern for the phenomenon to be preferentially located in the south or southern parts of buildings and spatially associated with hearths. Even in the case of Hoard 2039 where there was no extant oven nearby, the excavators note that Mellaart had truncated this space to below floor level removing all features in the process, and they suggest that an oven may have originally been located in that part of the building. At the risk of later contradiction, I will hazard that this is indeed the case and that the position of the hoards in the southern part of the building, where ovens and food preparation also seemed to occur, is an important part of their overall meaning and significance.

As the hoards are preferentially located in the southern half of buildings and under, or very near, objects and features used for food production there is at least a spatial link, and quite possibly a symbolic link between the two. This may be only to the extent that the southern part of a building was the

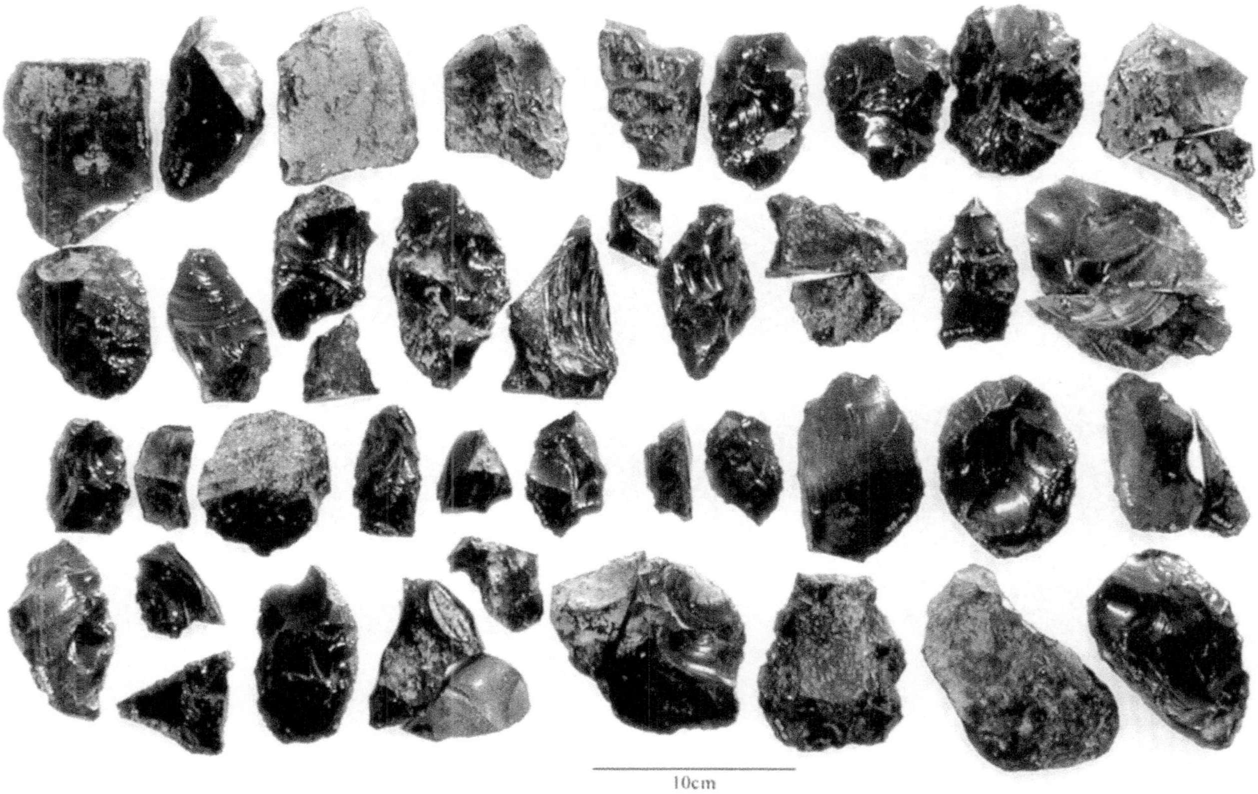

Figure 5.20: Hoard 4209.

appropriate place for hoarding surplus obsidian. Alternatively it may be for more ritualistic reasons, such as controlling of a 'wild' or powerful material (*cf.* Hodder 1992). This, however, places the obsidian at the centre of the interpretation. As an alternative, the apparent dichotomy between the wild obsidian and the transforming nature of the domestic oven — and the fact that some of the obsidian appears to have been buried without intention of retrieval — situates hoarding as a ritual practice linked to the foundation of the domestic area of the house. Perhaps obsidian, as a representative of the 'wild', may be being used to highlight or offset the power of the 'domestic'.

These ideas may all be slightly abstract, but at the minimum the evidence is suggestive that there were rules governing where obsidian hoards were located. In this respect, the burial of obsidian may be seen to be linked to ideas of control or transformation of wild resources. Although not all obsidian is buried in hoards, non-hoarded obsidian is continuously shaped by knapping; and as Mortensen (1988, 201) suggested earlier, this in itself may be seen as a process of transformation from a wild to domestic state, albeit through a different process.

5.4 Summary

The obsidian hoards at Çatalhöyük are technologically distinct from the non-hoarded obsidian and flint, and are found in contexts that seem to conform to well-defined spatial rules. Although the hoards may be seen as a product of storage of family wealth, the phenomenon may also be seen as a extension of the control and transformation of wild resources that takes place around the hearth area in Neolithic households. These interpretations are necessarily provisional, as we are only now beginning to gather a more comprehensive understanding of the depositional context than was previously available. It may be that as further data is gathered, the patterns that have here been used to interpret the phenomena will change and the interpretation will need to be modified.

The most recent information on the excavation can be found at: `http://www.catalhoyuk.com`

Notes

1. This excludes obsidian artefacts found with the dead; Mellaart separately lists three separate examples where 'unused and spectacular weapons' were found in burials (Mellaart 1964, 103).

2. Note that some were post-depositionally broken, and the total count of all objects is 65.

Figure 5.21: Hoard 4210.

3. Note that for the hoards 4280, 4276, 4138, 4209 and 4210 there are currently no available plans as, at the time of writing (15/05/00), the buildings are still under excavation. Plans will eventually be available at: `http://catal.arch.cam.ac.uk/catal/catal.html`

References

ATAMAN, K. 1988. 'A preliminary analysis of the upslion blade tools from Hayaz Höyük.' *Anatolica* 15: 81–5.

BALKAN-ATLI, N. & G. DER APRAHAMIAN 1998. 'Les nucléus de Kaletepe et deux ateliers de taille en Cappadoce.' *In* M. C. Cauvin *et al.* (eds.), *L'Obsidienne au Proche et Moyen Orient: du Volcan à l'Outil*, pp. 241–57. British Archaeological Reports International Series 738, Oxford.

BAR-YOSEF, O. & D. ALON 1988. 'Nahal Hemar Cave: the excavations.' *'Atiqot* (English Series) 18: 3–30.

BIALOR, P. 1962. 'The chipped stone industry of Çatalhöyük.' *Anatolian Studies* 12: 67–110.

BRADLEY, R. 1990. *The Passage of Arms.* Cambridge University Press, Cambridge.

CONOLLY, J. 1996. 'The knapped-stone.' *In* Hodder (1996a), pp. 173–98.

CONOLLY, J. 1999a. *The Çatalhöyük Flint and Obsidian Industry: Technology and Typology in Context.* British Archaeological Reports International Series 787, Oxford.

CONOLLY, J. 1999b. 'Technical strategies and technical change at Neolithic Çatalhöyük, Turkey.' *Antiquity* 73(282): 791–800.

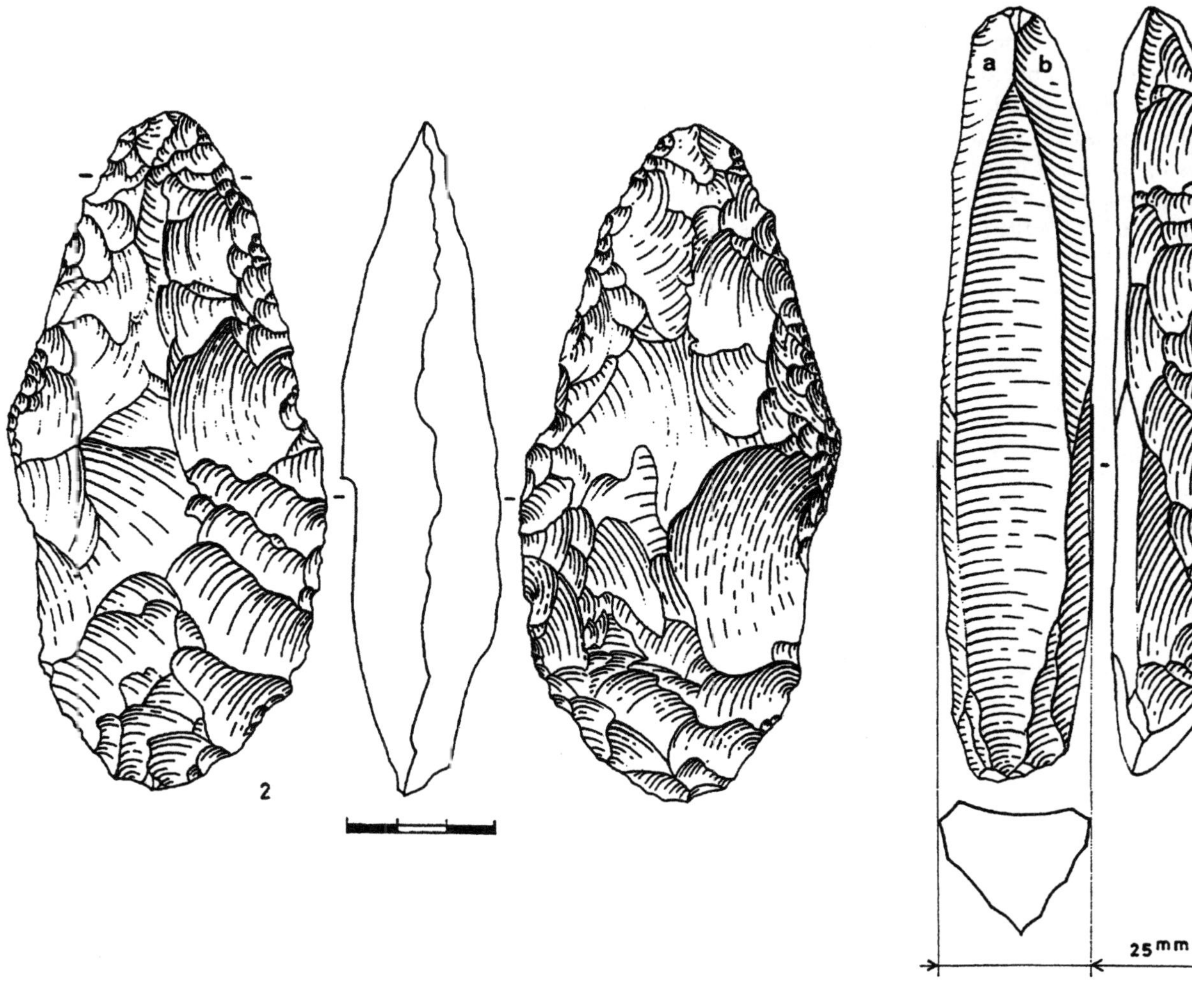

Figure 5.22: Cores and bifaces from Kaletepe.

HODDER, I. 1992. *The Domestication of Europe*. Blackwell, London.

HODDER, I. (ed.) 1996a. *On the Surface: Çatalhöyük 1993–95*. McDonald Institute for Archaeological Research and the British Institute of Archaeology at Ankara, Cambridge and Ankara.

HODDER, I. 1996b. 'Re-opening Çatalhöyük.' *In* Hodder (1996a), pp. 1–18.

LAST, J. 1998. 'A design for life: interpreting the art of Çatalhöyük.' *Journal of Material Culture* 3(3): 355–78.

LE ROUZIC, Z. 1927. 'Dépôts rituels de haches en pierre polie découverts dans la région de Carnac.' *Bulletin de la Société Préhistorique Française* 81: 89–96.

MATTHEWS, R. 1996. *The excavation of Building 1, North Area. Çatalhöyük 1996 Archive Report.* `http://catal.arch.cam.ac.uk/catal/Archive_rep96/rogmatthews96.html`.

MELLAART, J. 1963. 'Excavations at Çatal Hüyük, 1962, second preliminary report.' *Anatolian Studies* 13: 43–103.

MELLAART, J. 1964. 'Excavations at Çatal Hüyük, 1963, third preliminary report.' *Anatolian Studies* 14: 39–119.

MORTENSEN, P. 1988. 'A note on a small box with flint blades and arrowheads from Beidha — and its implications.' *In* A. Garrard & G. H. Gebel (eds.), *The Prehistory of Jordan: the State of Research*, pp. 199–207. British Archaeological Reports International Series 396, Oxford.

NIELSEN, P. O. 1977. 'Die Flintbeile der Frühen Trichterbecherkultur in Dänemark.' *Acta Archaeologica* 48: 61–138.

RECH, M. 1979. *Studie zu Depotfunden der Trichterbecher- und Einzelgrabkultur des Nordens.* Karl Wacholz, Neuminster.

RISSMAN, P. 1988. 'Public displays and private values: a guide to buried wealth in Harappan archaeology.' *World Archaeology* 20(2): 209–28.

WILKE, P. & L. QUINTERO 1994. 'Naviform core-and-blade technology: assemblage character as determined by replicative experiments.' *In* H. G. Gebel & S. K. Kozlowski (eds.), *Neolithic Chipped Stone Industries of the Fertile Crescent: Proceedings of the First Workshop on PPN Chipped Lithic Industries, Free University of Berlin, 29th March–2nd April, 1993*, Studies in Early Near Eastern Production, Subsistence, and Environment 1, pp. 33–60. Ex Oriente Free University of Berlin, Berlin.

James Conolly
Institute of Archaeology, University College London
31–34 Gordon Square
London WC1H 0PY
j.conolly@ucl.ac.uk

6

Flint Use in Later Bronze Age and Iron Age England? Some Criteria for Future Research

Jodie Humphrey and Robert Young

School of Archaeological Studies, University of Leicester

Abstract

In this paper we want to address the thorny question of the Later Bronze Age and Iron Age use of flint. We realise that this is a contentious area for debate and that many lithic specialists are sceptical of the notion of regular, post-Middle Bronze Age, utilisation of flint and other lithic resources. Indeed, one of the present authors (RY) was very wary of accepting the possibility of such a phenomenon until some recent work on lithic material in the east Midlands of England made him re-evaluate his position. Other workers, though, have made their opposition to the concept quite explicit.

6.1 Introduction

While admitting that there might be a case for *ad hoc* use of lithic resources in these later periods, Alan Saville has been quite forthright in his views on the issue of the regular use of flint tools. In 1981 he wrote that 'regular production and use of flint artefacts for everyday domestic activities declined and ceased altogether within the Later Bronze Age' (Saville 1981, 6). His stance has not changed since that time (Saville *pers. comm.*) and it reflects the commonly held view that a decline in flint use for tool manufacture was an economic phenomenon linked to the concomitant rise in the ubiquity of metal artefacts. This is an issue that will be addressed below.

6.2 'Residuality'

Most lithic material from Later Bronze Age and Iron Age sites is written off as 'residual'. By that archaeologists mean that it represents material that is out of context and left over from earlier human activity on the same archaeological site. Saville notes that this problem 'must always loom so large in any consideration of the presence of flintwork on later prehistoric sites' (1981, 8). We would not of course wish to argue that residuality does not occur on archaeological sites; certain locations were undoubtedly used by groups of people over long periods of time. We feel, however, that this problem needs to be given a much more detailed consideration than has previously been the case. The term should not be used simply as a 'catch-all' that might free the specialist from a detailed assessment of the difficulties of site and assemblage formation. In other words, we need to be quite clear about what activities flint and other so-called 'residual' material is actually 'residual' from. The frequent, and possibly unthinking, use of the term might allow for a massive oversimplification of a complex pattern of resource use on later sites. We think that a re-examination of our approaches to this phenomenon and to the problems and challenges of identifying and recovering contemporary lithic material from Later Bronze Age and Iron Age sites is essential.

It is easy to describe flint material as 'residual' if we have evidence for early, as well as later, human activity on a particular site. But what about those sites where flint occurs with no other earlier material? What about those sites where flint occurs in apparently sealed contexts, in association with nothing but Later Bronze Age or Iron Age material? These sites and contexts should, we argue, deserve some special attention as it is not logical to automatically consign the recovered finds to the dustbin of 'residuality'. We have examined some late sites where we think that the above conditions prevail and we believe that we can identify a series of regular, recurrent features in the flint assemblages which might allow us to suggest that they are contemporary with the Later Bronze Age / Iron Age occupation activity.

6.3 Some case-studies examined

At Wanlip, on the gravels of the River Soar in Leicestershire, excavations undertaken by Leicester University Archaeological Services (ULAS) have revealed a sequence of pits, post-holes and ditches. Some 1,168 pieces of flint were examined, with 430 coming from a series of sealed contexts (Cooper & Humphrey 1998).

The only other dateable material from these contexts comprised of two sherds of Late Bronze Age / Early Iron Age pottery and over 1000 pieces of Middle Iron Age pottery. There is no evidence at all for earlier activity on the site or from the surrounding area. Locally available, gravel flint pebbles and nodules were utilised, and all of the material examined was fresh and unabraded. Table 6.1 shows the ratio of lithic finds to pottery in the main Wanlip features.

Examination of the lithic material from the pit sequences showed that short, squat flakes with obtuse flaking angles and with a length:breadth ratio of 1:1 were being produced. Table 6.2 gives a typological breakdown of the lithic material from all sealed contexts.

All cores were irregular and did not fit into any of the standard core classification systems currently in use for earlier assemblages (*e.g.,* Clark 1960). The only clearly recognisable tools present were eleven very simple scrapers and seven piercers.

Analysis of bulbs of percussion suggests that direct percussion was applied generally throughout the knapping process, probably using a hard hammer. Flake shape, thickness, bulb size and the presence of much step and hinge fracturing along with multiple incipient cones on many pieces, implies that the knappers were either not skilled or not too concerned about the results of their work so long as broadly useful pieces were removed. The length:breadth ratios of flakes show that the majority are short and squat and only six examples exhibit blade-like proportions. On the basis of this data we think that a good case can be put forward at Wanlip for Middle Iron Age use of locally available flint.

Excavations by Clark & Fell (1953) at Micklemoor Hill, West Harling, Norfolk also produced a fresh flint assemblage, associated with Later Bronze Age / Early Iron Age pottery. In the past there has never been any debate about the contemporaneity of these two categories of finds, and most workers would accept the late date of this lithic assemblage. The raw material was locally available, nodular, surface flint from the gravel ridge on which the enclosures were constructed. Again, there was no evidence for earlier settlement on the site of West Harling, and a simple core/flake technology was dominant. The material showed little platform preparation and only one in four of the 200 primary flakes recorded showed evidence for a prepared striking platform.

The end product was short squat flakes with heavily obtuse striking angles. Seven very crude cores were recorded, all with irregular, multi-directional flaking. Fifteen scrapers were the only clearly discernible artefact types along with 15 very crude hammerstones (Clark & Fell 1953, 34–35) (Fig. 6.1).

At Birchanger in Essex, lithic material was recovered from a sequence of pits dated, by pottery associations, to the Early/Later Iron Age. A breakdown of the material is shown in Table 6.3.

The raw material used was, again, local flint, severely flawed. Austin (1994) suggested that the material had broken along natural fracture planes in the course of knapping. She argued that the lack of knapping ability was shown by the simple pattern of flake production. Flakes had large, plain butts and prominent bulbs of percussion. Several showed incipient cones from several unsuccessful attempts at flake removal. Many of the flakes were miss hits and Austin suggests that the pit deposits represent dumps of contemporary knapping debris.

Lithic material from the Iron Age / Romano-British enclosure at Warren Hill in Wiltshire should also be mentioned here (Stevens 1993). Some eight hundred pieces of flint were recovered from a well stratified ditch deposit associated with Early and Middle Iron Age pottery. No other earlier material was recorded on the site. Overall the industry was very crude and it certainly did not fit with standard classification systems. Indeed, Stevens had to devise a whole new set of terms with which to classify the finds. Table 6.4 shows the range of material recorded.

Hinge terminations were present on 18% of all complete flakes. More than 146 complete flakes showed evidence of being mis-hits, and over 50% of all flakes were broken. More than 70% of all recorded flakes showed evidence for multiple, irregular, dorsal scars, indicating a lack of control in the knapping process. Of the 803 pieces, 10 hammerstones and 37 of the utilisable flakes showed evidence for heavy abrasion and use wear.

Moving on to Hampshire, we would like to comment on one further site, at Winnall Down. Here, some 2816 pieces of flint were found in a variety of late contexts. Only 1% of all of the material consisted of finished tool forms, 3.5% were recognisable as cores, and 8% of the assemblage consisted of retouched or utilised flakes. Two thousand two hundred and twenty eight flakes were recorded overall.

Material recovered from phases 3, 4 and 5 of the excavation spanned the Early to Late Iron Age periods, and can be broken down as shown in Table 6.5.

Detailed analysis was carried out on the material from the sealed phase 3 ditch deposits, of Early Iron Age date. No obvious residual material was recorded from these contexts and the finds can be classified as is shown in Table 6.6.

As can be seen above, scrapers represented the only clearly definable tool type. Only one primary flake was recorded, and the secondary and inner flakes were generally broad and squat with obtuse flaking angles between 116° and 130° (Fig. 6.2). Winham (1985, 86) made the following comments on the material:

> Although this assemblage is rather small, the trend was clearly for large secondary flakes with a tendency for them to be broad rather than long. The small percentage of tertiary flakes, the almost total

	ditch	pit	post-hole	gully	cut feature
flint (F)	99	182	111	13	1
pottery (P)	272	674	339	34	3
ratio P:F	2.74:1	3.70:1	3.05:1	2.61:1	3:1

Table 6.1: Ratio of pottery to flint in cut features/sealed contexts at Wanlip (Leics.).

Artefact type	No.
complete flakes	68
complete flakes (retouched)	13
broken flakes	76
broken flakes (retouched)	12
blades	2
broken blades	2
broken blades (retouched)	3
'bladelets'	1
broken 'bladelets'	7
chips	170
chips (retouched)	1
chunks	40
cores	6
scrapers	11
?obliquely retouched piece	1
?microlith	1
serrated blade	1
serrated flake	1
'cutting' flake	6
piercers	7
?spurred tool	1
Total	430

Table 6.2: Breakdown of material from all sealed contexts at Wanlip.

Artefact type	No.
complete flakes	78
broken flakes	17
cores	9
naturally fractured lumps	6
hammerstones	2
fragmentary hammerstones	1
Total	113

Table 6.3: Breakdown of lithic finds from Birchanger (after Austin 1994, 43).

Artefact type	No.	%
preparation flakes	274	34.1
utilisable flakes	105	13.0
dressing chips	305	37.9
cores	55	6.8
nodules	54	6.7
hammerstones	10	1.2
Total	803	100

Table 6.4: Classification of lithic material from Warren Hill (Wilts.) (after Stevens 1993, 91).

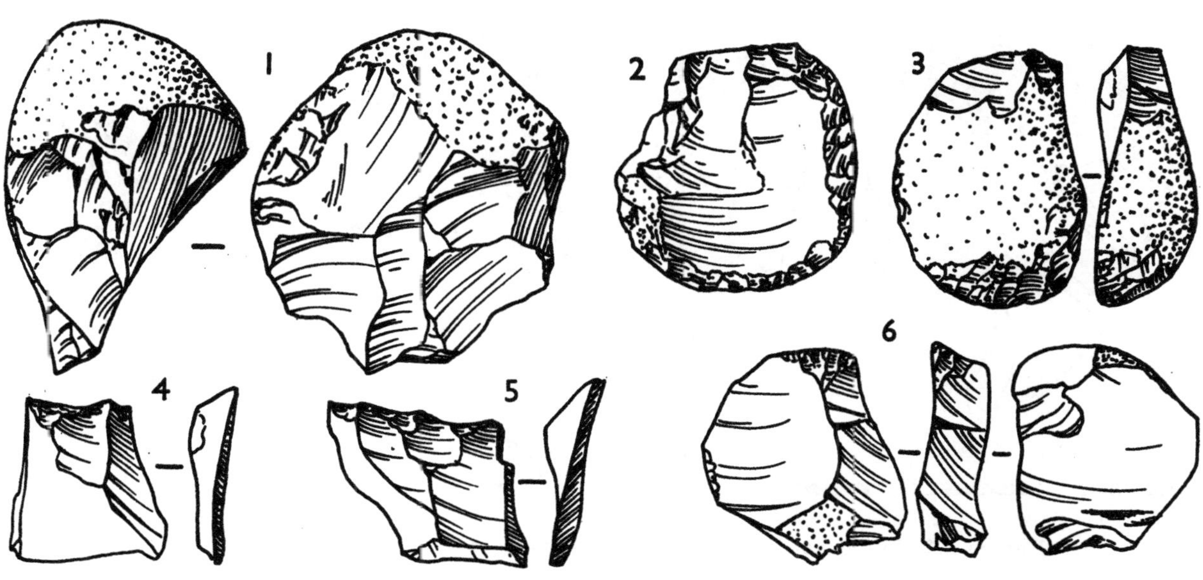

Figure 6.1: Lithic material from Micklemoor Hill, West Harling, Norfolk (after Clark & Fell 1953), no scale provided.

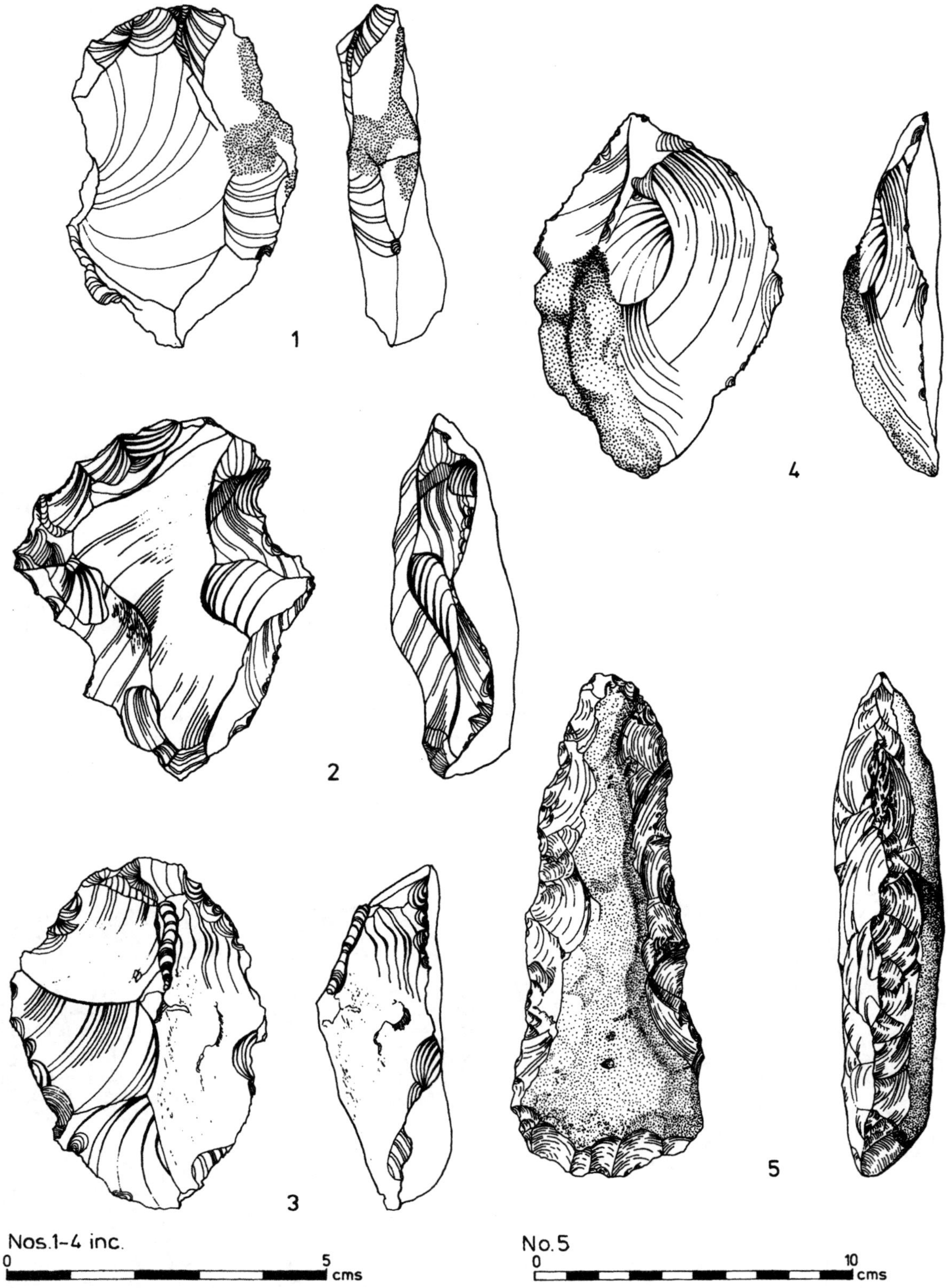

Figure 6.2: Lithic material from Phase 3, Winnall Down (after Winham 1985).

Artefact type	Phase 3 (EIA)	Phase 4 (MIA)	Phase 5(LIA)
flakes	707	527	36
retouched flakes	49	46	3
utilised flakes	19	21	1
cores	29	25	
chips	34	25	2
blades	2		
fragments	20	13	2
rejuvenation flakes	8	5	
scrapers	8	1	
borers	1	2	
points		2	
hammerstones	1		
Total	878	667	44

Table 6.5: Breakdown of lithic material from Phases 3, 4 and 5, Winnall Down (Wilts.), (after Winham 1985, 85).

Artefact type	No.
waste frags., chips and nodules	132
flakes	298
retouched flakes	23
utilised flakes	6
cores	14
core fragments	10
rejuvenation flakes	2
scrapers	6
Total	491

Table 6.6: Lithic material from Phase 3 ditch deposits, Winnall Down (Wilts.) (after Winham 1985, 84).

> lack of primary flakes and the small proportion of 'tools' compared with retouched flakes suggests the lack of formalised flint working procedure with detailed retouching. Any convenient flake, it seems was picked out to retouch, rather than knapping to produce flakes of a specific type for retouching to form tools. This may indicate that flakes were utilised for the job in hand and then thrown away rather than kept for further use.

He argued that the material from Winnall Down suggests that flint working was 'a crude industry in the Early Iron Age' (1985, 86) and also stated that:

> Comparisons indicate that this industry could be seen to continue the trend of deteriorating flint working from the Late Neolithic to the Early Bronze Age, but a lack of comparative data from other sites of the Early Iron Age and further Bronze Age sites makes any broad conclusions tentative. (Winham 1985, 86)

All of the above sites have, therefore, produced lithic material of Later Bronze Age / Iron Age date. We would suggest that there are no reasons for doubting the continued knapping and utilisation of flint at these locations. On the basis of this brief survey and our discussion below, we think that the features listed at the end of this paper might help us to identify potentially Later Bronze Age / Iron Age flint work in the field, and might equip us with criteria for the isolation of later material on sites where there is also obvious evidence for residuality.

6.4 Contexts of use, economic changes and social alternatives

Based on the above observations, we would suggest here, *contra* Saville (1981), that it may be precisely in the domestic sphere that lithic utilisation went on the longest, continuing well into the Iron Age in certain areas. In order to set this argument and our observations into context, however, it is necessary to consider the processes involved in the 'replacement' of stone tools by metal artefacts. It is undeniable that the scale of flint utilisation, and the range of flint tools manufactured, declined throughout the Bronze Age to the low, but still observable, levels that we have seen in the preceding discussion. We need to ask though, whether this decline was simply to do with economics and the broadening availability of new materials, or whether there may be more subtle and wide ranging social mechanisms at work here.

To get to grips with this question we need to think about what seems to have been happening in the Middle Bronze Age. This is the last chronological period in British prehistory when most researchers feel comfortable with the idea of regular flint utilisation. Certain trends are observable, including:

- Decline in knapping skills.
- Increased use of locally available raw materials.
- Non-curation of products.
- Growth in the expedient production of artefacts.
- Restricted range of recognisable 'tool' types.

The situation was neatly summed up by Edmonds when he said:

> Although stone clearly remained an important practical resource, many of these later assemblages suggest a further impoverishment of techniques and traditions that had persisted for many generations. In raw material terms there is very little to indicate that people used anything other than stone that was locally available. There is no evidence that raw materials were transported or exchanged over any distance and exotic artefacts are conspicuous by their absence in a majority of contexts. People simply made use of stone that was ready to hand to a far greater extent than before. (Edmonds 1995, 184)

As Herne has shown at Grimes Graves and on other Middle Bronze Age sites (1992), Middle Bronze Age flint working was characterised by simple core/flake technologies. Hard hammer, direct percussion, was universally employed and hammers were made from any suitably-sized flint nodules. Raw material does not seem to have been curated, but used expediently. Nor does controlled knapping seem to have been a feature of Middle Bronze Age assemblages. Thick, squarish flakes seem to have been the normal type of removal, and flake scars on the dorsal surfaces of these pieces are not regular, perhaps indicating that cores were simply rotated to find a good edge to use as a striking platform. The overall aim at sites both on and off the flint-bearing chalk seems to have been to generate as many potentially usable pieces as possible, without any emphasis on standardisation of end products (Herne 1992, 35–47) (Figs. 6.3–6.5).

Throughout the Bronze Age, Herne (1992, 67) suggests that the gradual decline in the use and range of flint and stone tools reflects the fact that these tools were required to fulfil fewer and fewer functions. As a result, he argues, the incentive to invest time and effort in flint procurement was lessened. At the same time and for the same reasons, less attention was paid to primary knapping or secondary flake modification. As a corollary, more attention was paid to expedient manufacture of flint tools.

It may be possible to argue, however, that the apparent decline in flint use has more important social implications than this straightforward economic picture suggests. By the end of the Middle Bronze Age, it would appear that the only recognisable regularly produced flint tools are scrapers, awls and some knife forms, all of which are invariably found in settlement contexts and are intimately bound up with the domestic mode of production (Herne 1992, 67; Edmonds 1995, 184–6).

It is now a commonly held view among prehistorians and others that tools and artefacts can have both utilitarian and symbolic or social functions. As many archaeologists have argued (Barrett 1994; Clark *et al.* 1985; Hodder 1986; Shanks & Tilley 1987; Tilley 1989) patterns of artefact use can be a medium of expression for particular social interests. In the early second millennium BC new techniques of flint working, such as invasive retouching, resulted in the manufacture of many highly formalised, 'fancy', lithic artefacts, including knives, daggers and arrowheads (Herne 1992, 72). Edmonds (1995, 122–183) has argued that the restricted distribution and consumption of these fancy items was integral to strategies of social demarcation. By the end of the Early Bronze Age, however, with the growing (if still restricted) circulation of metal objects among élites, there was a gradual diminution of the capacity of stone tools to stand for people and practices (Edmonds 1995, 187). The eventual outcome was a movement away from flint as a medium for the public constitution of ideas about self and society.

At the same time as this transition was going on throughout the Middle Bronze Age, Edmonds further suggests that the:

> ... metaphorical qualities of settlements and field systems may have come to play an increasingly important part in shaping people's understanding of their place within the community and within the broader social landscape. (Edmonds 1995, 187)

The contexts in which tasks were performed and the way in which labour was divided up and jobs shared out, rather than material display through artefacts, may have become the important media through which people and their social roles were defined. It is this seeming shift in terms of social emphasis — rather than any great change in economics or subsistence — which Edmonds suggests might better account for the treatment of stone by the end of the Bronze Age (Edmonds 1995, 187–8).

The erosion of the prestige-based role and the overtly social dimensions of stone had a material effect on traditional attitudes to procurement, production and consumption strategies. This was probably a gradual process, and:

> ... with each generation less and less importance may have been attached to the maintenance of traditional patterns of procurement ... where stone tools no longer served as important metaphors for people or their roles

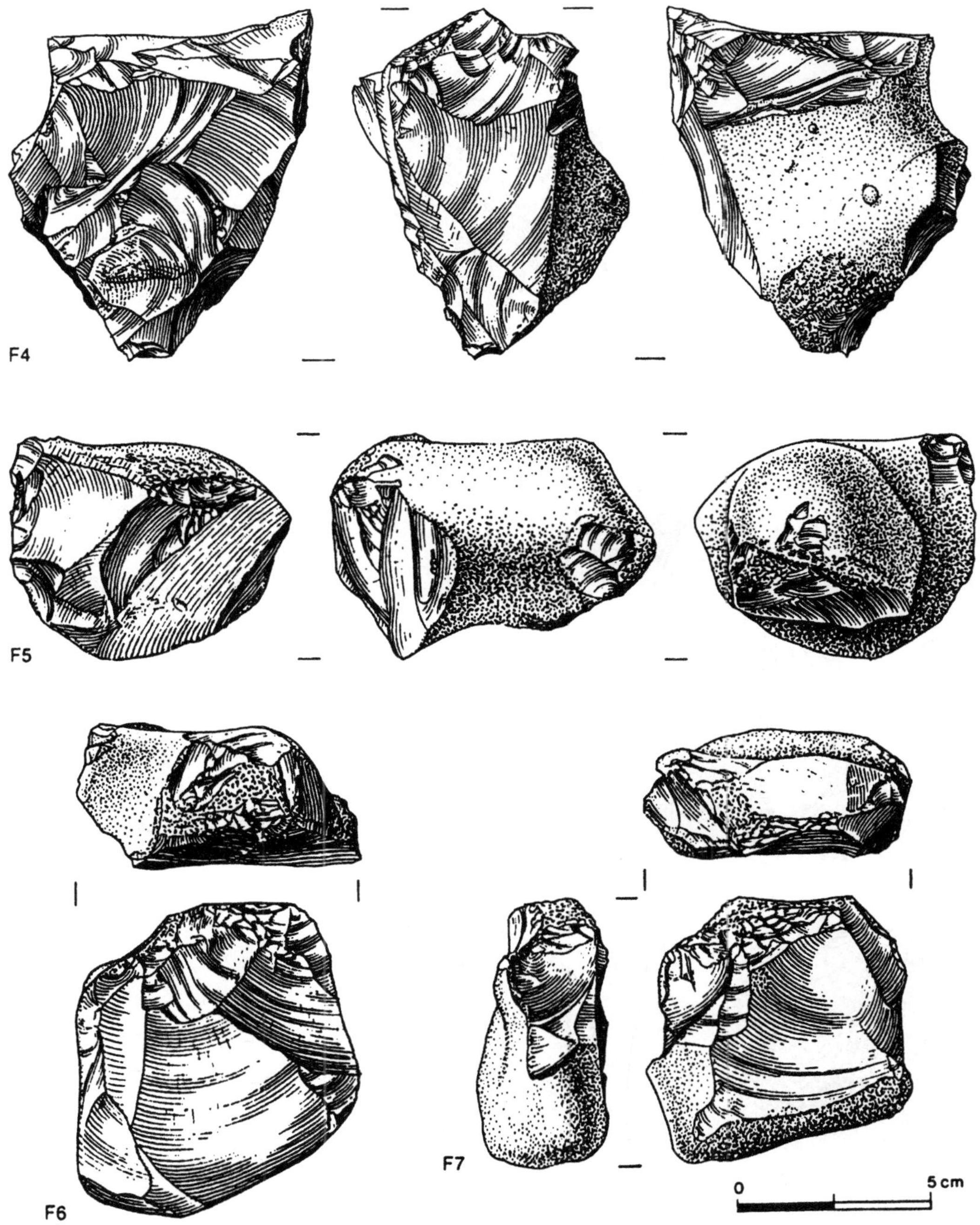

Figure 6.3: Floorstone and non-floorstone cores and core tools from Shaft X, Grimes Graves, Norfolk (after Herne 1992, 89).

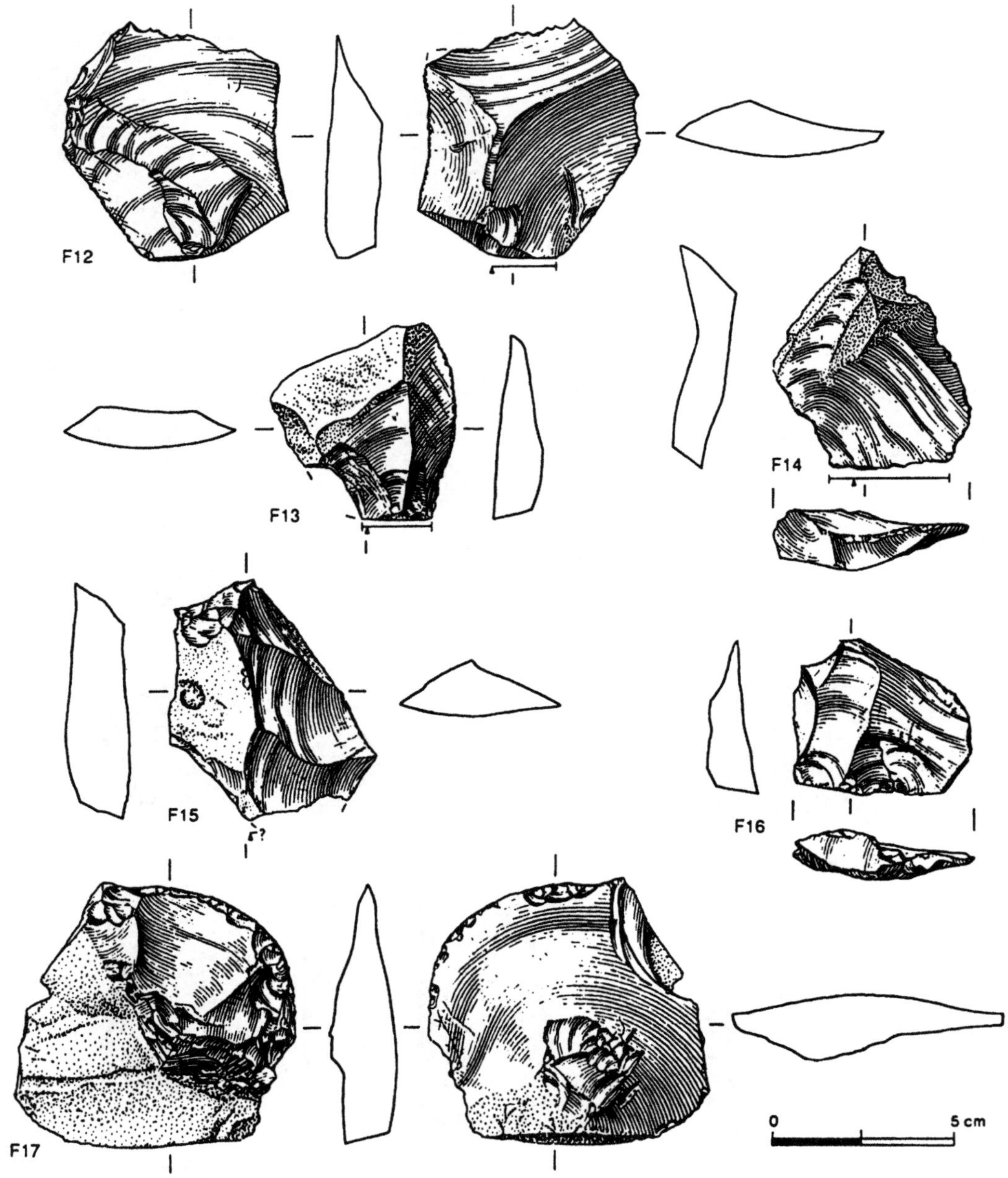

Figure 6.4: Flake tools from Shaft X, Grimes Graves, Norfolk (after Herne 1992, 84).

> and connections, the boundaries between formal artefact categories may have become increasingly blurred. Indeed the learning of complex knapping techniques may itself have ceased to be an important feature in the lives of many people. (Edmonds 1995, 188)

On this reading, flint and stone tool production and use may gradually have been located solely in the domestic sphere. The range of different, easily recognisable, tools gradually decreased until they became entirely functional and utilitarian.

6.5 Implications for the understanding of later flint use?

If all of the above holds good for the fate of flint and stone tools in the Middle Bronze Age, why can researchers not accept that similar processes may have continued on in time as the domestic mode of production became more important in certain areas (see *e.g.*, Hill 1994, 1995a,b, on Iron Age Wessex)?

It is at this juncture that we come up against one of the major stumbling blocks in a serious consideration of the potential later uses of flint: archaeologists' own accepted wisdoms and their own mind-set. No-one has a problem with Middle Bronze Age flint working and the gradual decline of knapping skills, or with the fact that the range of recognisable tools became severely restricted as flint use became more embedded in the domestic sphere. Once formal tools, (which can be easily assigned to typological categories and clear 'period' divisions) disappear, however, we are left with low quality, undiagnostic material like 'scrapers' and 'awls'. As a result, the archaeological recognition of later flint assemblages becomes a highly variable circumstance.

Here, we run the risk of getting caught up in a downward spiral: the range of easily classifiable and dateable tools decreases over time, and therefore workers do not expect to recover contemporary Later Bronze Age or Iron Age flint assemblages from excavations. As a result, excavators give up looking for such material and there is thus a tendency to write off everything from post-Middle Bronze Age contexts as 'residual'.

If, from the Middle Bronze Age onwards, lithic use is largely functional and utilitarian, then the rate and scale of flint utilisation on later sites will depend on a range of variables such as the availability of flint, its ease of procurement, resource perception on the part of local communities and regional variation in traditions of stone working. As a result it is probably unwise to talk in blanket terms about an end to flint working. It may have declined at different times in different places and it may become more difficult over time for us to tell the difference between *ad hoc* and regular use of flint. Maybe we need to alter our attitudes to take account of this and pay more attention to lithic material on later sites on a site by site basis. Particular attention should be paid to contexts of deposition and to other artefactual associations (as highlighted above), to the nature of any earlier settlement on sites, and also to the potential recycling of lithic material.

With these observations and the list of observable criteria set out below, we believe we can go on to tackle the difficult problem of flint use on complex sites like the Meare Lake Village and Glastonbury (Orme *et al.* 1983; Saville 1981; Smith 1981) but this is a topic for a future paper. This present attempt to isolate characteristics of potentially late flint assemblages is a statement of research in progress. In essence it has really been a plea for a reconsideration of accepted ideas about the use of flint in the 'metal age'. We would end this contribution by simply urging all excavators of Later Bronze Age and Iron Age sites who read this to pay more attention to material that might easily be relegated to the residual category. It might, as solicitors are wont to say, 'provide us with information that could be to our advantage'. We would suggest that the following are the main criteria to consider:

- Utilisation of highly localised raw materials — some of which may be very low quality.
- Low overall assemblage numbers.
- Simple core/flake technology — using hard hammer, direct percussion.
- Lack of skill in knapping, evidenced by:
  - obtuse striking angles
  - high instance of step or hinge terminations
  - thick, wide striking platforms
  - irregular dorsal flake scar patterns on flakes
  - short, squat flakes — L/B ratio 1:1
  - high instance of chips and chunks.
- Restricted range of formal tool types *e.g.*, scrapers, awls *etc.*
- Crude hammerstones.
- Predominance of secondary and inner flakes.

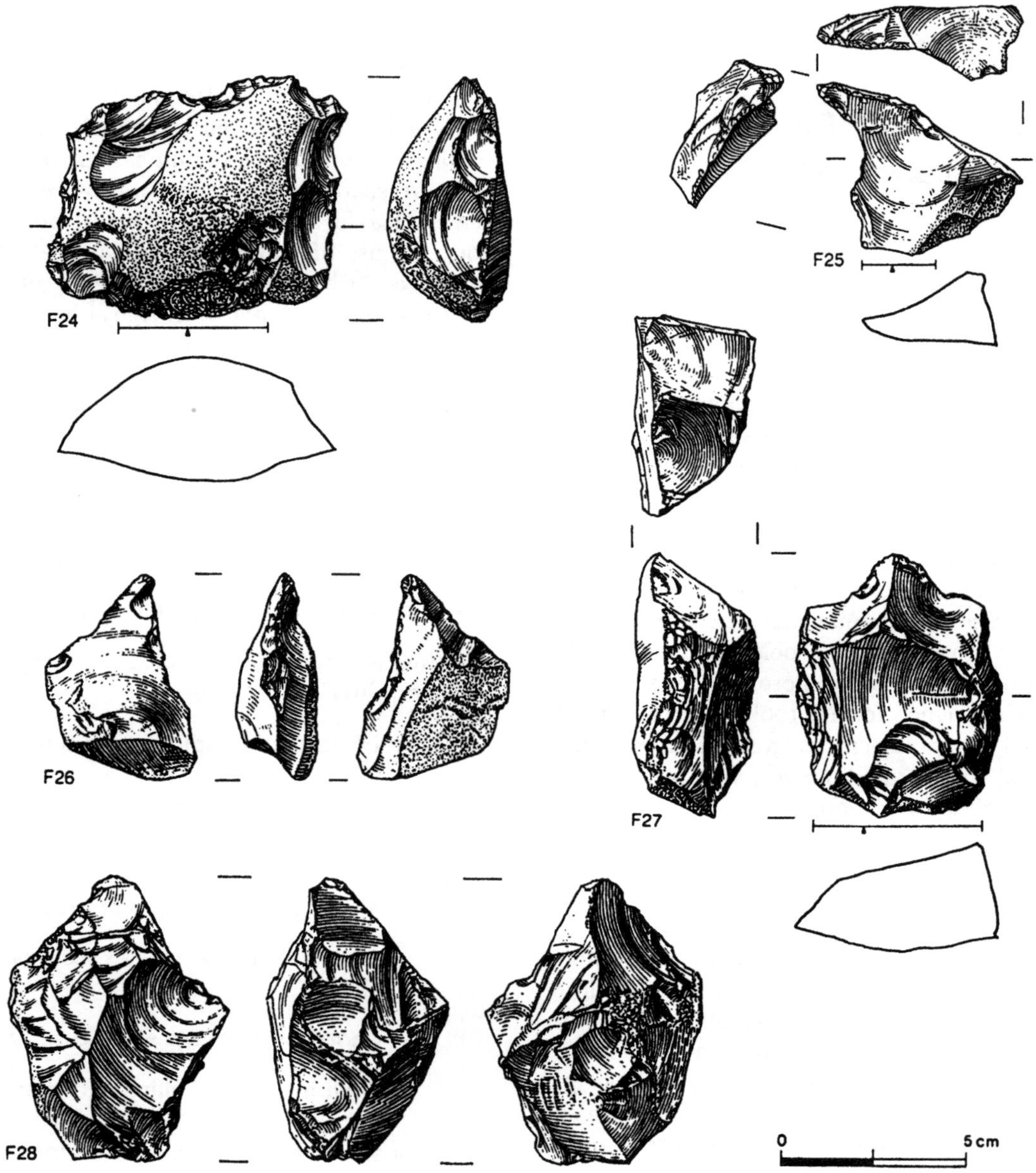

Figure 6.5: Flake and core tools from shaft X, Grimes Graves, Norfolk (after Herne 1992, 88).

References

AUSTIN, L. 1994. 'Flint report.' *In* M. Medlycott, *Iron Age and Roman material from Birchanger, near Bishops Stortford; excavations at Woodside Industrial Park, 1992*, pp. 43–4. *Essex Archaeology and History*, 25: 28–45.

BARRETT, J. C. 1994. *Fragments from Antiquity: an archaeology of social life in Britain, 2900–1200* BC. Blackwell, Oxford.

CLARK, D. V., T. G. COWIE & A. FOXON 1985. *Symbols of Power at the Time of Stonehenge.* HMSO, London.

CLARK, J. D. G. & C. I. FELL 1953. 'The Early Iron Age site at Micklemoor Hill, West Harling, Norfolk, and its Pottery.' *Proceedings of the Prehistoric Society* 19: 1–40.

CLARK, J. G. D. 1960. 'Excavations at the Neolithic site of Mildenhall, Hurst Fen, Suffolk, 1954, 1957 and 1958.' *Proceedings of the Prehistoric Society* 26: 202–45.

COOPER, L. & J. L. HUMPHREY 1998. 'The lithics.' *In* M. Beamish, *A middle Iron Age site at Wanlip, Leicestershire. Transactions of the Leicester Archaeological and Historical Society*, 72: 63–74.

EDMONDS, M. R. 1995. *Stone Tools and Society: Working Stone in Neolithic and Bronze Age Britain.* Batsford, London.

HERNE, A. 1992. 'The lithic assemblage.' *In* I. Longworth, A. Herne, G. Varndell & S. Needham (eds.), *Excavations at Grimes Graves, Norfolk, Fasicule III*, pp. 21–75. British Museum, London.

HILL, J. D. 1994. 'Why we should not take the data from Iron Age settlements for granted: Recent studies of intra-settlement patterning.' *In* A. P. Fitzpatrick & E. L. Morris (eds.), *The Iron Age in Wessex: Recent Work*, pp. 4–8. Association Française d'Etude de L'Age du Fer and the Trust for Wessex Archaeology, Salisbury.

HILL, J. D. 1995a. 'How should we understand Iron Age societies and hillforts? a contextual study from southern Britain.' *In* J. D. Hill & C. G. Cumberpatch (eds.), *Different Iron Ages: Studies on the Iron Age in Temperate Europe*, pp. 45–66. British Archaeological Reports International Series 602, Oxford.

HILL, J. D. 1995b. *Ritual and Rubbish in the Iron Age of Wessex: a study on the formation of a specific archaeological record.* British Archaeological Reports British Series 242, Oxford.

HODDER, I. 1986. *Reading the Past.* Cambridge University Press, Cambridge. Second edition 1991.

ORME, B. J., J. M. COLES & R. J. SILVESTER 1983. 'Meare Village East 1982.' *Somerset Levels Papers* 9: 49–73.

SAVILLE, A. 1981. 'Iron Age flint working — fact or fiction?' *Lithics* 2: 6–9.

SHANKS, M. & C. TILLEY 1987. *Re-constructing Archaeology.* Cambridge University Press, Cambridge.

SMITH, A. J. 1981. 'Flint.' *In* B. J. Orme, J. M. Coles, A. E. Caseldine & G. N. Bailey, *Meare Village West 1979. Somerset Levels Papers*, 7: 65–6 & Fiche F1.53–66.

STEVENS, J. 1993. *An Analysis of Flint From Warren Hill Enclosure, Salisbury Plain, Wiltshire.* Master's thesis, School of Archaeological Studies, University of Leicester.

TILLEY, C. 1989. 'Interpreting material culture.' *In* I. Hodder (ed.), *The Meanings of Things*, pp. 185–94. Cambridge University Press, Cambridge.

WINHAM, R. 1985. 'The flint.' *In* P. Fasham (ed.), *The prehistoric settlement at Winnall Down, Winchester: excavations of MARC 3 Site R17 in 1976 and 1977*, pp. 84–7. Hampshire Field Club in association with the Trust for Wessex Archaeology, Winchester.

Jodie Humphrey
School of Archaeological Studies
University of Leicester
University Road
Leicester LE1 7RH
England
`jlh11@le.ac.uk`

Rob Young
Northumberland National Park Authority
Eastburn
South Park
Hexham
Northumberland
`robert.young@nnpa.org.uk`

7
Later Prehistoric and Historic use of Flint in England

Hazel E. Martingell
Braintree, England

Abstract

My purpose in presenting this paper is to raise awareness that not all stone flakes found in the landscape are either prehistoric or natural. I propose therefore to comment on a continuing craft of flint knapping from its supposed demise in the late Iron Age up to the present day, and to provide criteria for some of the flakes and flaked stone pieces found in this period. The categories of struck flint discussed in this paper include: squat flakes, strike-a-lights, gunflints, threshing sledge flints (*tribulum*), flint used in architecture, craft pieces, replicas and forgeries.

7.1 Introduction

Early in 1998, I was field walking at East Harling, Norfolk, England, with the land owner James Chapman, looking for worked flint. I mentioned my interest in the continuing use of flint tools, even up to the present day. Mr Chapman replied that his father used sharp flints for cutting string while working in his fields and sheds, as it was much more convenient than using a penknife which would become rusted and blunt. This reminded me that during the 1980s a Mr Carter from near Chelmsford in Essex, demonstrated a method of flint knapping he had been shown in his boyhood. He held an oval river pebble in his left hand and with a similar pebble in his right hand, he struck flakes from one end rather in the manner of slicing salami. The resulting flakes had no platforms, some were edged all around with cortex and needed trimming to provide a sharp edge, but others fell away as sharp edged simple flakes with a positive bulb on the ventral surface opposed to a negative bulb on the reverse, dorsal surface. According to hearsay these sharp flakes obtained from this simple method of knapping have long been in common use by poachers of game, for skinning animals. Rabbits, for example, may be skinned and cut up on the spot, the sharp flakes then discarded become unrecognisable amongst all the other chipped stone in the landscape, a useful attribute of these tools should a gamekeeper appear on the scene. These flint flakes are of the simplest kind and require the minimum of skills to make but, where they occur in contexts with prehistoric flint artefacts, it is important to be able to distinguish the ancient from the recent. Recently knapped flint pieces have a dry-looking surface, unlike the glossy/waxy look of most prehistoric knapped flints. The flake edges may be very sharp and brittle and when in use cut with a tearing quality comparable to a modern day serrated knife. More research is required into the specific types of recently flaked artefacts.

7.2 Squat flakes

Another type of simple flake is of more interest. Here there seems to be an idea behind the knapping, which is to create a flake that is both strong and sharp, that can be struck from a casually picked-up surface flint or from the best quality mined material.

The distinguishing features of these flakes (Fig. 7.1, 1–3) are:

1. The width of the platform which may be greater than the length of the flake.
2. The marked obtuse angle of the platform to the ventral surface, which may also be the deepest part of the flake.
3. The prominent and large bulbs of percussion, which produce a wedge shaped profile, or a sinuous profile if the flakes are removed in sequence.

The cores are also distinctive:

1. They have a bold 'cusped' and sometimes denticulate appearance due to the negative impressions of the removed flakes.
2. The platform is large and flattish, which may be either a natural cleavage plane, a thermal split or a shallow faceted surface.
3. Occasionally the sharp curved edges at the junction of the platform and flaked surface show signs of heavy utilisation as notched pieces.

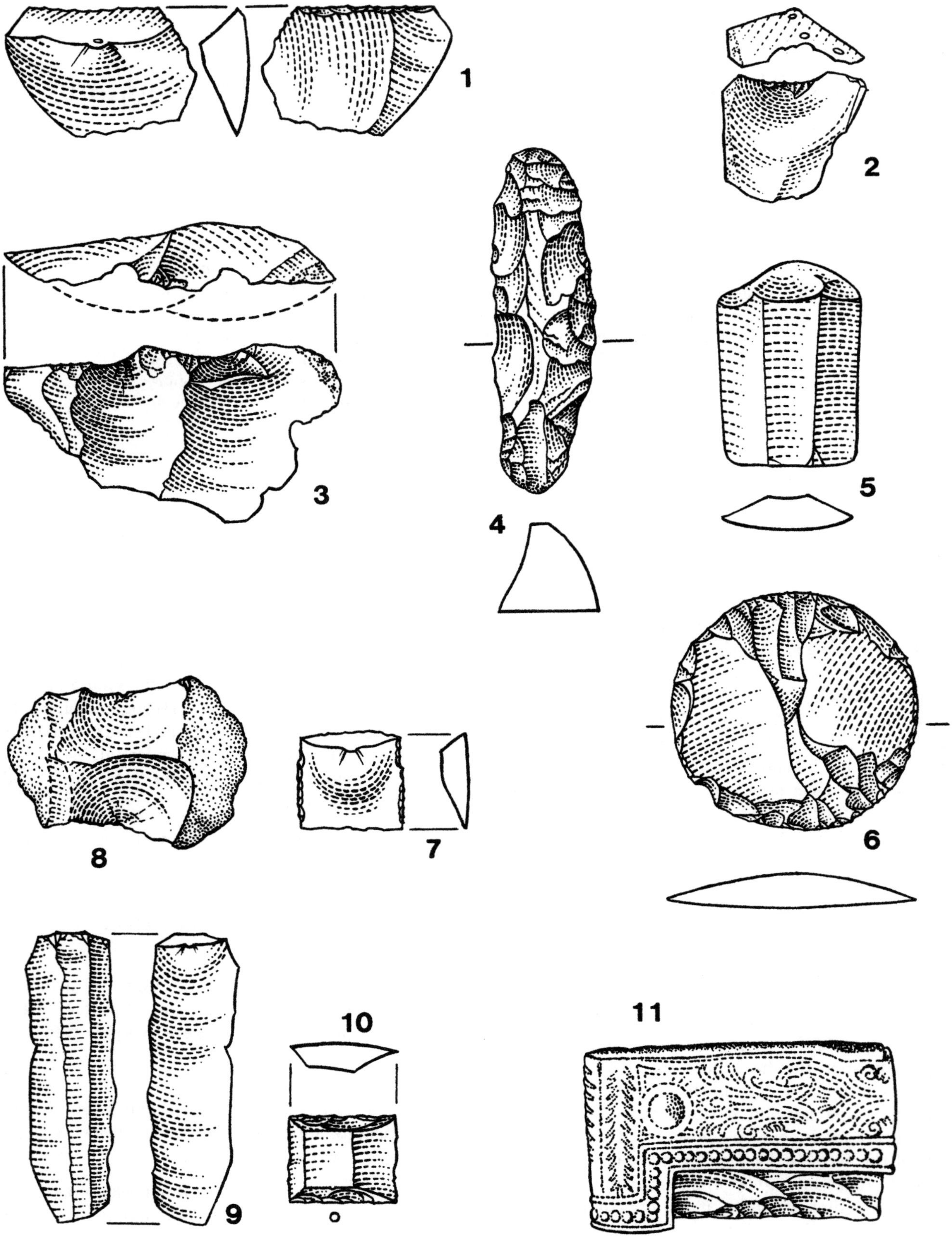

Figure 7.1: 1: Squat flake, schematic interpretation; 2: Squat flake; 3: Core with squat flake removal: 4: Strike-a-light, fabricator type; 5: Strike-a-light, scraper type; 6: Manufactured Strike-a-light; 7: Gunflint, wedge type; 8: Gunflint core, wedge type; 9: Gunflint blade for platform type of gunflints; 10: Gunflint, platform type; 11: Strike-a-light, bifacial and decorated lead holder.

The method by which these flakes are obtained appears consistently from the late Bronze Age, if not earlier, up to recent times (see also Humphrey and Young, this volume).

The squat flake was first recognised by Graham Clark in 1935 whilst studying the lithic assemblage from the Late Bronze Age site of Mildenhall Fen, West Suffolk (Clark 1936, 46–7). He noticed 'forms peculiar to this industry'. One group (d) he describes as:

> a group of flakes struck at a remarkable flaking angle and with trimmed striking platforms ... It will be seen that the angles formed by the intersection of the striking platforms and the primary flake surfaces are obtuse to a very marked degree. In consequence of this the flakes, of which there are a fair number from the site, are squat, being normally considerably shorter than the width of their platforms.

This appears to be the earliest assemblage to contain squat flakes in any number and is possibly the first report to define the type. Later, in 1953, Clark again recognised squat flakes amongst the assemblage from the early Iron Age site of Micklemoor Hill, West Harling, Norfolk (Clark & Fell 1953) and noted the similarity with those found at Mildenhall Fen. He writes:

> There can be no doubt that the Early Iron Age inhabitants made and used flint implements, though the range of these was restricted.
>
> 1. Of the 200 one in four was struck from the outer cortex of the parent nodule without the preparation of a striking platform.
> 2. In the case of those with striking platforms the angle between platform and primary flake surface was obtuse and as a rule markedly so.
> 3. In one out of ten the edge of the platform had been trimmed or strongly utilised so as to form either a convex scraping edge or one or more notches. It may be noted that the production of squat primary flakes ... has been observed at Late Bronze Age II settlements in East Anglia (Mildenhall Fen) and Wessex (Thorny Down) ...
> 4. Cores ... roughly made ... some from which squat flakes had been struck from an obtuse angle
>
> (Clark & Fell 1953, 34–5)

The first assemblages in which squat flake appear with a clear function is in context with the shale industries at Kimmeridge, Dorset (Davies 1937), originally dated within the Early Iron Age but now considered to be later, possibly Roman. The excavator noted:

> The appearance vividly suggests that here sat a worker, cutting the shale with sharp flint flakes, throwing down pieces that broke and the flint flakes as they blunted and striking off fresh flakes as he needed them.
>
> (Davies 1937, 212)

Many of these flakes possess the criteria for squat flakes, although some of the bulbs are not central to the wide platform but at one corner.

From the Iron Age on, and in the relevant literature, information regarding flaked stone tools becomes extremely sparse. It is as though they ceased to be used, which may account for the popular concept that the craft of flint knapping died out at this time. That this could not have been so has been disregarded. In studying prehistoric flint assemblages the emphasis has always been the cutting and scraping attributes of the artefacts. There is seldom comment about artefacts used to create a spark to light a fire or a lamp. These implements would have been essential from prehistoric times to the arrival of the lucifer (an early form of match) in the 19th century.

7.3 Strike-a-Lights

> The strike-a-light was the continuing tool from the Neolithic era to the First World War, when they were in demand owing to the match scarcity. The shaping of a flasher, strike-a-light, or gun flint required the same technique.
>
> (Forrest 1983, 8)

> The art of creating fire was known to Stone Age man and among his artefacts, flints which are thought to be strike-a-lights are occasionally found. These tools have edges which are polished from being struck repeatedly against crystals of iron pyrites to give a shower of sparks for kindling fire. (Mason 1978, 46)

Forrest here suggests that a strike-a-light and a gun flint are similar in form and Mason describes the edge-wear. With this information, and there is more, it should be possible to recognise any strike-a-lights in an assemblage of stone tools. But in practice this is not the case. First the shapes vary considerably: they may be rod-like (fabricators), scraper-like, circular discs, oblong or square and they may be unifacial or bifacial (Fig. 7.1, 4–6). Secondly, the edge-wear looks much like any smooth edged piece. To date, positive identification has to rely on the context from which they have been recovered and these are mostly burials.

One of the earliest references to a strike-a-light was the one recovered from the Bronze Age barrow of Green Low, Derbyshire. The excavators found 'a piece of spherical pyrites or iron ore and a flint instrument of the circular-headed form' (Bateman 1848, 55). And in later excavations at Simondston Cairn, Bridgend, Glamorgan, a Bronze Age cairn of about 1400 BC was opened and from the primary burials a 'flint fabricator whose rounded points show the smoothed surface characteristic of that tool' was recovered (Fox

1959, 82–3). A similar fabricator type of strike-a-light was found in the Bronze Age barrow at Rudstone, Yorkshire by the Rev. W. Greenwell. It is described as 'a narrow, thick, external flake the end of which has been trimmed to a semi-circular bevelled edge; a portion of one side has also been trimmed'. It was found with half a nodule of iron pyrites which lay above it. Evans illustrates both these pieces (Evans 1872, 284).

It is by no means certain that all rods/fabricators were used as strike-a-lights. As the name implies they may also have been used as punches in the manufacture of flint blades and some for working leather. It is probable that the preferred shape for a strike-a-light resembled a later prehistoric scraper, which makes identification extremely difficult in a large assemblage from mixed periods.

In the 1860s Evans purchased a:

> modern strike-a-light or gunflint ... at Pontlevoy, France ... made of a segment of a flake, one edge and the sides of which have been trimmed to a scraper-like edge and the other merely made straight. The resemblance between this and some of the ancient 'scrapers' is manifest.

He also purchased a modern strike-a-light from Treves, France, which was

> ... about $2 \times 1\frac{3}{8}$ inches, a flat flake, trimmed to a nearly square edge at the butt end and to a very flat arc at the point, both the trimmed edges being of precisely the same character as those of a scraper. By working such a flint and a steel/briquet together, much the same bruising of the edge is produced as that apparent on some of the old 'scrapers'.

Evans then concludes

> that a certain proportion of these instruments were in use, not for scraping hides like the others but for scraping iron pyrites and not improbably in later days iron or steel ... used for such a purpose they often present a bruising of the edge and an irregularity of form ... accounting for their great abundance.
> (Evans 1872, 282–3)

The Abbé Cochet, while excavating Merovingian interments in France in the 1870s, wrote to Evans that he had found implements 'resembling gunflints; one of these was apparently carried at the waist in a purse with money and other necessaries' (Evans 1872, 282). Similarly, in England at about the same time, a large number of Anglo-Saxon purse-mounts have been found in association with flints (Brown 1977).

While the town of Brandon in Norfolk is famous for the manufacture of gunflints, strike-a-lights were also made there, out of the large flakes the knappers had rejected as unsuitable for trimming into gunflints. These were usually circular discs about two inches in diameter; they might be unifacial or bifacial and some had fine flaking resembling retouch around the circumference. They can be very neatly made perfectly circular and thin in section.

The strike-a-light, although in time used with a tinder-box, underwent little change as a knapped oval or horseshoe-shaped flint. In Continental Europe however, the practice of carrying in the pocket a piece of flint and some prepared tinder continued into the 1870s at least. In 1887 an interesting paper was published, with illustrations, entitled *The Flint Knappers Art* in Albania. This describes strike-a-lights observed by the author. They are well flaked bifacial pieces square or oblong in outline and partially encased in ornamental lead sheaths studded with glass gems (Fig. 7.1, 11) (Evans 1887). These were obviously treasured possessions.

By contrast, in many regions, almost any piece of flint would have served to make a spark. But it should be borne in mind that as outcrops of flint are absent in many places, it has to be imported. These 'imports' needed to fulfil some concept of shape and size suitable for the job of making a spark. Some strike-a-lights remained in use until the 1920s at least (Forrest 1983).

7.4 Gunflints

Much information on gunflints is available in the many publications of Mr Seymour de Lotbiniere, who lived at Brandon Hall, Brandon in the 1970s. He gave many exciting talks, with demonstrations of firing flintlock pistols and muskets. Lotbiniere concluded there was no connection between the Neolithic scraper and gunflints (de Lotbiniere 1977). Skertchley in his report for the Geological Survey of England and Wales, 'argued the gunflint evolved from the medieval strike-a-light ... and this owed its genesis to the Neolithic scraper' (in Forrest 1983).

The earliest gunflints appear from about 1575. These were a do-it-yourself, horseshoe-shaped and wedge-profiled type of gunflint, very similar in appearance to squat flakes and made exclusively of surface flint. From 1660 to 1785 a craft-produced horseshoe-shaped and wedge-profiled gunflint, (Fig. 7.1, 7–8) known as 'Englishmen' was added to the do-it-yourself supply. In publications referring to the production of strike-a-lights and gunflints from workshops that were manufacturing commercially on a relatively large scale, the knappers are referred to as 'craftsmen'. This distinguishes them from individuals who casually knapped flints strictly for their own use. These craft-produced gunflints were described as 'flints ready cut'; they were regular issue for garrisons overseas and were supplied in lots of 5,000 or 10,000. The flint was obtained as a by-product from chalk quarries. The waste cores from this manufacture were large, thick-sectioned flakes, which after the removal of the gunflint flakes, became flaked flakes, a term also used to describe certain Clactonian artefacts, which can be very similar in appearance (McNabb & Ashton 1990). This waste, in general, remained in the chalk quarries or caves, but scatters or heaps of gunflint waste of flakes and cores have been found on various open sites. In a letter about gunflint manufacture at Greenhithe, there is mention of 'flint heaps between the river and the old main road' (de Lotbiniere 1980). What became of this waste is unknown.

From 1785 to 1980 a craft produced 'platform' type of gunflint, made from sections of blades became preferred (Fig. 7.1, 9–10). This different technique for the manufacture of gunflints appears to have been introduced from France in the mid-1780s and the gunflints were therefore known as 'Frenchmen'. Dr James Mitchel described this flint craft work as he observed it at Brandon, Norfolk (reported in Forrest 1983, 90). First the quarterer or 'cracker', broke up the large blocks of flint from the mines. Secondly the flaker struck off long blades with the skilful use of a pointed French hammer which

> ... when applied to a good running flint, resulted in flakes six to eight inches in length, either double or single backed with equal flanges.

These he then graded large, medium and small into different tins. Finally,

> ... the knapper uses a hammer which is merely a plate of steel extending two inches on each side of the handle and an inch in breadth and not above 1/6th of an inch in thickness. He takes into his left hand one of the flakes and lays it over the little anvil on the block, and with his hammer breaks it into three or four flints. All he has to do after that is to see which edge will be best for the flint and from the other he breaks a little off and the whole is complete.

The platform type of gunflint was ordered by the Board of Ordinance for Admiralty and Navy. It could be supplied in 23 different patterns for a variety of fire-arms. By 1880 the Board serviced territories that ranged from China to the Andes and included America, South America, India, the West Indies and Africa.

There are at least three other uses for simple flint flakes accounting for their presence in the landscape: as part of threshing sledges, as building materials and as craft pieces.

7.5 Threshing sledge flints: the *Tribulum*

The *tribulum* is a Roman threshing implement with, usually, flint 'teeth' (Fig. 7.2, 12). 'They are flat sledges of wood, five to six feet in length and two or three in breadth. The underside is pitted with a number of square or lozenge shaped holes and having in each hole a flake or splinter of stone' (Evans 1872, 256). Of one example in the Christy Collection, Evans states 'each flake is of cherty flint and has been artificially shaped'. He hypothesises that this is another cause why rough flakes of flint should accompany Roman remains, especially in the case of villas in country districts. It may be that the *tribulum* was never used in northern latitudes or, alternatively, the timber frames have just not survived. As it is, there is no evidence for their use in Roman times or later in the British Isles, whereas they continued in use in the Eastern Mediterranean into the 20th century (Ataman 1989). In 1994 the author visited the Folk Art Museum in Kyrenia, North Cyprus; one of the exhibits was a thresher with flints stuck into planks. The curator recounted seeing these used in the fields when he was a boy and was fairly sure an old farmer he knew was still working with one.

7.6 Architecture

The use of stone for decoration on the external surfaces of buildings is described in many publications, but is included here only in its relation to knapped flint. From AD 1300 onwards flints were in much demand for decorative surfaces to abbeys, castles, churches and manor houses. The descriptive terms for the patterns that adorned these buildings were called 'flushwork' and 'chequered panelling'. The outer surface of each small block of flint would usually be flat, more often a natural fracture plane of the flint, whereas the surface held in place by the cement was knapped to give grip. When these pieces are examined the knapping is often of such good quality that they could easily be mistaken for a prehistoric product were it not for the unusual oblong or heart-shaped outline (Fig. 7.2, 13). They were often knapped on the site of the building; a typical example is at Long Melford church, Suffolk where the flints adorning New Tower, built in 1903, were knapped in the churchyard (Forrest 1983).

At the domestic level, cottages in East Anglia are frequently decorated with flints, the flints being placed either 'random' or 'faced', the latter were rounded or squared. Sometimes flakes were also employed to decorate cottages, being inserted on edge into the unset mortar. This resulted in the sharp sides protruding and is known as 'Galleting'.

The building industry absorbed thousands of tons of flint waste from knapping yards. It was sometimes used to provide sound foundations for buildings and also railway tracks.

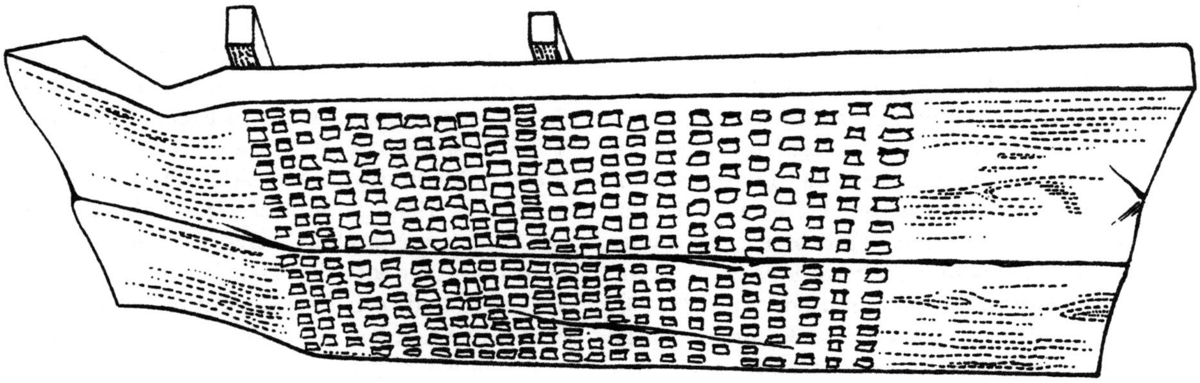

12

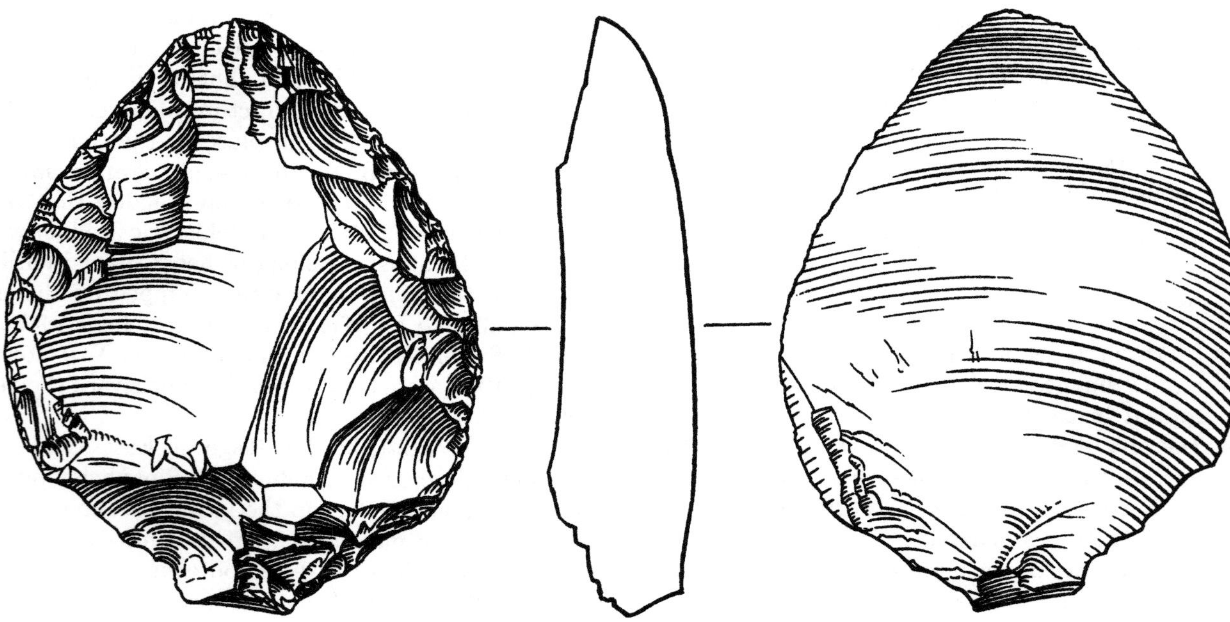

13

Figure 7.2: 12: *Tribulum*; 13: Architectural decorative flint.

7.7 Knapped flint craft pieces, replicas and forgeries

During the 19th century there was a growing interest in things antiquarian. A small trade developed in knapped stone artefacts, particularly flint arrowheads and handaxes. At first knappers replicated these ancient weapons for sale to collectors; some knappers became quite famous at recreating prehistoric artefacts, especially Mr Edward Simpson, known as 'Flint Jack'. Unfortunately, as by-and-large they were intended to dupe the collector, many of these pieces, being regarded as forgeries, were subsequently withdrawn from collections despite the admirable skill employed in making them (for a thorough account of this subject see Munro 1905). At Brandon during the 20th century craftsmen have knapped special pieces demonstrating a fascinating use of flint. These include the necklace and alphabet knapped by Bill Basham and the necklace and bangles of Fred Snare. Happily this craft continues to this day.

7.8 Conclusion

This paper intends to do no more than give a brief overview of the continuing use of knapped flint. The items discussed, strike-a-lights, gunflints, threshing sledge flints, the architectural use of flint and replicas, each have their own specialists. These specialists of the historic use of flint tend to work within a specific historic period; their interest in these pieces is related to the behaviour of people in the recent past. It is possible, therefore, for prehistoric and historic research to develop separately and this may account for the belief that flint knapping ceased with the arrival of the Romans, an idea the author found prevalent in the 1960s–70s.

> When the knappers of axeheads, knives, scrapers, borers, microliths, sickles and arrowheads deserted (the flint mines of) Grimes Graves in about 1500 BC, no-one could have known or guessed that ahead lay entirely new ages of flint craft. (Forrest 1983, 125)

References

ATAMAN, K. 1989. *The Chipped Stone Assemblage from Can Hasan III.* Ph.D. thesis, University of London.

BATEMAN, T. 1848. *Vestiges of the antiquities of Derbyshire and the sepulchral usages of its inhabitants from the most remote ages to the Reformation.* London.

BROWN, D. 1977. 'Firesteels and pursemounts again.' *Bonner Jahrbücher* 177: 451–77.

CLARK, J. D. G. 1936. 'Report on a Late Bronze Age site in Mildenhall Fen, West Suffolk.' *Antiqaries Journal* 16: 46–7.

CLARK, J. D. G. & C. I. FELL 1953. 'The Early Iron Age site at Micklemoor Hill, West Harling, Norfolk, and its Pottery.' *Proceedings of the Prehistoric Society* 19: 1–40.

DAVIES, H. F. 1937. 'The shale industries at Kimmeridge, Dorset.' *Archaeological Journal* 93: 200–19.

DE LOTBINIERE, S. 1977. 'The story of the English gunflint.' *Arms and Armour Society Journal* 9(1): 18–53.

DE LOTBINIERE, S. 1980. 'Gunflint enquiry.' *Kent Archaeological Review* 59: 198–201.

EVANS, A. J. 1887. 'On the flint-knapper's art in Albania.' *Journal of the Anthropogical Institute of Great Britain and Ireland* 16: 65–7.

EVANS, J. 1872. *The Ancient Stone Implements, Weapons and Ornaments of Great Britain.* Longmans, Green, Reader and Dyer, London.

FORREST, A. J. 1983. *Masters of Flint.* Dalton, Lavenham, Suffolk.

FOX, C. 1959. *Life and death in the Bronze Age : an archaeologist's fieldwork.* Routledge & Kegan Paul, London.

MASON, H. J. 1978. *Flint, the Versatile Stone.* Providence Press, Ely, Cambridgeshire.

MCNABB, J. & N. ASHTON 1990. 'Clactonian gunflints?' *Lithics* 11: 44–7.

MUNRO, R. 1905. *Archaeology and False Antiquities.* Methuen and Co., London.

Hazel E. Martingell
27 Sandwich Close
Braintree
Essex CM7 5RZ
England

8
Lithic Scatters and Dynamic Archaeology

Chris Barrowman

Department of Archaeology, University of Glasgow

Abstract

Lithic scatters are an important part of the prehistoric material record, and have been studied and interpreted in a variety of different ways. These studies generally start from the premise that scatters are an indication of prehistoric activity, but they can also be used to illustrate to a wider audience the fact that archaeology is not only about upstanding monuments or isolated to specific locales, it is part of our whole landscape. Even a ploughed field may hold a part of the past. Lithic scatters are created by our actions today; they represent a dynamic and fluid archaeology.

8.1 Introduction

> **lithic** /ˈlɪθɪk/ *adj.* **1** of, like, or made of stone ... [Gk *lithikos* f. *lithos* stone]
>
> **scatter** /ˈskætə(r)/ ... *v. & n.* ... **3** the extent of distribution of ... [ME, prob. var. of SHATTER]
>
> *The Oxford English Reference Dictionary* (Pearsall & Trumble 1995, 837 & 1293).

A surface lithic scatter within the context of a ploughed field has been created in the present. This discussion is concerned with the rôle surface lithic scatters can play in a social context today. It is in no way an attempt to clarify what scatters represent in terms of the prehistoric activities from which they derive; these themes are expressed elsewhere (see Stuart, this volume). A simple statement of the author's sentiments follow.

8.2 Lithic scatters

Let us assume that archaeology is represented by a vast array of material remains, ranging from upstanding monuments to buried flakes of flint. If a hierarchical system is created from these remains, we may assume that the biggest and best preserved of the monuments is considered the most archaeologically important in terms of information which may be gained by academics now and in the future, or in educating an audience who have a limited knowledge of prehistoric material culture. Little concern is given to the lower tiers of the hierarchy — and this is where lithic scatters lie due to their physical characteristics. Being described in the past as only signatures of a site (Schofield 1991a), the plough that creates a scatter is writing a *summary* of the archaeological story that lies beneath the soil. In some cases this may not be a true résumé as the surface collection of lithic scatters does not represent the whole and may give a misguided impression of what is in the rest of the topsoil and therefore any surviving archaeological remains underneath (Dunnell 1992; Shott 1995).

The result of the destruction of original archaeological deposits is a scatter of the material component which had a direct relationship to that original context, its spatial relationship having been changed by modern activity. I am assuming that the older the scatter, that is the more times the field in which the scatter is held is ploughed, the more the lithics will be decontextualised and spread throughout the topsoil or ploughzone (Boismier 1997). This obviously depends on the complexities of the agricultural regime, combined with natural processes, but for the benefits of this paper let us assume that 'lithic scatter' equals 'damaged archaeological deposits' or, that some elements of the material culture have been moved from their *original* context. I am also assuming that a lithic scatter represents the decontextualised lithic material which is uncovered through surface collection *i.e.*, those parts which are visible on the surface of the ploughsoil.

Through my fieldwork for the Scottish Lithic Scatters Project, I have had many dealings with scatters throughout South and Central Scotland, and can seriously say that the majority of our heritage lies within ploughed soil. As Wickham-Jones and MacKenzie state in a recent paper:

> Much current archaeological work is biased towards upstanding remains or corresponding site remnants (*i.e.*, crop marks). While not wishing to detract from their value, it should also be recognised that much of Scotland's past is not represented by upstanding

> material, nor by upstanding monuments and crop marks. Early sites are often unknown in advance of the work that may reveal them, and they may well be represented by lithic assemblages alone. It may be hard to identify them given the current pace and mechanisation of development. Unless we can identify them, however, and take steps to recognise their worth, we shall be losing significant elements of Scottish prehistory.
>
> (Wickham-Jones & MacKenzie 1996, 15)

It is clear that this archaeological resource is not recognised as being as important as other sites. Why is this?

One reason is due to the fact that lithic scatters are hard to see and so do not warrant the status given to upstanding monuments. Even when one can be seen, through walking a ploughed field and scrutinising the ground at every step, they are still not very interesting to look at. They do not work well as visual stimuli — a stone circle will evoke thoughts of past activity and the viewer may think about how it was built, and the meaning behind it. A ploughed field containing a lithic scatter is not quite as emotive. It does not grab your attention in quite the same way.

Another reason why surface lithic scatters are not recognised as being major archaeological resources is because they are accepted as being ***destroyed*** archaeology, as I mentioned above. The decontextualised surface material is not necessarily representative of all the lithics within the ploughzone, and without more intensive analysis it is considered unwise to base interpretations on such an ephemeral entity.

So what can we tell from surface scatters given that they are only a fraction of the lithic material in the ploughzone and represent damaged if not totally destroyed archaeological contexts? Much research has been undertaken over the past twenty years on artefact scatters and what they represent. Geographical models have been created which take scatters as indicators of sites within the landscape (Allen 1991; Wagstaff 1991). 'Distributed artefactual remains', as described by Foley (1981), have been used to illustrate 'off-site archaeology' and explain land-use patterns. Actualistic models have been created to study the effects of tillage and natural processes on the distribution of artefacts throughout the ploughzone (Boismier 1997; Clark & Schofield 1991), the results of which have been used to prove or disprove theories on the representation of surface scatters to the whole assemblage within the ploughsoil. In some cases lithic surface scatters are seen as being representative of the archaeology from which they are taken, in others the potential information gained from studying them is seen as being minimal and misguided. Interpretations of the artefacts which make up the scatters through macroscopic and microscopic investigation are numerous, and indeed seen as necessary if any information is to be derived from the scatter (*e.g.,* Clarke 1989). In this way simple biographies of the tool assemblage may be built up. In addition to this, recent examinations into the value of scatters have been discussed by Schofield (1995), which places the resource into the context of social theory.

All the above analyses concern the work undertaken on the surface material remains. Further investigation involving geophysical survey, test-pitting and excavation sometimes uncovers more information about the archaeology which the scatter is seen to represent (Barrowman 1998a,b; Barrowman & Stuart forthcoming; Bowden *et al.* 1991), and certainly helps to establish the relationship between surface and subsurface archaeology. I am not concerned with these aspects of surface scatter investigation in this paper.

We look at scatters from an archaeological viewpoint. Many problems are encountered when doing so because of the fragmentary nature of the resource. There is an apparent unreliability in knowing exactly what surface scatters represent. The truth of the matter is that they represent a very wide spectrum of past and present activities: past activities are desperately drawn out of the material residues; present activities are only looked at as a means on which to base formation process models (Schiffer 1987).

What exactly do scatters represent when it comes to present activities? How can they be made relevant to a wider public concern? Lithic scatters are created *today*, by activities occurring ***today***, by people living *today*. Lithic scatters are created by modern day activities which can be seen by people alive today. As a tractor ploughs a field it is re-structuring an element of the past, writing a new chapter of the biography of the lithics. The meanings they held as prehistoric debitage, tools or other social symbols has been changed, as it was when they were sealed within their archaeological contexts. This context has been destroyed, and a new one created. A lithic scatter has a modern day context, that is the ploughsoil of the ploughed field. The newly created surface lithic scatter results in a prehistoric material culture being redeposited into the dawn of the living. Lithic scatters bridge the gap to the past as they are an archaeological phenomenon which have been created today. They do not only represent prehistoric activity, they represent our activities on the landscape through time. They show that archaeology is about a past identity given new and changing values in the present. They represent a dynamic and fluid archaeology: dynamic because they are changing from one cultural value to another; and fluid because the material record is physically active, moving within and around the ploughsoil over time. The creation or appearance of a surface lithic scatter is not within archaeologists' control, unlike the exposure of material remains through excavation. Lithic scatters are part of a wider cultural system which encompasses how we work and live on the land today.

With this in mind, lithic scatters can be used to teach the public about the volume of history which our landscape has soaked up over the past few thousand years, as well as indicate the dynamic nature of the archaeological record, being influenced by what we do today. Lithic scatters can reveal many concepts which are normally hidden away and never dealt with through the usual public archaeology interface and teaching mediums. Museums deal with the exciting aspects of prehistory and illustrate their narratives with certain material objects. Popular archaeological texts have only recently started talking about them, such as Historic Scotland's publication *Scotland's First Settlers* by Caroline Wickham-Jones, which has a whole chapter dedicated to lithic scatter sites (1994, 62–74).

The presentation of scheduled monuments and other upstanding sites in the landscape completely ignores the presence of any lithic scatters around them, and it can be shown by the data contained in the Scottish Lithic Scatters database that there are frequently scatters close to such monuments (Barrowman & Stuart forthcoming). There are approximately 5,000 scatters across the whole of Scotland, mostly in arable areas. Walk along any river valley or across any field and you will be retracing the footsteps of someone in prehistory. Despite being 'destroyed archaeology', or, at best, a possible indication of damaged structural remains and what they may represent, lithic scatters still hold a value which, if released, may do wonders for teaching people about history of the landscape, creating a relevant link between the past and the way people negotiate their landscapes today (Bender 1998).

This 'weft of the past ... [feeding into] the warp of the present' (Shanks & Tilley 1992, 95) is the ideal tool for teaching and making people aware that the whole structure of the landscape is composed of traces of prehistoric cultures. Archaeology is present in the whole landscape; it is not only represented by what we can see, there is also an underlying presence almost everywhere. Prehistoric residues are with us today in every part of the surviving landscape, lithic scatters are the material proof. They do not only represent 'off-site' archaeology (Foley 1981), they represent the presence of humans over the whole landscape through time, a landscape which was incredibly important to the people who lived on, in and with it throughout prehistory. Public archaeology uses only a fraction of the archaeological record to create a narrative, the sites which have 'less value' in public terms are not included, but are equally as important. Richard Bradley, for instance, has recently shown that phenomenological studies which link archaeological sites and monuments in the landscape, and the natural features within those landscapes, are ideally placed within a wider public realm as ' ... landmarks, paths and viewpoints ... lead so many people to visit the countryside today' (Bradley 1996, 48). A wider landscape archaeology is immediate to the public who visit these places. Similarly, unrecognised locales marked by invisible scatters are part of this wider landscape concern and should be celebrated within a public archaeology. The social construction of space throughout prehistory has left material marks in the form of scatters across many parts of Scotland. These *spaces* are similarly constructed through the process of fieldwalking (Barrowman forthcoming).

If this concept is created within the public sphere, *i.e.,* that archaeology is everywhere and the landscape we live in has been inhabited since prehistory, then our habitat may be seen with fresh eyes. If people were to realise the time-depth of the landscape a fuller understanding of it may result. It may be possible to create a longer time-depth within peoples' minds than three or four generations. Prehistory may not only mean a stone circle or even a Channel Four television tea-time dip into a virtual past. Ploughed fields will take on new dimensions as people drive through the countryside or look out of the train window. Tom Greeves touches on this in his quote in the book *All Natural Things*:

> Many of those who shared the motorway with me would have perceived the countryside through which they passed as green in colour, natural in content and irrelevant to their lives except perhaps as a view. And yet I knew this was not always so, and need not be in the future. Moreover, I knew that landscape contained riches that could touch the lives of all of us, once recognised and shared.
>
> (Greeves 1992, 15)

An awareness of lithic scatters will give an immediacy to the prehistoric past, and an understanding of the cultural and historical extent of our landscapes. By incorporating the presence of lithic scatters into a wider interpretation of prehistory and including scatters in public archaeologies, an expansion of the public's awareness of the time-depth of the landscape is generated. By acknowledging that we create and appropriate scatters today, an immediate relationship to prehistoric material culture is achieved, enriching the experience of being in the landscape.

Acknowledgements

Thanks go to Rachel Harry, Eland Stewart and Michael Shott for their comments on drafts of this paper. Although the paper was initially criticised for being rather thin, I decided after much deliberation to keep it as a statement of my sentiments, rather than a more in-depth critique. Many of the themes mentioned in this paper are expanded upon in my doctoral thesis.

References

ALLEN, M. 1991. 'Analysing the landscape: a geographical approach to archaeological problems.' *In* Schofield (1991b), pp. 39–58.

BARROWMAN, C. 1998a. *Garvald Burn, Dunsyre*. Unpublished fieldwork report.

BARROWMAN, C. 1998b. *Weston: a Mesolithic Scatter*. Unpublished fieldwork report.

BARROWMAN, C. forthcoming. *Lithic Scatters as an Archaeological Resource in South and Central Scotland*. Ph.D. thesis.

BARROWMAN, C. & E. STUART forthcoming. *Scottish Lithic Scatters Project*.

BENDER, B. 1998. *Stonehenge, Making Space*. Berg, Oxford.

BOISMIER, W. 1997. *Modelling the Effects of Tillage Processes on Artefact Distribution in the Ploughzone: a Simulation Study of Tillage-induced Pattern Formations*. British Archaeological Reports British Series 259, Oxford.

BOWDEN, M., S. FORD, V. GAFFNEY & M. TINGLE 1991. 'Skimming the surface or scraping the barrel: a few observations on the nature of surface and sub-surface archaeology.' *In* Schofield (1991b), pp. 107–13.

BRADLEY, R. 1996. 'Ancient monuments and the modern public.' *In* P. Evans, P. Solway & D. Thackray (eds.), *The Remains of Distant Times; Archaeology and the National Trust*, pp. 38–46. Boydell for the Society of Antiquaries of London and the National Trust, Woodbridge.

CLARK, R. & A. SCHOFIELD 1991. 'By experiment and calibration: an integrated approach to archaeology of the ploughsoil.' *In* Schofield (1991b), pp. 93–106.

CLARKE, A. 1989. 'Corse Law, Carnwath, Lanarkshire: a lithic scatter.' *Proceedings of the Society of Antiqaries of Scotland* 119: 43–54.

DUNNELL, R. C. 1992. 'The Notion site.' *In* J. Rossignol & L. Wandsnider (eds.), *Space, Time and Archaeological Landscapes*, pp. 21–41. Plenum Press, New York.

FOLEY, R. 1981. 'Off-site archaeology: an alternative approach for the short sited.' *In* I. Hodder, G. Isaac & N. Hammond (eds.), *Pattern of the Past: Studies in Honour of David Clarke*, pp. 157–82. Cambridge University Press, Cambridge.

GREEVES, T. 1992. 'Reclaiming the land: why archaeology is green.' *In* L. Macinnes & C. Wickham-Jones (eds.), *All Natural Things: Archaeology and the Green Debate*, pp. 14–21. Oxbow, Oxford.

PEARSALL, J. & B. TRUMBLE (eds.) 1995. *The Oxford English Reference Dictionary*. Oxford University Press, Oxford & New York.

SCHIFFER, M. B. 1987. *Formation Processes of the Archaeological Record*. Albuquerque, University of New Mexico Press.

SCHOFIELD, A. 1991a. 'Interpreting artefact scatters: an introduction.' *In* Schofield (1991b), pp. 43–8.

SCHOFIELD, A. (ed.) 1991b. *Interpreting Artefact Scatters: Contributions to Ploughzone Archaeology*. Oxbow, Oxford.

SCHOFIELD, A. 1995. 'Settlement mobility and la longue durée: towards a context for surface lithic material.' *In* A. Schofield (ed.), *Lithics in Context. Suggestions for the Future Direction of Lithic Studies*, Lithic Studies Occasional Paper No. 5., pp. 105–14. Lithics Studies Society, London.

SHANKS, M. & C. TILLEY 1992. *Re-Constructing Archaeology: Theory and Practice*. Routledge, London and New York, second edition.

SHOTT, M. J. 1995. 'Reliability of archaeological records on cultivated surfaces: a Michigan case study.' *Journal of Field Archaeology* 22(1): 475–90.

WAGSTAFF, M. 1991. 'The archaeological site from a geographical perspective.' *In* Schofield (1991b), pp. 9–10.

WICKHAM-JONES, C. 1994. *Scotland's First Settlers*. Batsford, Edinburgh.

WICKHAM-JONES, C. & J. MACKENZIE 1996. 'An unusual lithic assemblage from Lunahead, Angus.' *Proceedings of the Society of Antiquaries of Scotland* 126: 1–16.

Chris Barrowman
Gregory Building
Department of Archaeology
University of Glasgow
Lilybank Gardens
Glasgow G12 8QQ
Scotland

9
Knowledge and Practice: The *Scottish Lithic Scatters Project* and Stoneworking in Prehistory

Eland Stuart
Department of Archaeology, University of Glasgow

Abstract

This paper has a single purpose expressed through two aims. The purpose is to address concerns that two scholars have put recently in print (Brown 1995; Schofield 1995b) about the seeming minimal value of lithic scatters to other archaeological studies of prehistoric society. What was the nature of working stone in prehistoric society, or how do lithic scatters represent prehistoric society? The question was repeatedly valid while gathering data for the Scottish Lithic Scatters Project (SLP), hence one aim is to briefly introduce the SLP and the other is to discuss lithic scatters as social products.

9.1 The Scottish Lithic Scatters Project (SLP)

The SLP was funded by Historic Scotland from 1995–98 and the final report is currently in preparation (Barrowman & Stuart forthcoming). The initial proposal was drafted by Dr Colin Richards of the University of Glasgow (1995) in consultation with a number of others including Mr Gordon Barclay of Historic Scotland, Dr Robin Boast, Mr John Humble, Dr John Schofield and Mr Bill Startin. As a sister project is also underway in England (see Schofield 1994) the information that we collected will provide part of a useful source of data for much of the British Isles. The English project aimed to consolidate a seemingly chaotic body of data (Schofield & Humble 1995) generated by numerous regional collection programmes which, upon fruition, present their data in isolation. Scotland differs in this respect. Fieldwalking has been widely employed in England using different systems of collection over a long period of time whereas, in Scotland systematic fieldwalking has only recently become an important component of archaeological survey and assessment (Richards 1995). Consequently the emphasis of the Scottish project was on re-assessment and informed planning for future work as well as consolidation of a disparate body of data.

The SLP had three principal aims:

1. to collect information into a single place from disparately located sources such as the National Monuments Record for Scotland (NMRS), museum records and those otherwise unrecorded but for journals and private sources;
2. to assess the resource;
3. to provide a foundation to manage that resource.

In tandem were a series of objectives which were:

1. to provide Historic Scotland with detailed information necessary for the evaluation of lithic surface scatters to impose preservation/protection policies;
2. to increase archaeological knowledge of lithic surface scatters and provide criteria for their assessment;
3. to produce a general methodology for the practice of systematic fieldwalking in Scotland;
4. to update the NMRS thereby developing an accessible database of considerable research potential;
5. to make available to local authorities and museums the results of the project.

To facilitate the aims and objectives a database was created. It fulfilled the first aim (collection of information) and formed the basis for the other aims and objectives. The results, which will fulfil the second aim — assessment of the resource — are still under analysis. Once this has been completed consultation may begin with Historic Scotland (HS) towards a management strategy (the third aim and the first three objectives), and will make objectives four and five (database development and accessibility) possible, although with reference to objective four, the product has already been examined by some scholars working on Scottish material (Finlayson & Warren 2000).

The database was designed to allow records of lithic scatters to be both classified and discriminated.

To do this a series of identical questions were asked of each scatter, each question forming the basis for a database field. The design of each field and its purpose followed that laid down for the English Heritage project and described by Schofield (1994). That information will not be duplicated here other than to say that some small differences were introduced into the Scottish database which do not much alter the fundamental design.

A field was added that describes the contents of scatters. Without this field interpretations of the period and function of scatters would have been isolated from the data that generated them. There was no built-in support for either 'when' or 'what' the scatters might relate to, or, in other words, no integral way for a database user to check or assess the interpretative claim. This said however, the information that we entered into the 'period' or 'function' field was not, in the main, of the judgement of project personnel, *i.e.*, Chris Barrowman and myself, but already recorded for a particular scatter. This raises a potential problem in that any mistakes or oversights made by others which have entered the archives will be perpetuated by our database. Where such omissions were detected, such as where microliths were found amongst, and recorded as, debitage, then we made the appropriate entries but it was not our primary purpose to check the quality or reliability of the statements of others. Thus many omissions may have gone undetected. Alan Saville recently pointed out this problem for Scottish Mesolithic data (Saville 1998, 213) and if anything our work, as a publicly accessible, centralised resource, could lay a foundation from which checks, should they be considered of importance by a researcher, could potentially succeed with greater ease.

The English project had a scoring method to relate sites one to another and assess their relative value. It was omitted from the SLP because most Scottish data was collected in the small scale activities of individual collectors, localised groups such as the *Lanark and District Archaeology Society* and the *Tain Archaeology Group*, or otherwise through serendipity. There has then been very little standardised collection which in turn has produced a piecemeal dataset so that a scoring method would have no statistical value (Richards 1995, 3).

Another problem encountered was that very few sites have any associated record of site function made by the collector. This is not a reflection of lack of effort or thought on the part of collectors but a problem generally with lithic data (*e.g.,* Shott 1995), and then not just surface lithic data. It is a general problem with lithic assemblages. For example, Ford (1987) employed a detailed method to examine lithic assemblages to infer their date and function. He had some success with period but site function remained elusive (Ford 1987, 72–8). The reason may not be the fault of the data; perhaps lithics are not a good index for the function of a site, in other words perhaps we may not always be asking the right questions of lithic data. If we want to know what the function of a site was, it is only if we are fortunate and there is sufficient material and only with great labour (*e.g.,* Schofield 1991c; Steinberg 1996) that we *might* provide an answer. Even then it is of the crudest terms, *e.g.,* 'This site was industrial', or 'This site was domestic' which are only loosely connected to the human subjects whose actions created the remains. As Schofield points out, many studies of surface remains seem short on interpretation (1991a, 3) and tell us much less about prehistoric occupation and landuse than we like to believe (1994, 90). This is a recurrent problem with lithic scatters as we on the project found during our own fieldwalking. Thus, we may ask 'Is it possible that lithic scatters, as a phenomenon, can be interrogated for other information, or thought about in other ways?' I believe that they can and, after briefly describing the other elements of the SLP, I shall address this question with reference to some Scottish material.

A small amount of fieldwork was done on the project to generate some primary data about fieldwalking and lithic scatters. Some fieldwalking, geochemical and geophysical testing and, in some places, test pitting was done. A handful of sites in several areas were located and briefly examined. Those areas selected were from across Scotland: in Caithness, North Angus, South Lanarkshire, The Isle of Bute and The Borders. Some details will be found in the SLP report (Barrowman & Stuart forthcoming) and other results are currently under analysis.

9.2 Social technologies of lithic working

In gathering data for the SLP, lithic scatters did not seem like the products of individual hands. In a recent paper Gamble (1998) acknowledges a very similar point with regard to Palaeolithic material. Lithic scatters seemed instead like dots on maps that may be used to depict an ancient human presence in a landscape but otherwise seemed removed from the practices that shaped peoples' lives. Schofield (1995b) expresses this concern for scatters in England when he considers that they make only a cursory appearance, if at all, in some notable recent prehistoric studies. Part of the concern of the Lithic Scatters Project in England was to address this issue (Schofield 1994, 90) and to increase the value of lithic scatter data to the study of prehistory. The SLP shares the same aspirations but it is unlikely that the existence of a database alone will be sufficient to foment the change; as Schofield points out the disparity is theoretical (1995b, 105). A pattern of lithic scatters across a landscape for a certain period of time creates a 'big picture' of human occupation covering a large area and a considerable time depth. In terms of human practice with stone that big picture was created through accumulated acts of human agency and the theoretical disparity concerns the operation of that agency. Thus rather than explore the 'big picture' I wish to turn to lithic scatters as products of people who made choices in their practices and in doing so left scatters across the landscape. The issue here is neither ways of viewing large patterns nor the interpretation of specifics of assemblages but to view lithic scatters as products of a knowledge that was historically situated.

The actions of individuals with stone only receive attention through refitting studies and the *chaîne opératoire*. Such work provides detailed accounts of a sequence of actions, that is short term events or the small moments in peoples' lives of the decisions that they made whilst knapping stone. They are vital to our understanding of human action in the more distant past but they do not, of themselves, provide any further understanding into any possible social nature of stone working practices. In such studies the sequence of actions is bounded; they consider stoneworking as an event rather than a practice. At the beginning is the lump of raw material and at the end finished products which are then put to work in the world. That other forces — social forces — are insinuated into the process of creation, in existence before and perpetuated by the process, but do not feature in the analysis. Thus we may ask 'How might the social feature in stoneworking, and how might we know?'

It would be easy to assume that people working stone in the past did so simply to produce the tools that they needed to procure the other necessities of life and that nothing else other than this guided those actions. On one hand it is true that people worked stone to make tools that could, in turn, be used for other material needs but, on the other, it is not true that nothing else guided those actions. There are two points here. First, that making tools is linked to the tasks for which they are intended. Gosden & Head (1994, 14) point out that most action implies future action that will take place at another time and possibly at a different place as well. The example that they use is knapping flint. The practice is embedded in further strategies of action that will involve the products. These strategies of action communicate socially or are, in other words, discourses. Sometimes these discourses were formal and ritualised practices such as burial (Barrett 1994), warfare (Edmonds & Thomas 1987) or other, perhaps more enigmatic, practices (Brown 1995) that communicated quite publicly. And often, they were humdrum, ordinary and everyday such as food procurement or preparation, or working with other materials. In many cases the domestic tasks communicate less publicly but are perhaps more fundamental to social life. These everyday tasks of repeated, routinised activities are no less powerful or compelling or, to ourselves, less interesting because of their everyday nature. In fact, quite the opposite. Precisely because those activities are repeated and routinised they shape peoples' social being less consciously and the order of the world becomes natural and obvious (Bourdieu 1977; Giddens 1984). In the past, as in the present, both public and domestic, perhaps open and closed 'fields of discourse' served to reproduce structure in prehistoric communities (Barrett 1988, 9–10). This introduces the second point which is that it is not true that making tools was guided by nothing other than making those tools. Making tools, or knapping stone, was linked by the tasks for which they were intended to the social identities and relationships that those tasks forged. In other words the relationship between activities or practices and the material things used in them is not straightforward but embedded in a social world (*e.g.,* Hodder 1986; Shanks & Tilley 1987a,b). People operate within social conditions and those conditions affect their actions, making some things possible and others not (Giddens 1984).

The amount of disturbance and movement at lithic scatters in ploughed fields casts doubt upon observations of material relations of association and avoidance (Odell & Cowan 1987, 477–9). But the way people structure material things is only part of the story. Another part is that the relationship between people and material culture is one of mutual ordering, we order it and this in turn orders us (Miller 1994, 397). Thus working lithics, as much as the products of it, played a part in shaping peoples' sense of who they were according to age, gender, labour and so on. These aspects were not always mapped spatially but mapped conceptually through the actual practice.

How do we know that this might be the case? It is often, but not always, possible to assign lithic scatters to a period of prehistory. The act of knapping flint produces lots of waste and lithic scatters often contain a mixture of waste and crafted implements. There is no independent method to date them so we rely upon inference from metrical analysis of those waste pieces and from diagnostic crafted implements, though, in many cases, neither is as straightforward as it sounds. Not all implements are diagnostic and not all assemblages are amenable to metrical analyses.

With diagnostic artefacts such dating is treated cautiously and we are right to be cautious because they can be misleading (Schofield 1987, 273, 1994, 91). For example, barbed and tanged arrowheads, associated with beaker pottery (Clark 1970) and therefore to the final Neolithic and Early Bronze Age, are often found at lithic scatter sites amongst material of other periods, *e.g.,* at Kinloch on the isle of Rhum (Wickham-Jones 1990, 26) and at Little Gight in Aberdeenshire (Baird & Finlayson 1994, 98) where most of the material was considered to be from the Mesolithic. These examples were quite easily detected due to other diagnostic forms such as microliths and the nature of the rest of the material present or, in other words, by looking at a mass of technological evidence. This takes us back to the use of waste pieces or metrical analysis; in these instances it indicated the manufacture of small blades which the authors used as evidence to suggest that the majority of the scatter was in fact Mesolithic. Few would have any real problems with this interpretation. But what is it based upon? It was a diagnosis of a period through the examination of technology or what people actually did when they worked the lithics. It could be argued that elements of an assemblage used in this way are used statistically to be diagnostic. There may therefore be little difference between the way that metrical analyses and diagnostic artefacts are used archaeologically. Any difference is due to the fact that the diagnosis is not from individual pieces such as projectile heads or scrapers, but statistically from elements of the assemblage which some scholars consider tend to occur together anyway (*e.g.,* Pitts 1978, 26). Thus metrical analyses may allow archaeologists to interpret the approximate date or period of lithic assemblages (Ford 1987; Ford *et al.* 1984; Pitts 1978; Pitts & Jacobi 1979). Crudely put, through time there was an increase in the ratio of flakes to blades (Ford 1987; Pitts 1978, 26). The observations point to tendencies within a whole assemblage rather than each and every piece in it.

However, there has been some objection to the use of technological or metrical analyses and therefore of their association with diagnostic artefacts (Schofield 1987). Schofield's objection is that diagnostic artefacts may not be accompanied by any such changes and that such technological change may not be chronologically sensitive but in fact relate to the availability of raw materials regardless of period. In turn Edmonds (1995, 37) rejects this because the ways that material was worked do not seem to respect the availability of raw material. This is partly true. In Scotland different raw materials are sometimes used where flint is scarce or absent altogether (Wickham-Jones 1979). Some of these materials were worked differently because their properties differ; such an example is the way that chert was used in southern Scotland (Finlayson 1990, 44). Chert splinters more readily than flint and so was worked slightly differently (Finlayson 1990, 46–8) and, as it is also common in southern Scotland, was consumed quite liberally (Finlayson 1990). This is discussed further below.

Whether or not we accept technological or metrical analyses, it is certain that some forms of artefact, such as forms of armature or arrowhead, were made over vast areas of space (Edmonds 1987, 156) and through long periods of time (Edmonds 1995, 20). Having said this we should also say that this general statement only applies to some forms. On the other side of the coin not all implements that are found are readily identifiable artefacts (Edmonds 1995). Indeed most assemblages have implements that fit no greater typological scheme and generally go unnamed. These objects are not flies in the ointment for interpretations about stone working in prehistoric society reached through observations that some artefacts have wide distributions and long shelf-lives. All of the materials are the products of human choices and we should not expect total homogeneity. This is because working lithics was embedded in the social. Archaeological explanation must be sensitive to these complexities through an understanding of human agency that allows us to see the remains as both the products of individual choices and, at the same time, structured by collective, social conditions (Giddens 1984). This notion of human agency is the paradoxical predicament of the human sciences.

It would seem to indicate that some objects, such as forms of projectile point, were tied in with wider social discourses that exerted an influence on the choices that people made when they sat down to make tools. Certainly in Scotland people sometimes adopted locally available materials where flint was scarce or unavailable (Wickham-Jones 1979, 36) and Saville considers that the use of lithics in Scotland is 'the story ... of ready adaptation to locally available materials' (Saville 1994, 57). That some forms of artefact were tied in with greater social discourses meant that, even where the best materials were not available, people still made some artefacts in ways that would have been recognisable by others many hundreds of miles away. Artefacts such as leaf-shaped arrowheads in chert from South Lanarkshire or quartz barbed and tanged arrowheads from the Western Isles and Shetland that Saville calls a triumph of skill over raw material (1994, 59). Clearly the choices exhibited by the makers were locked in with social constraints

That artefacts were being made in stone throughout prehistory that others in different places would be familiar with is not a claim that the meaning of these artefacts was universal. The meaning of things is generated in localised contexts of practice (Barrett 1994, 88) but there are two connected forces operating at different scales. One, as mentioned, generates localised meaning. The other generates the remarkable similarities shown by artefacts and monuments across vast regions of space at certain times. Such material homogeneity is not evidence of social homogeneity as explained by scholars within culture history since it is not possible that people far removed from each other, either temporally and geographically, knew of each others' practices. Groups of people do not generally exist in isolation and the relationships between distant areas only existed because of networks of local relations. These proliferating networks are the medium of culture but do not determine localised structures of meaning. It is unlikely that they made people across large regions of space directly accessible to other people in prehistory. They were not a medium that could be accessed but instead provide a way for archaeologists to understand the patterns of similarity of archaeological material. That we can have localised meaning and broad scale patterns is not contradictory but a duality which enables us to address the complexities of difference and similarity across prehistoric Britain

Technologies, such as the way to work lithics, were not created from scratch. People were born and socialised into a world where they already existed as a necessity of life. Thus behind any individual knapping flint was a body of knowledge that was sensitive to both the physical practicalities of working with the material - manifest in the skilled handling of the material, and to the social practicalities (Giddens 1984, xxiii) — manifest in certain artefacts and technologies that operated through long periods of time and were, at the same time, peculiar to those periods and hence, chronologically sensitive.

The practice of working stone and teaching other members of the community to work stone reproduced those structures of knowledge so that they became institutionalised into prehistoric communities as traditions. Peoples' practices reproduced those traditions through action but traditions are a set of conditions for action, they do not determine it. Human beings make choices within tradition so there is an inbuilt capacity for invention and change. Those traditions were also tied to other activities and discourses through the created products. Thus when the significance and details of those activities changed, such as when people changed their economic relationships with animals from hunting to husbandry at the end of the Mesolithic, then conditions were created where people also began to change the way that they worked stone. At a big scale, for example across Scotland, the mechanism of change looks impossible to find. But it was lived through in peoples' choices in their course of action. They surely were not aware of the consequences, that is our privilege, but it was through choices, perhaps relating to other concerns, that gave rise to the story of humanity as we see it.

9.3 Conclusions

The chronologically sensitive, diagnostic artefacts and possibly also the results from metrical analyses indicate the social nature of stone working practice. Not all assemblages are amenable to metrical analyses since diagnostic pieces may be absent or metrical analyses inconclusive. Despite this, assemblages that appear to lack chronological sensitivity were still created by knowledgeable practice. It is simply that the technology employed cannot be used archaeologically to locate them within the long-term narratives of prehistory. The practice was bedded into the flow of life in the past. It implied future acts and it served to reproduce an aspect of vital everyday knowledge. It is unlikely that the products of knapping stone determined cultural change in the past but then that is true of all cultural phenomena regarded singularly. Thinking about how human societies reproduce traditions or institutions through practice provides a starting point for the changes we observe through time in material residues, but not the forms taken by those residues. The point here is that the knowledge of the ways to work lithics was reproduced through the acts of working lithics. It was a form of social discourse and thus some forms display a homogeneity and a longevity. This grounds lithic scatters in human practices through time. The result is that lithic scatters are not isolated products which are explicable only in their own terms but products of knowledgeable action situated in the passage of time as lived through by people engaged in routine activities. They are not the neutral traces of communities eking out a living, each generation isolated from the other. They are instead the products of practices that shaped, in a small but not insignificant way, the lives of prehistoric people that in turn left very durable traces for the archaeologist.

When we look at a pattern of lithic scatters across a landscape we are looking to see how human groups organised themselves and their activities through a particular time. We then look to see how that was different to times before and after and produce accounts of the evolution of the human occupation in an area. This is a panoramic history spanning perhaps thousands of years but there is another history available in this data. Lithic scatters are also places where knowledge and tradition fused and were kept alive by necessity. I would argue that if we also consider this agency and lithic scatters as the products of situated practices, then the histories that we write are no longer the histories of things but the histories of human beings who were situated in, and acted within, conditions that those actions served to reproduce.

Schofield recently wondered in print why lithic scatters seems to tell us so little (1995b). The same question was evident working on the SLP. In this paper I have introduced the SLP and attempted to address some of the issues that the question raises. I hope that in doing so I have raised yet more issues.

Acknowledgements

Thank you to those with whom I discussed aspects of this paper and those who made comments upon earlier drafts. Chris Barrowman, Rae Harry, William Kilbride, Gavin MacGregor and Colin Richards. Thank you to my anonymous reviewer who helped with streamlining and to Norah Moloney for the opportunity to contribute and subsequent patience. All faults are mine.

References

Ashton, N. & A. David (eds.) 1994. *Stories in Stone.* Lithic Studies Society Occasional Paper No. 4. Lithics Studies Society, London.

ASHTON, N., F. HEALY & P. PETTITT (eds.) 1998. *Stone Age Archaeology: Essays in honour of John Wymer.* Lithic Studies Society Occasional Paper No. 6. Oxbow, Oxford.

BAIRD, D. & B. FINLAYSON 1994. 'A Mesolithic and later flint scatter at Little Gight, Grampian region.' *Proceedings of the Society of Antiquarties of Scotland* 124: 95–102.

BARRETT, J. C. 1988. 'Fields of discourse: Reconstituting a social archaeology.' *Critique of Anthropology* 7: 5–16.

BARRETT, J. C. 1994. *Fragments from Antiquity: an archaeology of social life in Britain, 2900–1200* BC. Blackwell, Oxford.

BARROWMAN, C. & E. STUART forthcoming. *Scottish Lithic Scatters Project.*

BOURDIEU, P. 1977. *Outline of a Theory of Practice.* Cambridge University Press, Cambridge.

BROWN, A. G. 1995. 'Beyond stone-age economics: a strategy for a contextual lithic analysis.' *In* Schofield (1995a), pp. 7–36.

BROWN, A. G. & M. R. EDMONDS (eds.) 1987. *Lithic Analysis and Later British Prehistory: some problems and approaches.* British Archaeological Reports British Series 162, Oxford.

CLARK, D. L. 1970. *Beaker Pottery of Great Britain and Ireland.* Cambridge University Press, Cambridge.

EDMONDS, M. R. 1987. 'Rocks and risks: Problems with lithic procurement strategies.' *In* Brown & Edmonds (1987), pp. 155–79.

EDMONDS, M. R. 1995. *Stone Tools and Society: Working Stone in Neolithic and Bronze Age Britain.* Batsford, London.

EDMONDS, M. R. & J. S. THOMAS 1987. 'The Archers: An everyday story of country folk.' *In* Brown & Edmonds (1987), pp. 187–99.

FINLAYSON, B. 1990. 'Lithic exploitation during the Mesolithic in Scotland.' *Scottish Archaeological Review* 7: 41–57.

FINLAYSON, B. & G. WARREN 2000. 'The Mesolithic of Eastern Scotland.' *In* R. Young (ed.), *Mesolithic Lifeways: Current Research from Britain and Ireland*, Leicester Archaeology Monographs No. 7, pp. 133–41. University of Leicester, Leicester.

FORD, S. 1987. 'Chronological and functional aspects of flint assemblages.' *In* Brown & Edmonds (1987), pp. 67–83.

FORD, S., R. BRADLEY, J. HAWKES & P. FISHER 1984. 'Flint working in the metal age.' *Oxford Journal of Archaeology* 3: 157–73.

GAMBLE, C. 1998. 'Handaxes and palaeolithic individuals.' *In* Ashton *et al.* (1998), pp. 104–9.

GIDDENS, A. 1984. *The Constitution of Society.* Polity Press, Cambridge.

GOSDEN, C. & L. HEAD 1994. 'Landscape — a usefully ambiguous concept.' *Archaeology Oceania* 29: 113–6.

HODDER, I. 1986. *Reading the Past.* Cambridge University Press, Cambridge. Second edition 1991.

MILLER, D. 1994. 'Artefacts and the meaning of things.' *In* T. Ingold (ed.), *Companion Encyclopaedia of Anthropology: Humanity, Culture and Social Life*, pp. 396–419. Routledge, London.

ODELL, G. H. & F. COWAN 1987. 'Estimating tillage effects on artifact distributions.' *American Antiquity* 52(3): 456–84.

PITTS, M. W. 1978. 'On the shape of waste flakes as an index of technological change in lithic industries.' *Journal of Archaeological Science* 5: 17–37.

PITTS, M. W. & R. M. JACOBI 1979. 'Some aspects of change in flaked stone industries of the Mesolithic and Neolithic in Southern Britain.' *Journal of Archaeological Science* 6: 163–77.

RICHARDS, C. R. 1995. *Proposal to Historic Scotland for a Scottish Lithic scatters Project, revised version.* Unpublished mss.

SAVILLE, A. 1994. 'Exploitation of lithic resources for stone tools in earlier prehistoric Scotland.' *In* Ashton & David (1994), pp. 57–70.

SAVILLE, A. 1998. 'Studying the Mesolithic period in Scotland: A bibliographic gazetteer.' *In* Ashton *et al.* (1998), pp. 211–24.

SCHOFIELD, A. 1991a. 'Interpreting artefact scatters: an introduction.' *In* Schofield (1991b), pp. 43–8.

SCHOFIELD, A. (ed.) 1991b. *Interpreting Artefact Scatters: Contributions to Ploughzone Archaeology.* Oxbow, Oxford.

SCHOFIELD, A. (ed.) 1995a. *Lithics in Context. Suggestions for the Future Direction of Lithic Studies.* Lithic Studies Occasional Paper No. 5. Lithics Studies Society, London.

SCHOFIELD, A. 1995b. 'Settlement mobility and la longue durée: towards a context for surface lithic material.' *In* Schofield (1995a), pp. 105–14.

SCHOFIELD, A. J. 1987. 'Putting lithics to the test: Non-site analysis and the Neolithic settlement of southern England.' *Oxford Journal of Archaeology* 6: 269–86.

SCHOFIELD, A. J. 1991c. 'Artefact distributions as activity areas: Examples from south-east Hampshire.' *In* Schofield (1991b), pp. 117–28.

SCHOFIELD, A. J. 1994. 'Looking back with regret; looking forward with optimism: Making sense of surface lithic scatters.' *In* Ashton & David (1994), pp. 90–8.

SCHOFIELD, A. J. & J. HUMBLE 1995. 'Order out of chaos: Making sense of stone age finds.' *Conservation Bulletin* 25: 9–11.

SHANKS, M. & C. TILLEY 1987a. *Re-constructing Archaeology.* Cambridge University Press, Cambridge.

SHANKS, M. & C. TILLEY 1987b. *Social Theory and Archaeology.* Polity, Cambridge.

SHOTT, M. J. 1995. 'Reliability of archaeological records on cultivated surfaces: a Michigan case study.' *Journal of Field Archaeology* 22(1): 475–90.

STEINBERG, J. M. 1996. 'Ploughzone sampling in Denmark: isolating and interpreting site signatures from disturbed contexts.' *Antiquity* 70: 368–92.

WICKHAM-JONES, C. R. 1979. 'Flaked stone technology in northern Britain.' *Scottish Archaeological Forum* 11: 36–42.

WICKHAM-JONES, C. R. 1990. *Rhum: Mesolithic and later sites at Kinloch excavations 1984–6.* Society of Antiquaries of Scotland Monograph Series No. 7. Society of Antiquaries of Scotland, Edinburgh.

Eland Stuart
Gregory Building
Department of Archaeology
University of Glasgow
Lilybank Gardens
Glasgow G12 8QQ
Scotland

Part II

Analytical Techniques in Lithic Analysis

10

A Refitter's Paradise: On the Conjoining of Artefacts at Maastricht-Belvédère (The Netherlands)

Dimitri de Loecker[1], Jan Kolen[1], Wil Roebroeks[1] and Paul Hennekens[2]

[1]*Faculty of Archaeology, University of Leiden;* [2]*Maastricht, The Netherlands*

Abstract

Although refitting analysis has been known as a valuable methodological tool for more than a century, it is only seldom explored on a systematic basis for the interpretation of stone age sites. This article reviews the results of a long-term refitting programme for the Middle Palaeolithic site complex at Maastricht-Belvédère, The Netherlands. On the basis of the Belvédère example, the potential value of such refitting programmes for the reconstruction of Palaeolithic site taphonomy as well as early human behaviour will be discussed. Specific attention is paid to the impact of post-depositional processes on artefact scatters, the reduction processes of Middle Palaeolithic core technologies, the complex life histories of single stone tools, the use of space by early humans and the spatial dynamics of lithic technologies.

10.1 Introduction

> This wonderful feat was first performed by my friend Mr F. C. Spurrell of Dartford. On first thoughts the thing seems utterly impossible, and it is obvious that no flake can possibly be replaced upon an implement unless one lights on the exact spot where the implement was made, and finds both implement and flakes in position. (Smith 1881, 582)

Until the late 1970s the Middle Palaeolithic archaeological record of The Netherlands was made up exclusively of lithic surface collections and diagnostic stray finds (Roebroeks 1980, among others). In the Dutch Southern Limburg area these surface scatters were located in the higher landscapes (outside the river valleys) in particular and investigations concentrated mainly on typo/technological interpretations. It was only from the 1980s onwards, and especially as a result of the detailed long-term excavations carried out at Maastricht-Belvédère, that information on a river valley became available as well. The emphasis shifted towards the excavation of stratigraphically 'sealed' and well-preserved settlement remains in sedimentation areas, *e.g.*, the river valleys with a fine-grained fluvial matrix, and also the loess covered plateaux. The well excavated artefact scatters at Belvédère documented a number of different 'on-site' activities and provided a better understanding of early human behaviour in a small segment of a 250,000 year old riverine landscape (Roebroeks 1988; Roebroeks *et al.* 1992, 1993).

Together with this shift of interest, a wide range of new questions were introduced into Dutch Palaeolithic archaeology along with a series of analytical tools to answer them. As a result, the excavated levels at Belvédère have generated new information on absolute dates, stratigraphy, palaeoenvironment, well-preserved spatial distributions of artefacts, food procurement and in particular on the technological strategies used (de Loecker forthcoming; Roebroeks 1988; Roebroeks *et al.* 1992, 1993; van Gijn 1988, 1989; Vandenberghe *et al.* 1993). In addition to the use-wear studies (van Gijn 1989) and a very detailed lithic description of the flint assemblages (de Loecker forthcoming; Roebroeks *et al.* 1997; Schlanger 1994, 1996; Schlanger & de Loecker 1992), conjoining studies were the most instrumental at Belvédère for making inferences on early human behaviour.

It has often been stated that refitting of lithic artefacts proved essential to the (re)construction of reduction strategies. As well as providing typological and technological documentation, the method can be used in the investigation of site formation processes (both human and non-human). Combined with distribution maps, refitting proves very useful in locating areas where artefacts were made, used and discarded, and in determining which artefacts were transported. By doing so refitting can inform us about the spatial and temporal characteristics of 'sites' (Cahen *et al.* 1979; Hofman 1981; Roebroeks 1988; Roebroeks *et al.* 1997). Although refitting has a history of more than a 100 years, its analytical importance has only been recognised during the last three decades. In particular, the investigations and refitting results by Leroi-

Gourhan & Brézillon (1966) and Cahen *et al.* (1979) at the Upper Palaeolithic sites of Pincevent and Meer II respectively and the 'Big Puzzle' symposium organised at Monrepos (Neuwied) in Germany in September 1987 (Cziesla *et al.* 1990) turned out to be significant milestones in European stone age archaeology. Moreover, they inspired many researchers to go on, or start, using the method of conjoining artefacts.

The aim of this article is twofold; it intends to deal with the methodology of conjoining artefacts and investigate the significance of refitting as a major analytical tool. Moreover, conjoining data from the Dutch Middle Palaeolithic sites at Maastricht-Belvédère will be used to illustrate its informative value for archaeology. Different research topics will be tackled by using information from different excavated areas (sites). The conjoined artefact compositions from Sites K and J are used for monitoring vertical and lateral post-depositional disturbances. Reconstruction of lithic reduction processes, the identification of the recycling and reworking of flakes and tools and the typological transformations during the uselife of these tools are illustrated by the refitting results of Sites A, G, C, K and J. Furthermore, the conjoined material from Site K is used to identify artefacts which were produced elsewhere and were subsequently transported to the excavated surfaces. In addition, site K is relevant for the identification of activity areas, such as loci of manufacturing, use, resharpening or discard and the question of temporal relationship between different loci within the same 'site'.

10.2 Refitting: a short historical review

The method of conjoining artefacts goes back to the 19th century. One of the earliest refits was reported from the Palaeolithic site at Brixham Cave, Devon, England (Evans 1897, 512–6; Lubbock 1872; Pengelly 1867, 1874; Prestwich 1873). During the August/September 1858 excavations, conducted by Pengelly and his colleagues, two pieces of a handaxe were recovered at different times and from different areas of the cave but from the same geological layer. According to Evans' description of Brixham cave (1897) some time afterwards (possibly at the end of 1858 or the beginning of 1859) Falconer discovered that the two fragments could be rejoined so that the true character of this tool was seen (Fig. 10.1), although no further inferences were made (Berridge & Roberts 1990). In the late 19th century other 'restorations' of broken tools were established at the Upper Palaeolithic site of Solutré near Mâçon, France (de Mortillet & de Mortillet 1881, 1903) with the conjoining of broken Solutrean leaf points (*feuille de laurier*) found some distance from each other. The first successful attempts to systematically refit flint artefacts were undertaken at the Middle Palaeolithic site of Crayford in Kent, England, where Spurrell refitted a number of flakes onto a flint handaxe and was able to reconstruct the original form of the flint nodule and the flint knapping technology employed (Smith 1881; Spurrell 1880a,b). However, he did not go beyond the observation that the nodule had been reduced on the spot. In the words of Arts & Cziesla (1990) he was in fact only 'restoring' the artefact's original form. Spurrell's refitting skill is referred to again in a monograph on the Egyptian site at Medum (Petrie 1892, 18). Here, he was able to conjoin a sequence of 17 well made blades in their original order of production and reconstruct their mode of manufacture.

In the late nineteenth century, a very systematic and detailed refitting analysis was also done by Worthington G. Smith (Evans 1897, 598-600; Smith 1892, 1894). In *Man, the Primeval Savage* (Smith 1894, 126–8) he described the elaborate conjoining and 'replacing' of the Acheulian flint assemblage from Caddington, England. A total of more than 500 artefacts were refitted in about three years (1890–1893). Not only was his working method very similar to the one we are still using today but he also interpreted some of the results in terms of human behaviour.

> It is remarkable that some of the cores found by me are of a certain colour, or naturally marked in some peculiar way, and that no flakes of a similar colour or marking have been found. I assume that the flakes were struck off these cores for some special purpose, and carried to some other position not lighted on by me. Again, some flakes are of a peculiar colour, or naturally marked in special way quite distinct from any core; these flakes, I suppose, must have been struck off elsewhere, and brought to the spot examined by me.
> (Smith 1894, 128)

In addition to these interpretations on artefact transportation he also made inferences on recycling, (re)sharpening and modes of flake and tool production/manufacture. In fact his analysis was a reconstruction of reduction schemes *avant la lettre*. Smith also undertook refitting, but with less success, at the Palaeolithic 'floors' of Stoke Newington Common in North London (Smith 1883, 1884, 1894) and Round Green in Luton, England (Smith 1916).

Another remarkable early conjoining of artefacts was established by de Munck (1893). At the Lower/Middle Palaeolithic quarry site of St-Symphorien-Hardenpont, Spiennes, Belgium, he managed to refit an almost complete Levallois core; only the larger prepared flakes could not be refitted as they had probably been transported to another location. As Otte noted (1995) this is probably the first true description and interpretation of a Levallois core in Belgium. In its early days refitting was not only confined to Palaeolithic material; for example at a Neolithic findspot near Dundrum Bay, Ireland (Archer 1881), part of a flint pebble, consisting of 22 flakes,

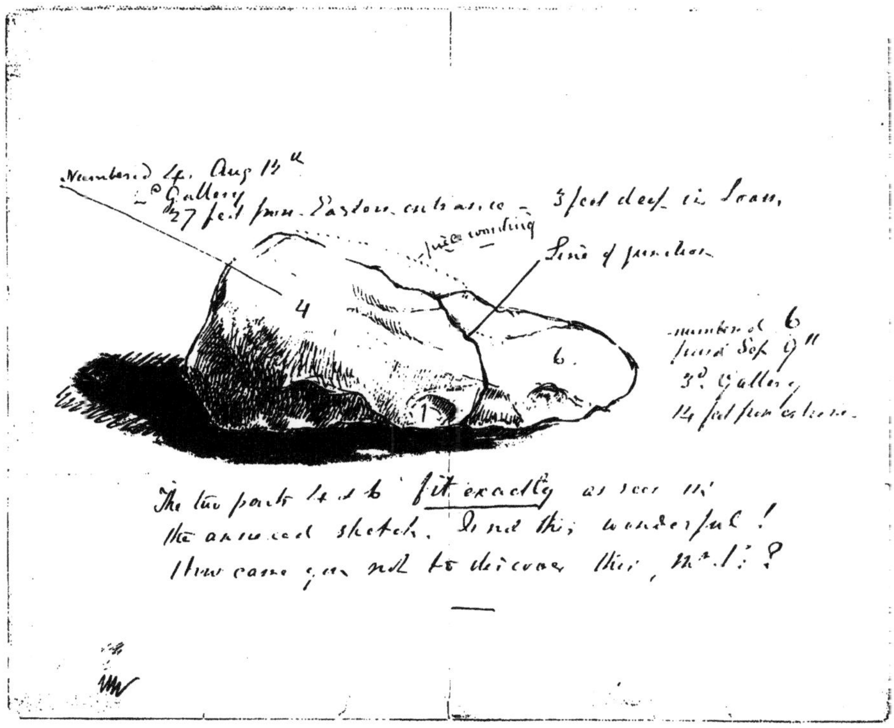

Figure 10.1: Brixham (Windmill Hill) cave. Falconer's sketch of an ancient broken and refitted handaxe (round-pointed lanceolate form). Note the text written underneath the sketch: 'The two parts 4:6 fit exactly as seen in the annexed sketch. Is not this wonderful! How came you not to discover this Mr. P?' (after Falconer's sketch, late 1850s–early 1860s, and used with the kind permission of the trustees of the British Museum).

was restored. Refits are also reported from the Neolithic flint mining industry at Spiennes Belgium (Cels & Depauw 1885–1886). Moreover, the publications of Henri-Martin 1907, 1909 show that the conjoining of faunal remains was not unusual either.

Refitting was used sporadically during the 19th and early 20th centuries, and until the 1960s mainly for technological reconstructions. From the 1960s onwards new excavation methods involving more precise documentation of finds were introduced for Palaeolithic sites. Through the innovative use of refitting at Pincevent (Leroi-Gourhan & Brézillon 1966) it became clear that the potential of the method exceeded reconstructing procedures for flake or blade manufacture and so refitting became crucial in behavioural and site analysis. Today, the conjoining of artefacts (bone and/or stone) is considered as a standard procedure in the analysis of stone age sites. The growing popularity of the method is nicely shown by a cumulative diagram of international publications (1880–1988) on the subject which was published by Arts & Cziesla (1990, 652, Figure 2; see also Hofman 1981, 691). Recent examples where refitting analyses have been successfully employed are the Upper Palaeolithic sites of Meer and Rekem in Belgium (Cahen *et al.* 1979; de Bie 1998), Pincevent in France (Leroi-Gourhan 1984) and the Lower and Middle Palaeolithic sites of Boxgrove in England (Austin 1994; Bergman & Roberts 1988; Bergman *et al.* 1990), Rheindahlen in Germany (Thieme 1983, 1990) and last but not least, the findspots at Maastricht-Belvédère in the Netherlands, some samples of which are discussed in this article (de Loecker 1994; Roebroeks 1988; Roebroeks *et al.* 1997).

10.3 Introduction to the Maastricht-Belvédère site complex

The Maastricht-Belvédère gravel and loess pit is situated on the left bank of the river Maas (Meuse) about one kilometre northwest of Maastricht, The Netherlands (Fig. 10.2). In the 1980s an interdisciplinary project studied the exposures of loess and river deposits in the quarry and between 1981 and 1990 the Institute of Prehistory of Leiden University undertook several (rescue) excavations on a yearly basis in three important archaeological levels. These levels can be roughly dated to the Saalian and the Weichselian.

The most important archaeological levels at Belvédère are situated in fine-grained fluviatile sediments dating from a warm, humid intra-Saalian interglacial. Several independent dating methods indicate that these sediments were deposited some 250,000 years ago and they can be correlated to Oxygen Isotope Stage (OIS) 7 (*cf.* Roebroeks 1988; Vandenberghe *et al.* 1993). The interglacial variation of biotopes around the Maas river valley at Belvédère attracted a large number of animal species (Duistermaat 1993; Meijer 1985; van Kolfschoten 1993) remains of which were recovered in association with Middle Palaeolithic lithic artefacts (for more details on the local geology, palaeontology and dating see Roebroeks 1988; van Kolfschoten & Roebroeks 1985; Vandenberghe *et al.* 1993).

The large quantities of flint artefacts, recovered from various findspots, indicate that the Belvédère

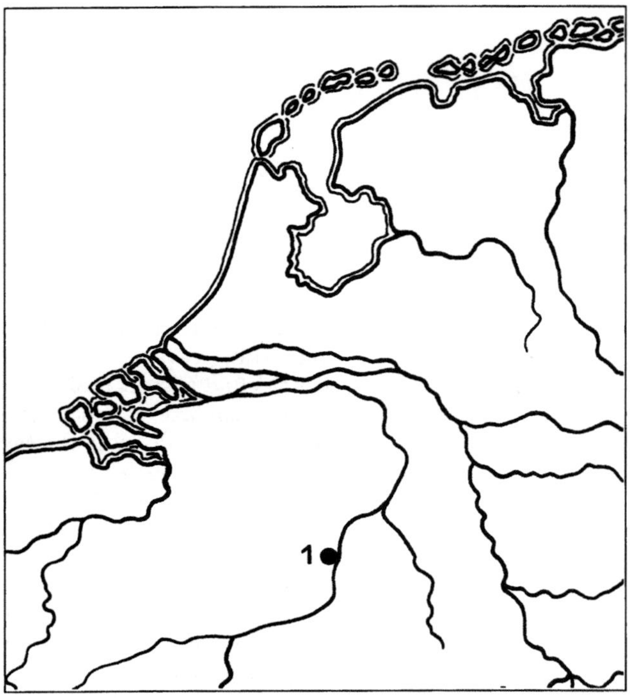

Figure 10.2: Location of the Maastricht-Belvédère pit (1) (after Roebroeks *et al.* 1997).

landscape was frequently visited by early humans. As a result of the long-term project, we are relatively well informed about their activities (Roebroeks 1988; Roebroeks *et al.* 1992). At Belvédère the excavated surfaces are traditionally being coined 'sites' (Site A, B, C, *etc.*, according to the chronology of research). However, it is evident by now that site terminology does not offer an adequate framework to understand the nature of the patterns encountered. Clearly these are not pin-pointed occurrences as recent excavations show that large parts of the interglacial valley bottom originally must have been littered with artefacts. This large scale and continuous artefact distribution, interpreted by Roebroeks *et al.* (1992) as a 'veil of stones', shows internal variations in artefact density, composition and refitting potential. At one end of the density scale are the 'low density scatters' consisting of scattered flakes, worn out tools and relatively few dorsal/ventral refits (Sites G and N), while at the other end are the 'patchy' occurrences or 'high density scatters', with a striking dominance of flint knapping debris which turned out to be a refitter's 'paradise'. Some of these patches, for example at Sites C, F and K (de Loecker 1992, 1994; Roebroeks 1988), were so well preserved that extensive refitting studies allowed detailed reconstructions of former reduction schemes to be produced (de Loecker forthcoming; Schlanger 1994).

10.4 A characterisation of the site data

As mentioned above, analysis of the Belvédère site complex shows that two kinds of find distributions can be identified in the intra-Saalian landscape: 'high' and 'low' density distributions. In order to better understand the site data it is important to stress that in all probability these find distributions can be interpreted as findspots created under similar environmental conditions over a relatively short period of time. The sites are probably contemporaneous in terms of having been formed in the same climatic optimum of an interglacial period. Apart from this 'contemporaneity' we can assume that no significant changes in the availability of raw material occurred during the formation of the archaeological assemblages. Moreover, the sites have been documented in a small area of approximately six hectares which allows us to infer that they were formed in comparable biotopes (Roebroeks 1988). The methods used to document the archaeological occurrences could not be chosen freely; they were always a compromise between the interests of the commercial exploiter of the pit and research aims. Since 1986, the emphasis has been on documenting large surfaces instead of focusing on a very detailed documentation of small areas.

10.4.1 The high density find distributions: Sites C, F and K

One of the find distributions with a 'patchy' appearance is Site C. Here mainly flint artefacts, but also burnt flints, badly preserved faunal remains, two clusters of (small) charcoal particles and some haematite dots were recovered from an area of $264m^2$ (Roebroeks 1988; Roebroeks & Hennekens 1990). The flint assemblage consists of 3,067 pieces predominantly less than 30mm in size (87.1%) and included only five tools *sensu stricto*, all scrapers and backed knives, and 18 flakes with macroscopic signs of use. The assemblage is to a large extent the product of prepared core technique, with several 'classical' Levallois flakes and the products of a *débitage Levallois récurrent* (*cf.* Boëda 1986). Most flakes are relatively thin and many striking platforms are facetted. The flint from Site C could be attributed to six different Raw Material Units (RMU) suggesting the presence of at least six different flint nodules with distinct core reduction sequences and spatial patterns. As a result of refitting, 21.5% ($n = 659$ artefacts) of the total assemblage could be conjoined. Flint knapping was evidently a main activity at the location, as the majority of the (refitted) lithic assemblage consists of small chips, unretouched flakes and four waste cores.

A slightly different kind of high density distribution was excavated at Site F (Roebroeks 1988), where except for some charcoal particles only flint artefacts were found. In total 1,177 pieces of debitage were recovered from an area of only 42m^2, of which 87.7% were less than 30mm. Artefact and tool density is much higher than the mean densities at Site C. The Site F assemblage consists of eight tools, including a scraper, a borer and six pieces with macroscopic signs of use. The scatter consists predominantly of unretouched flakes without significant dorsal preparation and mainly without facetted striking platforms. The flint artefacts obviously represent the waste of different flint knapping episodes, but probably a different core reduction strategy was used than that at Site C. Only two cores were recovered; one disc and one miscellaneous. Of the total artefacts 13% ($n = 153$ artefacts) could be refitted. It is highly possible that the excavated Site F surface formed part of a much larger scatter resembling Site K in terms of both densities and typo/technological composition.

An area with a 'patchy' nature is most evident at Site K where some badly preserved faunal remains, a few charcoal particles and a large quantity of flint debitage and tools were excavated over a surface of 370m^2 (de Loecker 1992, 1994). The flint assemblage includes a total of 10,912 artefacts, most less than 30mm (71.1%). Mean artefact and tool densities are the highest for any site at Maastricht-Belvédère. The lithics include 111 tools *sensu stricto*, 26 pieces with macroscopic signs of use and 91 cores. Various types of scrapers dominate while disc and discoidal cores, usually very irregular in appearance, are the most common core type. Furthermore, 617 artefacts were identified as burned. At Site K all stages of production and technology are represented. Cores show a reduction strategy with limited attention to core preparation; there are few traces of striking platform or butt preparation and negatives of earlier flakes in the reduction sequence are usually used as striking platforms. The large number of cores, the presence of waste from all core reduction sequences and refitting studies imply flint knapping as the main activity responsible for the patchy nature of the distribution. The elaborate refitting analysis resulted eventually in the conjoining of 16.8% ($n = 1,828$ artefacts) of the total assemblage.

10.4.2 The low density find distributions: Sites G and N

It is clear that the dense patches of artefacts are archaeologically highly visible and represent the most frequently excavated surfaces (the 'classical' sites). However, it should be mentioned that they are rather local phenomena; between the large clusters of artefacts such as Sites C, F and K, stray finds have been found all over the pit which appear to have been discarded as isolated objects or in a small group of one up to a few dozen (*cf.* Isaac 1981).

A section of this suggested 'all over' low density distribution was excavated at Site G (Roebroeks 1988). At this scatter, 75 flint artefacts were recovered in association with faunal remains and burnt flints over an area of 50m^2. Most artefacts (70%) are $\geqslant$30mm and raw material studies showed that the larger flakes were probably struck from a least three different flint nodules. The flints were fairly evenly distributed among a few bone fragments. Eight artefacts are tools and include a scraper, two backed knives and pieces with macroscopic signs of use. One backed knife is made on a large flake that had been struck from the lateral side of a prepared core (*éclat débordant* as defined by Beyries & Boëda 1983). The rest of the assemblage consists of small flakes and chips which show traces of striking platform or butt preparation. Sporadically these artefacts form small groups of refits ($n = 25$ or 33.3% of all artefacts). The refitting analysis indicates that minimal flint knapping took place inside the excavated area as only some very fine knapping debris could be refitted to two larger flakes. More than half the conjoins are of fractures. Cores are absent.

Although refitting evidence excludes the possibility that bones and stones were washed together by fluvial activities or by other natural deposition processes, it does not provide any decisive arguments for assuming a human involvement in the spatial distribution of Site G faunal remains. However, the large backed knife mentioned above was recovered from a concentration of rhinoceros remains (Roebroeks 1988). Microwear analysis indicates that the knife was probably used to cut the hide of an animal with thick skin, such as that of a rhino or elephant, possibly during butchering (van Gijn 1988, 1989) which seems to give us the best possible archaeological evidence at Belvédère for translating spatial associations between bones and stones into behavioural terms.

A shift in interest in recent research has highlighted the importance of the thin 'background noise' in artefact distributions for the interpretation of the Belvédère site complex. The decision was made to fully concentrate on the 'low' density patterns to gain an insight into the nature of the archaeology surrounding the high density patches. Therefore *c.* 100m to the south of Site K an area of 765m^2 was excavated (Roebroeks *et al.* 1992). As with Site G, the Site N surface formed part of one continuous and widespread scatter of artefacts and a few faunal remains. Only 450 flint artefacts, mostly ⩾30mm (80.2%), were found and bone material was virtually absent. A rather high number of tools ($n = 26$) was recovered. Twelve tools *sensu stricto* (mainly scrapers — sometimes heavily reduced and broken), 14 flakes with macroscopic signs of use and (unretouched) backed knives were recovered. The artefact distribution shows no evidence for substantial primary flint knapping and tool production. There was only one (exhausted) disc core and cortical flakes are clearly under-represented. Furthermore the tools:waste ratio of 1:16 is the lowest of all Belvédère sites. Further evidence for this assumption is found in refitting; 16.2% ($n = 73$) of all artefacts could be refitted and represent mainly fractures. Many flakes show careful butt preparation and the Facetting Index values are among the highest at Belvédère. Raw material study indicates that the artefacts were probably made from at least eight different flint nodules. Technologically and typologically this tool-dominated assemblage differs in a number of aspects from that of Site K for example (see de Loecker forthcoming; Roebroeks *et al.* 1992). It may be that a selection of tools and blanks was taken to this area, which was probably visited repeatedly during a number of 'short' unrelated events. The Site G and N scatters, which may be representative of the overall background noise in this kind of interglacial river valley setting, may have been produced while using technology in direct resource-procurement, while the rich sites certainly represent discard from activities essentially aimed at maintenance of technology such as flake and tool production

10.5 The Weichselian flint assemblage of Site J

As the Weichselian Site J (Roebroeks 1988; Roebroeks *et al.* 1997) revealed unique refitting data on the reuse and recycling of primary flakes and the production and (re)sharpening of tools, it is important to incorporate these results in this article. Site J is one of the few well excavated Middle Palaeolithic scatters on the plateau edges in Southern Limburg and it surroundings. Dating evidence suggests an Early Weichselian age, possibly in the range of OIS 5c to 5a (see Roebroeks 1988, 100–2 for details). No absolute dates could be established. In contrast to the intra-Saalian artefact assemblages, which were all recovered from a meandering river valley setting, the loess covered Site J was geomorphologically located in the higher areas of the Southern Limburg landscape. In the Early Weichselian the river Maas had cut its bed at least ten metres deeper and therefore Site J was situated originally at the edge of a terrace plateau overlooking the river valley which became braided (for a detailed description on the dating evidence, stratigraphy and palaeoenvironment see Roebroeks 1988; Roebroeks *et al.* 1997; Vandenberghe *et al.* 1993).

Although the findspot was partly destroyed by commercial quarrying, an area of at least 210m^2 was excavated revealing a 'high' density find distribution with many flint artefacts, a few scattered charcoal fragments and some poorly preserved molar fragments, probably of *Mammuthus primigenius*. The excavation uncovered a total of 2,863 artefacts (most ⩾30mm, 74.4%) including 134 complete and fragmented tools predominantly scrapers, denticulates and flakes with macroscopic use retouch. Furthermore, evidence for the use of prepared (Levallois and/or disc) cores is absent. Most cores are irregular and polyhedral indicating a rather unsystematic reduction strategy. The majority of unretouched flakes show no significant dorsal or butt preparation. In general it seems that all stages in core reduction and tool production (including resharpening) are well represented suggesting that flint knapping was one of the main activities at this findspot. This is further indicated by the refitting results; 39% ($n = 945$) of the artefacts could be conjoined and the number could have been much higher, as part of Site J had been destroyed before the excavation. Interestingly Site J does not seem to be present against the background of thin 'low density' scatters as previously reported for the intra-Saalian interglacial patches. Instead the concentration was of a more isolated character, or in the words of Roebroeks *et al.* (1992, 1997) the assemblage seems to have been 'parachuted' onto an archaeologically quite sterile landscape

10.6 The refitting programme used at Maastricht-Belvédère

From the first Belvédère excavations onwards much time and energy was invested in systematically refitting the flint material. As a result of Cziesla's publications (1986; 1990) the refitting programme, and especially the visualisation of the results, became even more standardised in the second half of the 1980s. Consequently the conjoining results of Sites J, K and N are recorded slightly differently from Sites C and F which were excavated earlier.

During the excavation of Site K it became apparent that the flint assemblage possessed a certain refitting potential as many artefacts (both small and large) from the same square metre could be conjoined in the field. Therefore the decision was made to undertake detailed refitting analyses. Because conjoining artefacts is a time-consuming activity, several different strategies can be followed. Often the raw material is first divided into different flint types and subsequently a refitting analysis is undertaken on every group (*cf.* Rensink 1992). At other sites it is more feasible to divide the excavated material into different technological stages or types. For instance all decortication flakes or blades and blade fragments can be selected first; next a refitting analysis is undertaken on every group, after which one can endeavour to find mutual conjoinings. At Site K a third formula was used. All artefacts ⩾20mm were put on laboratory tables according to the square metre in which they had been found (roughly the spatial configuration of the assemblage). Artefacts were first conjoined 'within' the square metres and subsequently adjacent square metres were involved in the refitting analysis. In contrast to the previous refitting analysis at Maastricht-Belvédère (see Roebroeks 1988), at Site K we were able to take advantage of the 'Cziesla approach' (Cziesla 1986, 1990, *cf.* Fig. 10.3c and 10.3d); because stone artefacts can undergo various stages of 'reduction', Cziesla emphasises the importance of distinguishing between several types of refits during reconstruction. He makes a distinction between the following types of conjoinings (Cziesla 1986, 253–4, 1990, 15):

Aufeinanderpassungen (refitting of production-sequences) refer to the refitting of all products from 'basic' or 'primary' production/reduction. It concerns only dorsal/ventral conjoins *e.g.*, flake series in a reduction sequence.

Aneinanderpassungen (refitting of broken artefacts) indicate the reconstruction of 'basic products' like flakes, blanks and tools. It mainly concerns the refitting of broken flake/blade fragments, but also the conjoining of 'split bulbs/cones' (*accidents de Siret*).

Anpassungen (refitting of modifications) refers to the refitting of all products resulting from the modification of a blank into a tool or resharpening of a tool (*e.g.*, Cornford 1986).

As Cziesla (1986, 1990) noted, all flint artefacts originate in a 'basic' or 'primary' production (*Aufeinanderpassungen*), but they can change to other classifications such as fractures (*Aneinanderpassungen*) and/or 'secondary' modifications (*Anpassungen*). In addition to these three types of refits a fourth class was introduced (Cziesla 1986, 1990), namely *Einpassungen* (inserts). This group concerns the refitting of fragments produced by natural processes (frost- and heat-damage).

In the refitting analysis of Site K (and other Belvédère sites) the Cziesla approach was used mainly in its original design. Only one adaptation to the approach was made. If a broken flake (*e.g.*, consisting of two parts) was conjoined dorsal/ventral to a complete flake, Cziesla uses two *Aufeinanderpassungen* that go from the complete flake to both broken parts. Because of the quantity of fits at Site K which could make the conjoining maps too complex, the option was chosen to connect the complete flake only with the proximal or most proximal part of the broken flake. Of course, as in Cziesla's methodology, both broken parts stay connected with an *Aneinanderpassung* (see Fig. 10.3e).

10.7 Computer applications

As well as refitting itself, the recording and subsequent visualisation of the spatial results can be very time-consuming. It is almost impossible to make a manual three-dimensional model of all conjoining results for a single flint-rich site. To overcome this problem the information revealed by refitting can be computerised with the aid of a software packages such as SITEFIT (Lindenbeck 1990), ANALITHIC (Boekschoten & Stapert 1993, 1996; Stapert 1992) or AUTOCAD. Initially at Site K, SITEFIT (specifically designed to assist the refitting analysis of stone age sites) was used. The programme operates with dBASE files and allows size-, location- and orientation-proportional representation of the spatial distribution of finds. The program is able to make two- or three-dimensional models that can be rotated, tilted or magnified in every direction. The only information it requires is three-dimensional recording of artefact location and the kind of refit established according to the Cziesla approach. If the artefacts are measured (length, width and thickness) a three-dimensional cube of fixed dimensions of every object can be visualised. In addition, the artefacts (cubes) can be enlarged and specific technological information can be added to them. This information can be pictured in an excavation grid (Lindenbeck 1988, 1990). The SITEFIT package is also equipped with SITEPLOT for graphic printouts and SITECHCK for basic statistics.

Although SITEFIT is a successful programme for visualising and computing conjoined artefacts and has been tailored on the Cziesla approach, use was also made of other flexible commercial programmes such as dBASE, AUTOCAD, MAPINFO and SURFER. We had a few special dBASE IV programmes made to create and maintain information about artefact compositions. AUTOCAD was eventually used to store and manipulate graphic representations of refits and a limited statistical analysis was done with a SPSS.

10.8 Describing and visualising the refitted reduction sequences

During a refitting analysis (dorsal/ventral fits) an attempt is made to refit all stages (flakes) in their chronological order, working from young to old; in fact

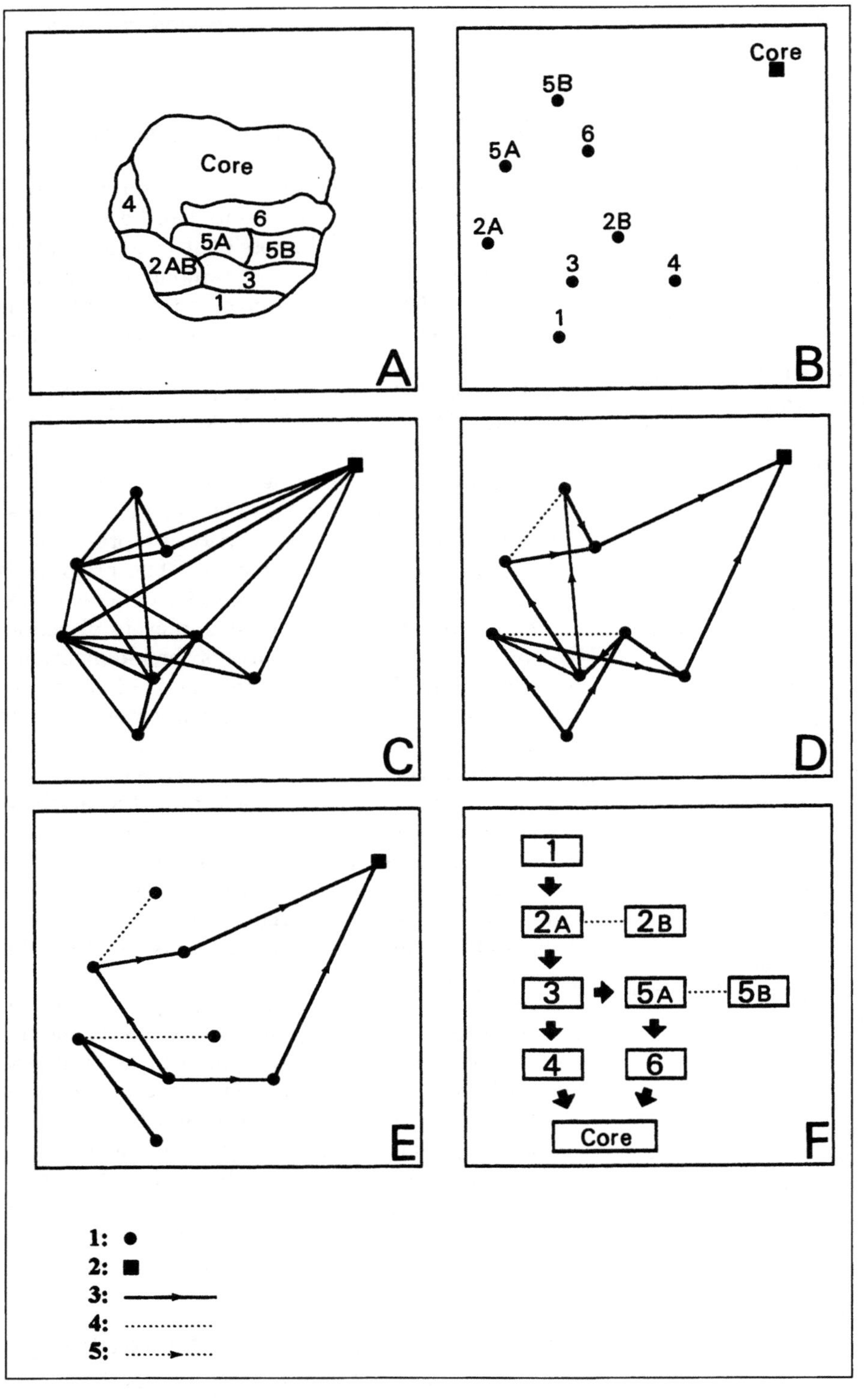

Figure 10.3: Models used for the graphic representation of the Maastricht-Belvédère refitting data after Cziesla 1986, Figure 7; see also Roebroeks 1988, Figure 45. **A:** Hypothetical example of a flaked core showing the conjoined elements (view of the striking surface). **B:** Spatial distribution of the refitted elements. c: core, 1: flake, 2a/2b: flake broken into two pieces [2a: proximal part, 2b: distal part], 3 and 4: flakes, 5a/5b: broken flake (split cone, accident de Siret), 6: flake. **C:** All contact surfaces linked by lines. The 'pre-Cziesla approach' (*cf.* Roebroeks 1988). **D:** The 'Cziesla approach'. Dorsal/ventral refits are traced back to the core, following the reduction sequence, as indicated by the arrows. Broken artefacts are indicated by dashed lines. **E:** The adapted 'Cziesla approach' used at Site K. Here a complete flake is only connected with the proximal or most proximal part of a broken flake. Both broken parts stay connected through a dashed line. **F:** Reduction sequence of the conjoined flakes put in a Harris-matrix. The numbers refer to the individual finds. Number 1 is the highest flake in the 'stratigraphical' sequence. In other words the oldest one or better the first artefact flaked. **Key:** 1. Flake; 2. Core; 3. *Aufeinanderpassung* (production sequences); 4. *Aneinanderpassung* (broken artefacts); 5. *Anpassung* (modifications).

a description of the technological reduction sequence is made. A Harris matrix sequence can be used to visualise such a reduction sequence (Fig. 10.3f). When we describe and visualise a reduction sequence several action stages, or parts in it, can be identified, representing information on what the French call the *chaîne opératoire* (Pelegrin *et al.* 1988) and as a result several technological actions or decisions made by the flintknapper can be distinguished. It is therefore useful to make a distinction between these stages of the *chaînes opératoires*. At Maastricht-Belvédère the refitting evidence of Site J (Roebroeks *et al.* 1997) and in particular Site K (de Loecker 1994) led to the development of a framework which focuses on three main reduction sequences (or three parts of one bigger reduction sequence):

1. Raw Material Unit reduction sequence: If a (large) flint nodule enters a site it may be split into smaller blocks as is shown by the Site K refitting (de Loecker 1994, and below) which can in turn be used as cores. The raw material may be split by means of large, thick flakes or along natural cracks caused by frost action (*cf.* Stapert 1976). As there is no hierarchy in the reduction sequence, when flint is fractured along natural cracks, a spatial distribution map of the smaller units (flaked cores) is used to visualise the raw material unit reduction sequence (Fig. 10.4a).

2. Primary reduction sequence: Once the raw material has been fractured, the smaller units can be used in the primary reduction sequence, *i.e.*, all flakes and cores (including blanks on which tools were made) which are produced/discarded during the basic reduction of a core. To visualise the primary reduction the Cziesla approach is used. Therefore, only *Aufeinanderpassungen* and *Aneinanderpassungen* will be found in this reduction sequence (Fig. 10.4b).

3. Secondary reduction sequence: During or after the primary reduction flakes can be selected for secondary modification and/or use. These flakes can be retouched into tools *sensu stricto* or secondarily used as cores (flaked flakes, *cf.* Ashton *et al.* 1992). This stage of reduction may also include resharpening of tools (*cf.* Cornford 1986). The resulting debitage is classified as part of the secondary reduction sequence within which we find only *Anpassungen* and *Aneinanderpassungen* (Fig. 10.4c).

When the primary (and secondary) reduction sequence is represented as a Harris matrix all flakes and cores are placed in their hierarchical order and a secondarily reduced flake (retouched, resharpened or flaked flake) will be presented as a separate action within a reduction 'stratigraphy' (Fig. 10.4d).

10.9 Putting a technology back together

Through the very detailed analysis and visualisation of the conjoining results much behavioural data was gained at Maastricht-Belvédère. In the following sections some of the refitting examples from the Belvédère sites, together with their implications for the reconstruction of early human behaviour, will be discussed in the context of the research topics mentioned above. Moreover, this summary highlights the information value of conjoining artefacts in general.

At Site C and K refitting proved to be especially important in reconstructing the *chaînes opératoires* and the spatial articulations of the reduction sequences. The Weichselian Site J conjoinings exemplify a rather paradoxical core reduction demonstrating both economizing and wasteful behaviour. Moreover this scatter clearly shows the production and a constant resharpening of tools.

10.9.1 A Levallois recurrent core approach: Site C

On the basis of the specific flint properties — texture, cortex, inclusions and colour, most of the 3,067 flint artefacts recovered from Site C were attributed to at least six different Raw Material Units (RMUs) *i.e.*, six different flint nodules (Roebroeks 1988). The large amount of small flaking debris together with a considerable quantity of refits, including small and large flakes and cores, clearly indicate on-site knapping activities. The technology at Site C shows that some cores were well prepared. Moreover, this preparation appears to be oriented towards larger Levallois flakes (Roebroeks 1988; Schlanger 1994, 36–59, Chapter 2). Most refitted groups at Site C are 'small' sequences of flakes and broken fragments of flakes although some large compositions were also determined. Conjoining of RMUs 3, 4 and 5 revealed spectacular results; technologically the elaborate refitting programme demonstrated that some RMUs were the products of a discoidal and Levallois reduction sequence. Besides this technological variety, the refitted groups show a 'biographical' variety (only specific stages in core reduction are represented). The initial (decortication) flaking stages are missing from some flint nodules, while other RMUs are represented by larger flakes which could not be refitted to the bulk of the other debris.

As most of the Site C refitting data has already been published (Roebroeks 1988; Schlanger 1994) it is sufficient to focus here on RMU 4 conjoins which technologically (and spatially, see below) are probably the most interesting at Site C. The core must have entered the excavated area in a prepared form as no cortex flakes could be refitted to it (Fig. 10.5a). Subsequently a series of flakes was produced from a single striking surface. While most flakes were refitted onto the striking surface of the core only a few

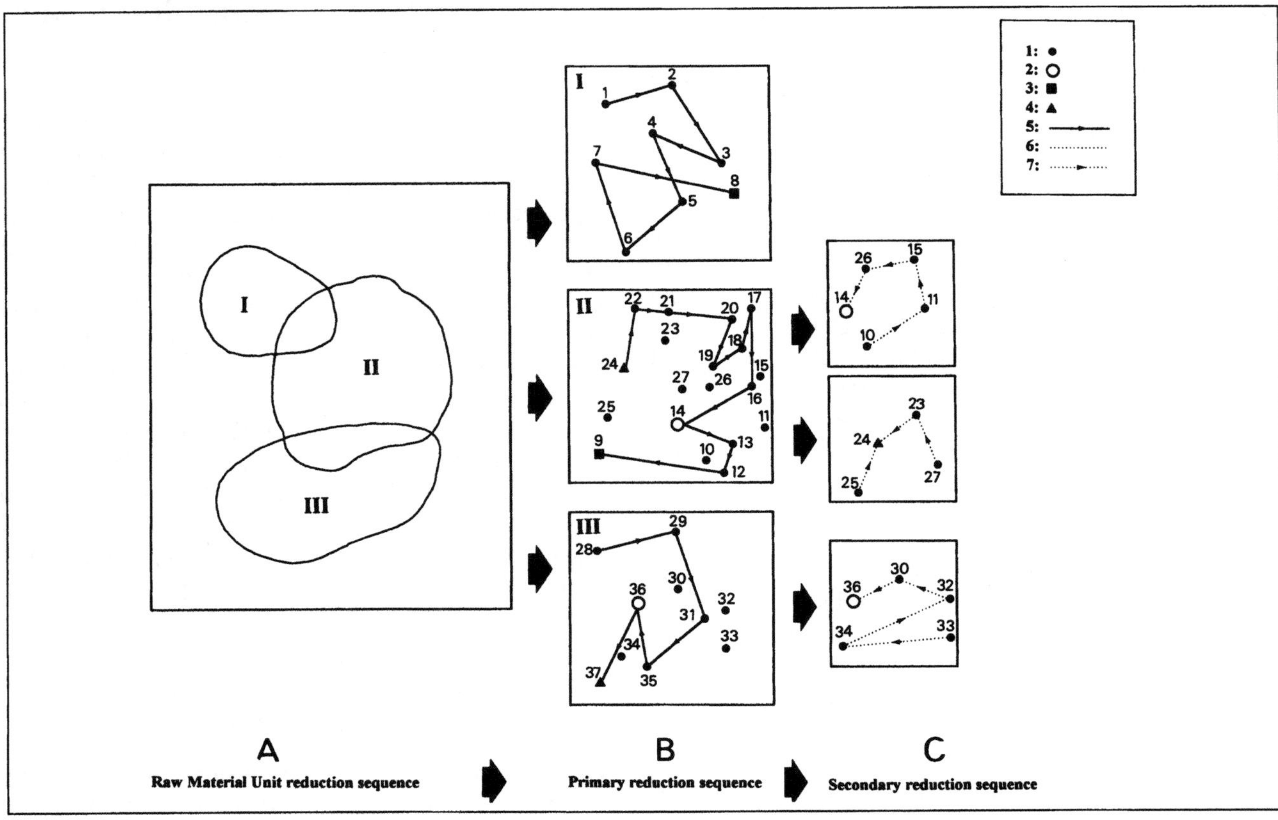

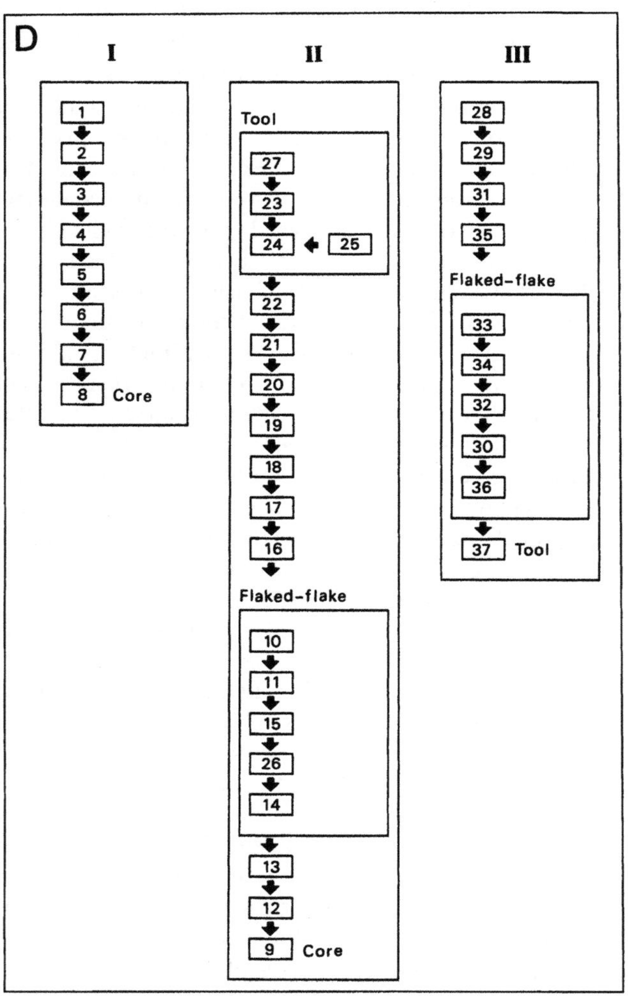

Figure 10.4: Framework which focuses on 'three' main reduction sequences, based on refitting, used at Maastricht-Belvédère. **A:** Raw Material Unit reduction sequence: because sometimes no hierarchical placing is noticed, a spatial distribution map of the units is used to visualize the raw material unit reduction sequence. **B:** Primary reduction sequence: primary reduction refers to all flakes and cores which are produced/discarded during the basic reduction of a core. The Cziesla approach is used to visualise the data. **C:** Secondary reduction sequence: secondary reduction refers to all debris produced during secondary modification and/or use. The 'Cziesla approach' is used to visualise the data. **D:** Harris-matrix including the Raw Material Unit reduction sequence, Primary reduction sequence and the Secondary reduction sequence. **Key:** 1. Flake; 2. Flaked-flake; 3. Core; 4. Tool; 5. *Aufeinanderpassung* (production sequences); 6. *Aneinanderpassung* (broken artefacts); 7. *Anpassung* (modifications).

could be conjoined to the 'continuous' striking platform. The core produced a rather regular alternating of smaller 'preparation' flakes and larger 'Levallois' flakes. Figure 10.5b shows this cyclical pattern of distinctive phases (see also Roebroeks 1988, 48, Fig. 56). The RMU 4 reduction sequence clearly shows that technology was not directed towards the production of a single Levallois *sensu stricto* flake (Boëda 1986; Bordes 1961), but towards a series of prepared Levallois flakes. The Levallois flakes produced during (at least) these seven phases of reduction show some distinctive qualitative features: part of the dorsal side consists of centripetal preparation while the other part displays one scar of a previous larger flake. In general this type of reduction, which is based on a careful preparation of the convexity of the core's working surface, can be described as *débitage Levallois récurrente* (Boëda 1986, 1993, 1994). At Site C the repeated sequences of preparation and production continued until the 'Levallois' core was exhausted and eventually discarded on the spot. It should be mentioned that a few larger flakes are missing from the refitted sequence while a number of larger flakes could not be conjoined. Seven of the latter show usewear traces, while none of the refitted flakes show signs of use.

In general the Site C assemblage is the result of a prepared core technique which resulted in disc(oidal) and Levallois cores. Moreover, it includes the products of a *débitage Levallois récurrente* core approach. Site C, then, represents a locus where technology was primarily maintained. However, some curated cores, flakes and tools entered and left the area as well.

10.9.2 From the supply of raw materials to the production of flakes, using a disc(oidal) core approach: Site K

The rather sophisticated form of *Levallois récurrente* documented for Site C contrasts with the non-prepared core reduction strategy evident in the rich assemblage from the larger Site K. Here dorsal flake patterns do not indicate careful core preparation; facetted butts are virtually absent and the few flakes that do have facetted butts might have been imported. The large quantity of chips, flakes and cores recovered close together, the large number of conjoined artefacts (large and small), the relatively short refit distances and the presence of two flint hammerstones again point to on-site core reduction. Refitting (1,828 conjoins) shows that most of the reconstructed nodules were reduced using a very uniform reduction strategy of which all stages are well represented. Most conjoined sequences, together with the high percentage of cortex on both cores and flakes, indicate that large flint nodules were introduced to the area without any, or hardly any, preparation or decortication (Fig. 10.6a). The abraded cortex indicates that the raw materials were probably collected in nearby river deposits. Protruding areas of the nodule (pre-core) were removed in an initial rough shaping achieved by the removal of several large, thick, flakes which were sometimes secondarily used as cores. Subsequently, the large nodules were split into smaller blocks or cores. The exploitation of natural fissures (frost cracks) in the nodules played a major part in the initial splitting of the raw material. These natural imperfections initially simplified the splitting of nodules but certainly posed problems during further reduction, possibly seen in the discard of the high backed discoidal cores at an early stage of reduction. Moreover the natural fissures could indicate a non-selective choice, lack of better quality raw material or that the nodules had not been tested before their introduction into the area. The smaller blocks (cores) were then roughly shaped to remove irregularities after which a long uninterrupted series of flakes were produced. Continuous flake production was made possible through a constant turning and twisting of the core but was not aimed at the production of specific predetermined flakes as at Site C. Furthermore the production of a series of flakes from a striking surface was only interrupted or changed to another striking surface to rejuvenate or maintain a good working edge angle. In a sense this could mean that the same striking surface and striking platform of a core were used as long as possible for the production of flakes.

The high percentage of disc and (high backed) discoidal cores at Site K is therefore the result of a continuous, sometimes radial, removal of flakes (Fig. 10.6b). For disc cores one main striking surface is exploited with minimal preparation of a second one, while for discoidal cores two striking surfaces are exploited. Refitting showed that the first flake is at the same time the dorsal preparation (or butt 'preparation') for the second one, the second for the third etc. This manner of core reduction resembles *le débitage discoïde* as described by Boëda (1993). Moreover the reconstructed cores at Site K demonstrate a unifacial disc(oidal) approach in which each working surface holds its function throughout the reduction sequence (one working surface is used as striking platform and one as striking surface) as well an interchanging bifacial discoidal approach (in which the function of the two working surfaces can be changed at any moment within the sequence). The differentiation between both discoidal approaches is only recognisable when the reduction sequence (or parts of it) can be reconstructed through refitting.

10.9.3 An economizing and a 'wasteful' raw material strategy within one and the same flint composition: Site J

According to the lithic description and refitting the Weichselian Site J assemblage is very similar to Site K. At Site J lithic raw materials seem to have been

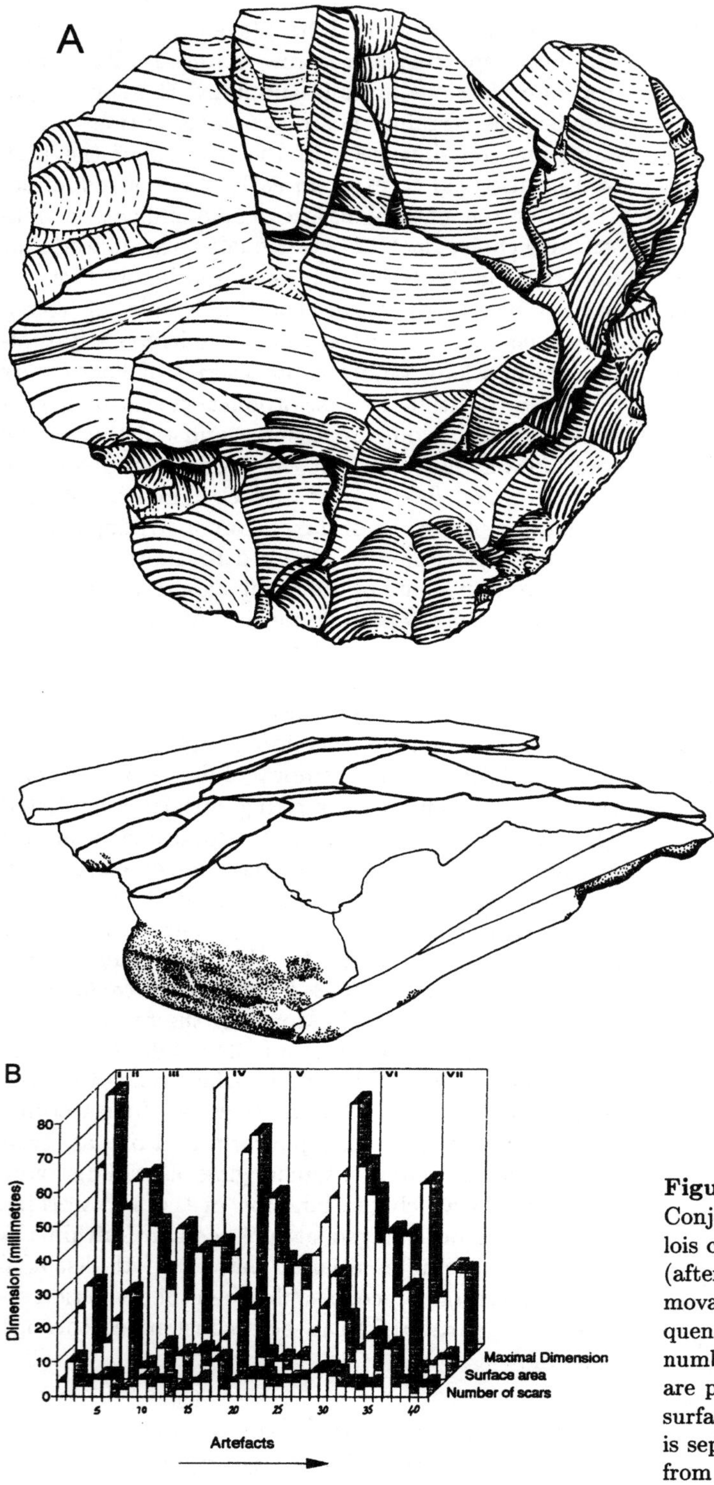

Figure 10.5: Maastricht-Belvédère Site C, RMU 4. **A:** Conjoined group of artefacts ($n = 29$), including a Levallois core. The nodule has a maximum dimension of 94mm (after Roebroeks 1988). **B:** The succession of flake removals from the RMU 4 core and their properties. The sequence is given on the horizontal axis, with Levallois flakes numbered and highlighted in white. On the vertical axis are presented, for each artefact, the number of scars, the surface area, and the maximum dimension. The sequence is separated by vertical lines into seven phases, numbered from I (first) to VII (last) (after Schlanger 1994, 1996).

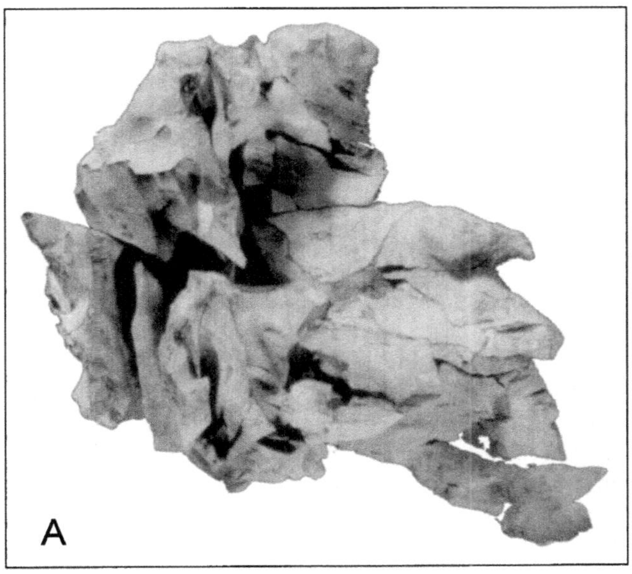

Figure 10.6: Maastricht-Belvédère Site K. **A:** Refitted block which consists of 146 flakes and cores. The cortex covered nodule has a maximum dimension of about 350mm. **B:** Refitted composition which consists of 47 artefacts and has a maximum dimension of 225mm. The whole core edge was used to remove radial orientated flakes in a circular direction ('anti-clock wise') from one and the same striking surface.

locally procured and all stages of core reduction, production and use of tools are well represented. Moreover, the reduction sequences as reconstructed by refitting all fit into one scheme (Roebroeks *et al.* 1997). Initially the flint nodules were divided into smaller blocks, mostly along natural fissures and a long uninterrupted series of flakes were produced in which specific activities such as decortication and correction of striking platforms are poorly defined. As with Site K this continuous flake production was made possible through a constant turning and twisting of the core, *i.e.*, through multidirectional flaking. However, unlike Site K only few flakes were produced in a continuous sequence from a single striking platform and from the same striking surface; as a rule new negatives were used as platforms for each new flake removal. This was done in such a regular and consistent way that Roebroeks *et al.* (1997) suggest that the complete reduction process was extremely systematic although in a non-classical manner. Refitting furthermore shows that initially rather large, thick flakes were produced (up to 200mm), some of which were used as cores (flaked flakes *cf.* Ashton *et al.* 1992) or were modified into tools, while many others were discarded on the spot. Moreover, a few larger cores were rejected at an early stage of reduction while others were completely reduced by means of very small flakes. In this sense the *chaîne opératoire* seems to display a somewhat paradoxical character regarding raw material. While the presence of large unused but suitable flakes indicates that earlier reduction phases were quite wasteful, the number and size of flakes scars on most of the exhausted conjoined cores point to a more economising behaviour during the final phases of core reduction. It must be emphasised that both modes of raw material behaviour appear within the same flint nodule (Fig. 10.7). This persistent (economising) reduction can also be seen in the constant modification and reuse (resharpening) of tools. Conjoining studies clearly showed that secondary working of flakes, tool production, repair of working edges and eventually discard of worn-out tools all took place within the excavated area. In some cases refitting even demonstrated such a degree of modification that tools changed from one typological category into another.

In conclusion the method of conjoining artefacts used at Maastricht-Belvédère proved to be a valuable analytical tool in the reconstruction of the different *chaînes opératoires*; the *Levallois récurrente* approach (Site C) and the splitting of large flint nodules along natural fissures (Site K and J) would have been missed had we not refitted. This also applies to the differentiation between the unifacial disc(oidal) approach and interchanging bifacial discoidal approach at Site K and the rather paradoxical raw material behaviour as described for Site J.

10.10 Recycling and its typological implications

It has been demonstrated by both regional research and site-oriented analyses that typological variations occur in and between Middle Palaeolithic assemblages. Frequently these typological variations relate to practices of reuse and/or resharpening of tools. In a broad sense this is known as the 'Frison effect' (Jelinek 1976). In his model Dibble (1987a; 1987b; Dibble & Rolland 1992) explains Middle Palaeolithic assemblage variability to a large extent as a function of tool scraper reduction through continuous resharpening and (re)modification of the working edge. During this process the use life of tools is extended intentionally, often leading to typological transformations. Accordingly, Dibble defines a reduction sequence from single to double, convergent and pointed side-scrapers. Steeply retouched transverse side-scrapers may also be viewed as the final stages in strategies of continuous reuse and remodification. In this process scrapers diminished in size and secondary working became more elaborate, sometimes leading to forms displaying Quina-like retouch. It can therefore be argued that assemblages consisting of large numbers of simple scrapers reflect a less intense utilisation (and reduction) of tools whereas assemblages with large numbers of double, convergent and transverse scrapers reflect a more intensive use of the lithics (Dibble 1987a,b). Several typological and raw material studies indicate that there is a relationship between intensity of retouch and distance of transport (Geneste 1985, among others); retouched flakes are generally discarded at much greater distances from the raw material source than non-retouched flakes and cores (Roebroeks *et al.* 1988). It follows, therefore, that sequences of reuse and typological transformation often have spatial significance insofar as they occur in the process of transportation of tools and tool kits.

At Site A, G and K few tool trimming elements were found (Fig. 10.8). In the case of Site K a Long Sharpening Flake (LSF, *cf.* Cornford 1986) was removed from a parent tool that was not present in the excavated part of the site. The fact that it cannot be conjoined onto an implement may indicate that the actual tool was subsequently transported away from the area. Identical scenarios are suggested for Sites A and G. In these last cases, however, it concerns Transverse Sharpening Flakes (TSF, *cf.* Cornford 1986). Judging from the variety of raw materials and the refitting data, many of Site N artefacts were introduced to the site as finished pieces. Among them is a double concave-convex side-scraper which shows a conspicuous dorsal flake scar on its left proximal side; this negative indicates that a secondary flake, knapped in the same direction as the actual flake, had partially removed the working edge (Fig. 10.9). The scar possibly originates from an LSF resembling a burin-spall. The macroscopic (use) wear traces on the newly created working edge probably indicate that the scraper was used again as well as introduced to the Site N area after resharpening but whether it was used in-or outside the excavated area is unclear.

Also at Site J refitting evidence points to the frequent (re)use and (re)modification of tools (Roebroeks *et al.* 1997). As various LSFs and TSFs could be conjoined to their parent tools, it is clear that the reworking of tools was practised on the spot. Moreover, the occurrence of intensively retouched tools coincides with the use of local raw materials suggesting that no substantial importation of finished tools was involved — an assumption confirmed by refitting. At Site J, then, we are dealing with the remnants of an expedient technology (cf Geneste 1985) in which LSFs and TSFs created new functional edges (Figs. 10.10a–b). The presence of some macroscopic use retouch and/or microwear on the new working edges implies that the tools were reused after rejuvenation, and that we are indeed dealing with resharpening and not with the final flakes of a first stage of tool manufacture. The refitting studies furthermore indicate that tools were modified for the purpose of reuse to such a degree that they were transformed from one typological category to another. Some tools have been transformed from a piece with a scraper edge to a denticulate, from denticulate to notch of so that virtually nothing remained of the former tool.

Refitting studies at Maastricht-Belvédère clearly document an early human behaviour that involved the resharpening of tools. For the Saalian Sites A, G and K this was exclusively performed on transported items, while at the Weichselian Site J resharpening was clearly part of an expedient *chaîne opératoire*. In addition to the evidence for the resharpening of unbroken tools, Sites G and K show evidence for the recycling of broken tool fragments. At both findspots a blade-like flake was introduced, probably as a finished item (Fig. 10.11). Within the excavated areas the flakes were broken into two conjoinable fragments (*Aneinanderpassungen*). As the broken parts show retouched edges it can be suggested that the breakage was a result of use. However, the proximal frag-

Figure 10.7: Maastricht-Belvédère Site J. Refitted group of large flakes and tools. One flake was completely reduced in at least 37 steps, while others were retouched into a scraper and Mousterian point. Onto the latter two resharpening spalls could be conjoined. Some of the flakes show traces of use. Scale 2:3 (after Roebroeks et al. 1997).

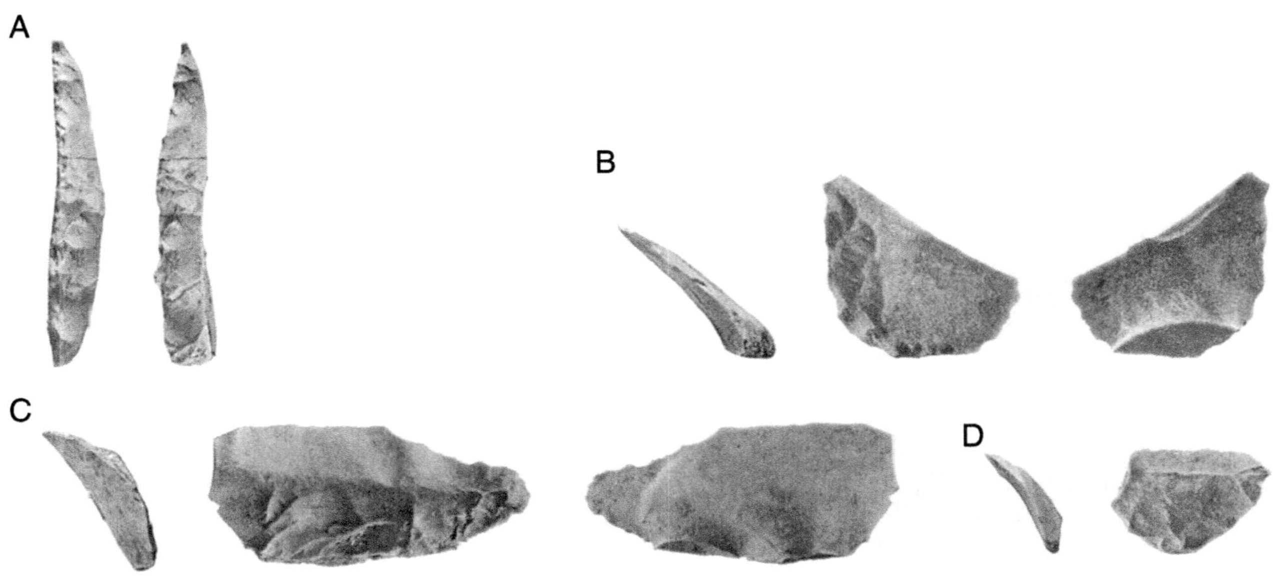

Figure 10.8: Maastricht-Belvédère. Different types of resharpening flakes. Scale 2:1. **A:** LSF from Site K. **B–D:** TSFS from respectively Sites A, G and K.

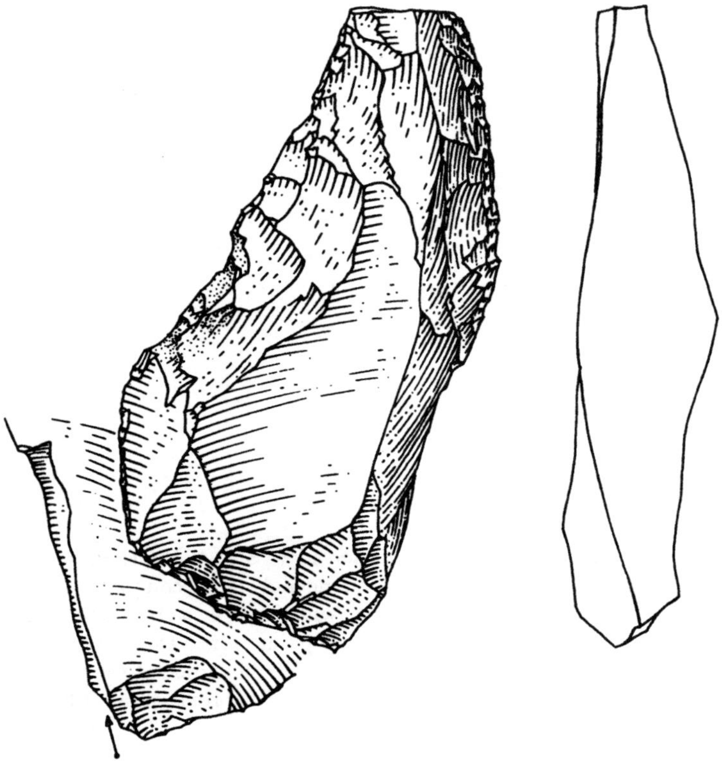

Figure 10.9: Maastricht-Belvédère Site N. Double concave-convexe side-scraper which shows a conspicuous dorsal flake scare on its left proximal side. The scar possibly originates from an LSF resembling a burin-spall. Scale $\frac{2}{3}$ (after Roebroeks *et al.* 1992).

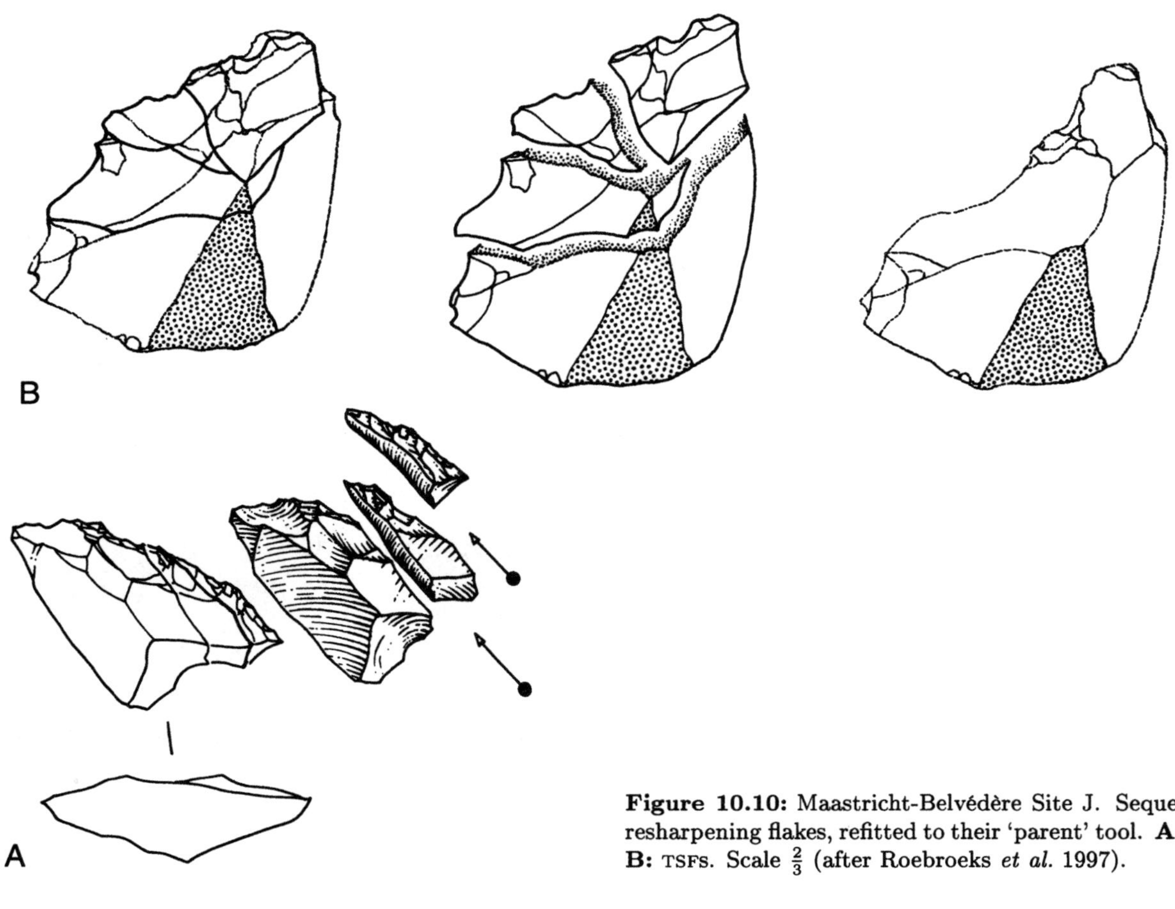

Figure 10.10: Maastricht-Belvédère Site J. Sequence of resharpening flakes, refitted to their 'parent' tool. **A:** LSFs. **B:** TSFs. Scale $\frac{2}{3}$ (after Roebroeks *et al.* 1997).

ments show signs of more intensive use (retouching) than the distal parts indicating their continued use after the flakes had been broken. In an alternative scenario it can be suggested that the blade-like flakes were used (?), broken and recycled outside the excavated area and that both parts were subsequently introduced to the site. Recycling parts of broken (transported) tools is also recorded at Wallertheim D, a Middle Palaeolithic site in the German Rhineland (Conard & Adler 1997). Without refitting this kind of information would have been lost.

10.11 Transportation of lithic technologies

The Saalian high and low density scatters at Maastricht-Belvédère do not form separate and clearly defined spatial units; in fact, all finds lie somewhere on a sliding scale between high density areas at Sites K and F and low density ones at Sites N and G. Nevertheless, there are some striking general differences between both types of scatters (see de Loecker forthcoming; Roebroeks *et al.* 1992, 11, Table 2). For example the percentage of tools within low density distributions (Site G: 10.7%, Site N: 5.6%) is much higher than for the high density patches (Sites C and F: 0.7%, Site K: 1.3%). Two factors seem to be responsible for these differences. The scatters and scatter types represent different trajectories within the life histories of Middle Palaeolithic flake technologies, *i.e.*, of cores and flakes. At one end of the continuum are Sites K and F where reduction sequences began and the degree of importation is low, except for some scrapers and points (especially at Site K); at the other end are the low density scatters of Sites G and N where flake technologies ended. Meanwhile, cores had been much reduced and transported flakes had been transformed into retouched and worn out tools. Stated differently, the high and low density scatters reflect different places in the spatial organisation of technologies. It may well be, however, that this was related to the execution of different activities. It is suggested by Roebroeks *et al.* (1992, see also Isaac 1981) that while high density scatters predominantly reflect the maintenance of technology (*i.e.*, the preparation of new cores, flakes and tools), low density scatters might relate to the actual use of these technologies in direct food procurement. The butchery of the young rhino at Site G fits this hypothesis well (see Roebroeks 1988; van Gijn 1988, 1989). Most of the Belvédère scatters can be perceived as belonging to a technological strategy that involved a regular transport of flakes, flake-tools and cores.

At Site K several classic Levallois flakes are present in the debitage, especially among the tools (scrapers). This is remarkable as no clear Levallois *sensu stricto* cores (Boëda 1986, 1993, 1994; Bordes 1961) were described in the assemblage. Moreover, the raw material of most of these tools looks somewhat different (*i.e.*, exotic and fine grained) as compared to the other artefacts. Therefore it can be assumed that most of the Levallois flakes, and especially the tools, entered the site as end products and were discarded on the spot, an assumption supported by refitting. The negative results of conjoining debitage to Levallois *sensu stricto* flakes also shows that previously assumed scenarios (*cf.* de Loecker 1992) such as transport of Levallois cores to the excavated area, production of Levallois flakes at Site K and subsequent transport of the cores away from the findspot, or a transformation of Levallois cores into other types such as disc and/or discoidal nuclei (*cf.* Boëda 1993, 393; Vynckier *et al.* 1988, 135), can be excluded.

As we have seen before, scrapers clearly dominate the Site K tool assemblage. In general these scrapers consist of three major classes (*cf.* Dibble 1987a,b); single, double and convergent types made, among others, on Levallois blanks. On the basis of Dibble's reduction models (1987a; 1987b, see above) convergent and pointed side-scrapers are expected to be shorter than the single and double-edged side scrapers. It is therefore remarkable that within the Site K tools the complete convergent side-scrapers through double-edged side scrapers are on average about 15mm longer than the single-edged side scrapers. If the complete reduction sequence had taken place on the spot, then, in the case of Site K, Dibble's model does not work. However, raw material (especially that of convergent side-scrapers) appears somewhat different to the rest of the assemblage and these tools may have entered the site as end products whose transformation could have taken place outside the excavated area. Further support for this assumption is that none of the convergent and double-edged side-scrapers could be refitted to the rest of the Site K material. Many of these scrapers, too, are produced on Levallois *sensu stricto* flakes. It is suggested that the Site K assemblage forms a link between a Levallois component produced on transported raw materials and the occurrence of convergent (including Mousterian points) and double-edged side-scrapers (*cf.* Geneste 1985).

The Site K data provide sufficient evidence to indicate early human transportation of lithic material either to or from the excavated area. It can be assumed that a 'toolkit' with flakes and scrapers, and partly made on exotic Levallois *sensu stricto* flakes, entered the excavated area perhaps for subsequent use or maintenance. Part of the toolkit was discarded on the spot while another part, possibly supplied with newly-made tools, was removed. Transportation of lithic material (to or away from the locus) could suggest an interaction between the Site K patch and other findspots within the same cultural system. As Roebroeks *et al.* (1992) have already mentioned, in this case Isaac's distinction between locations where technology was maintained (*e.g.*, Site K) and locations where technology was used in non-maintenance activ-

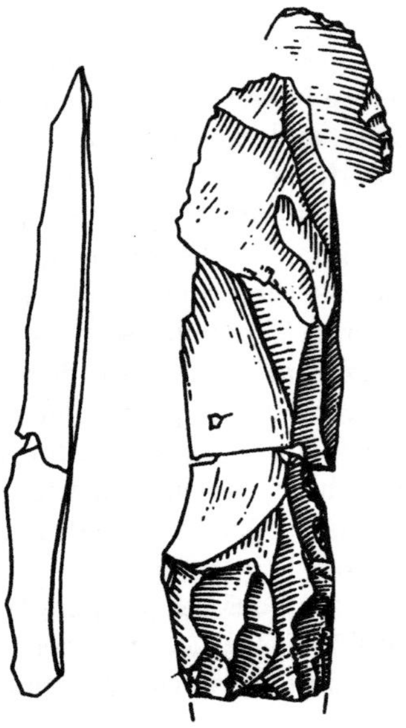

Figure 10.11: Maastricht-Belvédère Site K. Ancient broken blade-like flake of which the proximal fragment shows signs of a more intensive use (retouching) than the distal parts. The proximal part indicates a continued use after the flake was broken. Scale $\frac{2}{3}$.

ities would make sense (Isaac 1981).

10.12 Refitting evidence and site-formation processes

In order to make inferences about the spatial patterns of flint assemblages in terms of human behaviour it is necessary to investigate the possible role of natural site-formation processes first (*cf.* Schiffer 1972, 1987). In this section we will focus mainly on the intra-Saalian Site K and the Weichselian Site J results. Both areas were excavated using a similar excavation strategy. Site K probably represents an area with minimal horizontal post-depositional disturbance while at Site J some reworking of the original scatter was apparent

10.12.1 Horizontal disturbances of artefact distributions

Most of the intra-Saalian interglacial Belvédère assemblages were embedded in a loamy, fine sandy matrix which may reflect a quiet fluviatile deposition of sediments (Roebroeks 1988, 79). In such settings it is very likely that disturbance by embedding processes was minimal and that artefacts are still in primary archaeological context. This actually seems to be the case at Site K. Size distribution shows that most artefacts consist of small flakes (chips and spalls). Refitting shows that many conjoined artefacts, both small and large, were found close to each other.[1] Within an area of about three metres, a group of very small burin-spalls refitted to a burin (see de Loecker 1992, 456, Figure 5), and the conjoined burnt artefacts (potlids onto their parent pieces) provide good examples of spatially clustered material. The same refitting study also shows that the products of different technological stages (parts of the same *chaîne opératoire*) appear to be spatially clustered (de Loecker forthcoming). These arguments indicate that Site K has suffered minimal horizontal post-depositional disturbance, and that the spatial configuration probably can be used to make behavioural inferences.

The Site J excavation revealed a rather high density distribution of artefacts with a main western concentration of material within an area of *c.* 5 × 10m. Most artefacts, 52.7%, are less than 20mm (Roebroeks *et al.* 1997), nearly identical to Site K and again small artefacts were found in spatial association with larger ones. Based on this and the many refits which had already been established during excavation, our first impression was that here too we had a primary context distribution (*cf.* de Loecker 1988). However systematic conjoining resulted in numerous refits (39%) from relatively large conjoining distances; between flakes in a reduction sequence (*Aufeinanderpassungen*), between broken artefacts (*Aneinanderpassungen*) and of resharpening flakes to their parent tools (*Anpassungen*). Moreover, the refitted products of different technological stages were not spatially clustered. (see Roebroeks *et al.* 1997, for some of the spatial refits). Compared to the refit patterns at Site K, and most other areas excavated in the lower, fluviatile OIS 7 deposits (*cf.* Roebroeks 1988), it can be concluded that some horizontal displacement of the original Site J scatter(s) took place — probably limited but certainly enough to exclude any meaningful spatial interpretation of the finds.

10.12.2 Vertical disturbances of artefact distributions

At both Sites K and J, the artefacts were vertically dispersed to some extent, up to 40cm (Site K) or 45cm (Site J), and the finds were concentrated around a minor erosional level (see de Loecker forthcoming; Roebroeks *et al.* 1997). In general, the larger artefacts seemed to lie near the lower margin of the vertical distribution, while artefacts less than 20mm display greater vertical distribution. Refits reflect a similar vertical displacement. This applies to refits between flakes in a reduction sequence (*Aufeinanderpassungen*) as well as for refits of broken pieces (*Aneinanderpassungen*). Furthermore, the degree of vertical displacement of conjoining elements at Site K corresponds more or less with that at other Belvédère sites in a similar loamy, fine sand matrix (Roebroeks 1988). It is, however, not possible to point to one agent as primarily responsible for this vertical dispersal of artefacts. As Roebroeks (1988, 58) has mentioned regarding Site C, post-depositional vertical movement of artefacts was probably caused by a cumulative effect of several agents: weight of the artefacts, frost action on the site matrix, alternate wetting and drying of sediments, earthworm activity, plant roots *etc.*, and perhaps some trampling activity (Hofman 1986).

Generally the archaeological evidence (especially refitting) at most of the Maastricht-Belvédère findspots suggests that post-depositional displacement of lithic material must have been minimal, although some differences in preservation occur (*cf.* Roebroeks 1988).

10.13 Spatial differentiation

The excavation of buried archaeological levels in the Belvédère pit has generated new insights into the technological strategies used by early humans (Roebroeks 1988; Roebroeks *et al.* 1992, 1993, 1997). However, relationships between artefacts and faunal remains generally remained difficult to specify (Roebroeks 1988, 139–48, for Maastricht-Belvédère) and spatial overlap had to be viewed as a contingent result of human repetitive behaviour. The reconstruction of 'isolated' events or single episodes of site use turned out to be illusory even within the most fine-grained and well-preserved find scatters. The scatters with low densities of flint artefacts and bone fragments (the intra-Saalian Sites G and N), for instance, could not be discriminated from the high density clusters (Sites C and K) in terms of time scales and resolution. Both scatter types could have been formed within a few days, but equally in the process of many repeated visits spread over hundreds or thousands of years.

Early human activities behind the low density scatters are still difficult to assign despite the exceptional preservation of the distribution patterns. The scatters could, for instance, relate to activities of relatively large groups, each individual using his/her own transported tool kit consisting of a core and a few flake tools. But they might as well relate to activities of individuals or small groups operating in (temporary) isolation from the group. If this last option is true, the low density scatters excavated at Belvédère must be considered palimpsests formed over longer time periods. Independent of this, parts of the scatters may represent intervening episodes of tool maintenance or the implementation of technologies in food procurement leaving behind few visible traces.

Refitting contributed more substantially to the reconstruction of spatial behaviour on the local scale at Site K. This site shows some internal differentiation due to a structured use of space, whereas at Sites C and F no such spatial structure is evident at all. It appears from the distribution of the various artefact classes that an internal division into separate activity areas is present at Site K (see de Loecker 1992, 458, Fig. 6). A cluster with retouched tools was found in the central part of the excavated surface where overall artefact densities are relatively low. This centre of tool discard is surrounded by a zone with higher artefact densities, mainly consisting of cores, debitage and burned flints. Most of this waste material clusters within the southeastern part of the area.

A further spatial differentiation is suggested by the refitting data. A good example of this is provided by the reduction of a large block of flint (Fig. 10.6a). This large block was divided into at least eight pre-cores in the southern area. Five units were completely reduced on the same spot (Fig. 10.12, 1), as was a sixth, except for a scraper which was taken to the central area and abandoned (Fig. 10.12, 2). After some large decortication flakes had been removed from the seventh core it was taken 5m to the northeast where further reduction took place (Fig. 10.12, 3). Likewise, the eighth core was moved 8m to the north after initial working in the southeast area, large flakes were struck from it before it was carried a short distance away where smaller flakes were removed and the core abandoned. If we look at the spatial distribution of the complete nodule, the 'movement' of all eight parts (clusters of conjoined artefacts) seem to repeat spatially; there appears to be a close relationship between production and consumption zones. Because the actions and movements were repeated for different reduction sequences from different nodules (*cf.* de Loecker forthcoming) it can be suggested that an internal structuring in the use of space was preserved at Site K. Here, we may be dealing with activities which are emphatically spatially structured.

10.14 Conclusion

Although refitting analysis has been known as a valuable tool for site analysis for more than a century it is only seldom explored on a systematic basis for the interpretation of stone age sites and technologies. In most research projects such interpretations are based solely on lithic analysis, a fairly detailed typological and technological description of artefacts, or, when used, refitting is only applied to a small sample of the

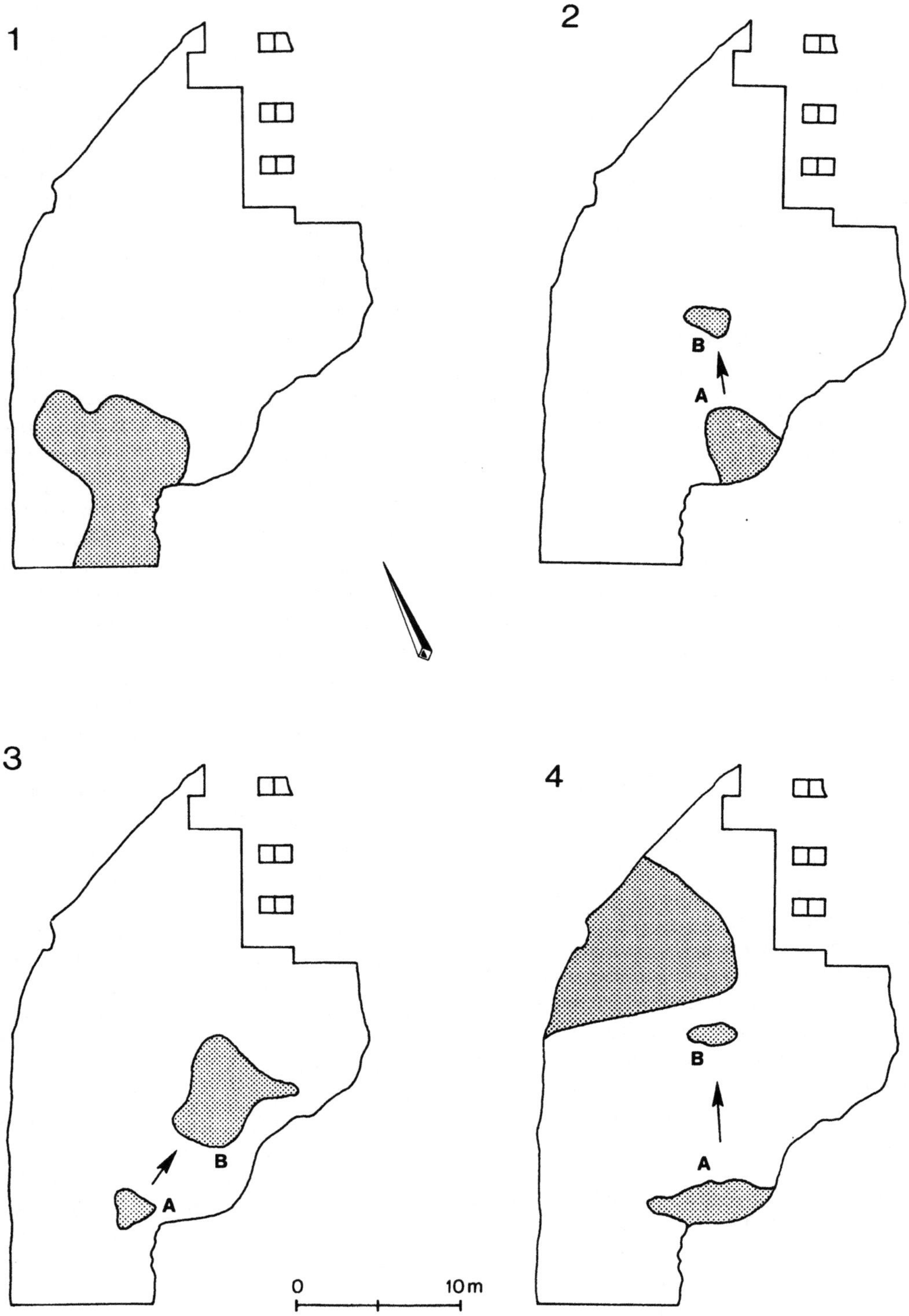

Figure 10.12: Maastricht-Belvédère Site K. Distribution maps of several parts of the refitted group shown on Fig. 10.6, A. **1:** Map with the spatial distribution of five cores that were reduced and discarded at the place where the nodule was split; **2:** Distribution map of a refitted group containing 13 flakes; **A:** the area where this part of the large block was reduced completely; **B:** the place where a fabricated simple-side scraper was discarded; **3:** Spatial distribution of a composition which consists of 23 artefacts; **A:** the area where several large, partly decortication flakes were removed; **B:** the place where further reduction took place after on-site transport of the core; **4:** Spatial distribution of a part of the nodule consisting of 59 artefacts; **A:** the locus where a number of large decortication flakes was removed; **B:** the area where after further transport some large flakes were produced; **C:** the area where the core was completely reduced by the production of smaller flakes.

assemblage. As conjoining programmes are time consuming, and as recent archaeological projects are increasingly being designed to minimise time and costs, the implementation of conjoining analysis may even be considered less favourable in future stone age research. Where refitting covers integral assemblages, however, its value for reconstructing both site taphonomy and human behaviour is well attested. It may even be argued that refitting is a must for reconstructing prehistoric lithic technologies. If lithic analysis alone is used, the processes of production, use and reuse remain largely hidden, many technological details and idiosyncrasies will not be observed, and the spatial dynamics of technologies, both on the site and at inter-site level, will be overlooked. The great potential of refitting is well illustrated at Maastricht-Belvédère where it has shed new light on the following subjects:

- the reduction processes of Middle Palaeolithic core technologies, including choices made by early humans when confronted with irregularities in raw materials and flaking;
- the often complex life histories and typological transformations of single stone tools in the process of reuse and recycling;
- the use of space by early humans on the local level resulting in both high and low density artefact scatters;
- the spatial organisation of technology when viewed from an inter-site/regional level;
- the taphonomic histories of Middle Palaeolithic artefact distributions, including the post-depositional horizontal and vertical displacement of lithic materials.

The results imply that, although refitting is time consuming, it should be applied where possible to improve the quality of interpretation.

Acknowledgments

The authors wish to thank Nick Ashton (British Museum) for providing Figure 10.1. We thank Henk de Lorm for Figures 10.2 and 10.12, Jan Pauptit for part of the photographic work (Figs. 10.5a, 10.6, 10.7 and 10.8), Gilles Leroy and Henk de Lorm for drawings of the Maastricht-Belvédère flint artefacts (Figs. 10.5a, 10.9, 10.10 and 10.11).

Notes

1. It should be noted that most artefacts in Site K and Site J were collected per square metre only (Roebroeks 1988, in contrast to Site C,). It is therefore impossible to assign exact distances to the refits, except in the three-dimensionally documented parts of the sites. For both sites most artefacts have been ascribed random coordinates within their square to permit visualisation.

References

ARCHER, F. 1881. 'Replacing flint flakes.' *Nature* 25: 8–9.

ARTS, N. & E. CZIESLA 1990. 'Bibliography (1880–1988) on the subject of refitting stone artefacts.' *In* Cziesla *et al.* (1990), pp. 651–83.

ASHTON, N. M., J. COOK, S. G. LEWIS & J. ROSE (eds.) 1992. *High Lodge: Excavations by G. de G. Sieveking, 1962–68 and J. Cook, 1988.* British Museum Press, London.

AUSTIN, L. 1994. 'The life and death of a Boxgrove biface.' *In* N. Ashton & A. David (eds.), *Stories in Stone*, Lithic Studies Society Occasional Paper No. 4, pp. 119–26. Lithics Studies Society, London.

BERGMAN, C. A. & M. B. ROBERTS 1988. 'Flaking technology at the Acheulian site of Boxgrove, West Sussex, England.' *Revue Archéologique de Picardie* 1–2 (numéro spécial): 105–13.

BERGMAN, C. A., M. B. ROBERTS, S. COLLCUTT & P. BARROW 1990. 'Refitting and spatial analysis of artefacts from Quarry 2 at the Middle Pleistocene Acheulean site of Boxgrove, West Sussex, England.' *In* Cziesla *et al.* (1990), pp. 265–81.

BERRIDGE, P. & A. ROBERTS 1990. 'Windmill Hill cave, Brixham: setting the record straight.' *Lithics* 11: 25–30.

BEYRIES, S. & E. BOËDA 1983. 'Etude technologique et traces d'utilisation des 'éclats débordants' de Corbehem (Pas-de-Calais).' *Bulletin de la Societé Préhistorique Française* 80: 275–70.

BOËDA, E. 1986. 'Le débitage de Biache-Saint-Vaast (Pas-de-Calais): première étude téchnologique.' *Bulletin Association Française pour l'étude du Quatérnaire* suppl. 26: 209–18.

BOËDA, E. 1993. 'Le débitage discoïde et le débitage Levallois récurrente centripète.' *Bulletin de la Société Préhistorique Française* 90(6): 392–404.

BOËDA, E. 1994. *Le Concept Levallois: Variabilité des Méthodes.* Monographie de CRA. CNRS, Paris.

BOEKSCHOTEN, G. R. & D. STAPERT 1993. 'Rings and sectors: a computer package for spatial analysis; with examples from Oldeholtwolde and Gönnersdorf.' *Helinium* 33(1): 20–35.

BOEKSCHOTEN, G. R. & D. STAPERT 1996. 'A new tool for spatial analysis: "rings and sectors plus density analysis and trace lines".' *In* H. Kamermans & K. Fennema (eds.), *Computer Applications and Quantitative Methods in Archaeology CAA95*, pp. 241–50. Institute of Prehistory, University of Leiden, Leiden. Analecta Praehistorica Leidensia 28.

BORDES, F. 1961. *Typologie du Paléolithique Ancien et Moyen.* CNRS, Paris.

CAHEN, D., L. H. KEELEY & F. VAN NOTEN 1979. 'Sone tools, toolkits, and human behavior in prehistory.' *Current Anthropology* 20: 661–83.

CELS, A. & E. DEPAUW 1885–1886. 'Communication de Mm. Cels et Depauw. Considérations sur la taille du silex, telle qu'elle était pratiquée à Spiennes à l'âge de la pierre polie.' *Bulletin de la Société d'Anthropologie Bruxelles* 4: 246–52.

CONARD, N. J. & D. S. ADLER 1997. 'Lithic reduction and hominid behaviour in the Middle Palaeolithic of the Rhineland.' *Journal of Anthropological Research* 53: 147–75.

CORNFORD, J. M. 1986. 'Specialized resharpening techniques and evidence of handedness.' *In* P. Callow & J. M. Cornford (eds.), *La Cotte de Ste. Brélade, Jersey. Excavations by C. M. B. McBurney 1961–1978*, pp. 337–51. Geo Books, Norwich.

CZIESLA, E. 1986. 'Uber das zusammenpassen von geschlagener Steinartefakte.' *Archäologisches Korrespondenzblatt* 16: 251–65.

CZIESLA, E. 1990. 'On refitting stone artefacts.' *In* Cziesla *et al.* (1990), pp. 9–44.

CZIESLA, E., S. EICKHOFF, N. ARTS & D. WINTER (eds.) 1990. *The Big Puzzle: International Symposium on Refitting Stone Artefacts.* Holos, Bonn.

DE BIE, M. 1998. *Intra-Site Analysis of the Federmesser Camp at Rekem (Belgium). A contribution to the study of Late Palaeolithic settlement sites in NW Europe.* Ph.D. thesis, University of Leuven.

DE LOECKER, D. 1988. *Ruimtelijke Analyse Site J.* Unpublished typescript, Institute of Prehistory, University of Leiden.

DE LOECKER, D. 1992. 'Site K: A Middle Palaeolithic site at Maastricht-Belvédère (Limburg, The Netherlands).' *Archäologisches Korrespondenzblatt* 22: 449–60.

DE LOECKER, D. 1994. 'On the refitting analysis of Site K: a Middle Palaeolithic findspot at Maastricht-Belvédère (The Netherlands).' *Ethnographisch Archäologischen Zeitschrift* 35(1): 107–17.

DE LOECKER, D. forthcoming. *Beyond the site: the Saalian archaeological record at Maastricht-Belvédère (The Netherlands).* Ph.D. thesis, University of Leiden.

DE MORTILLET, G. & A. DE MORTILLET 1881. *Musée Préhistorique.* Reinwald, Paris.

DE MORTILLET, G. & A. DE MORTILLET 1903. *Préhistorique, Paris.* Reinwald, Paris, second edition.

DE MUNCK, E. 1893. 'Observations nouvelles sur le Quaternaire de la région de Mons-Saint-Symphorien-Spiennes. Présentation de pièces: Nuclëi de l'époque paléolithique sur lequels se rappliquent plusieurs éclats.' *Bulletin de la Société d'Anthropologie Bruxelles* 11: 198–210.

DIBBLE, H. L. 1987a. 'The interpretation of Middle Paleolithic scraper morphology.' *American Antiquity* 52: 109–17.

DIBBLE, H. L. 1987b. 'Reduction sequences in the manufacture of Mousterian implements of France.' *In* O. Soffer (ed.), *The Pleistocene Old World. Regional Perspectives*, pp. 33–43. Plenum Press, New York and London.

DIBBLE, H. L. & N. ROLLAND 1992. 'On assemblage variability in the Middle Paleolithic of Western Europe.' *In* H. L. Dibble & P. Mellars (eds.), *The Middle Paleolithic: Adaptations, Behaviour and Variability*, pp. 1–28. University Museum of Philadelphia Museum Monograph 78, Philadelphia.

DUISTERMAAT, H. 1993. 'The interglacial mollusc fauna from Maastricht-Belvédère Site G.' *Mededelingen Rijks Geologische Dienst* 47: 61–7.

EVANS, J. 1897. *The Ancient Stone Implements, Weapons and Ornaments of Great Britain.* Longmans, London, second edition.

GENESTE, J. M. 1985. *Analyse lithique d'industries Moustériennes du Périgord; une approche téchnologique de comportement des groupes humaines au Paléolithique Moyen.* Ph.D. thesis, Université de Bordeaux I.

HENRI-MARTIN, H. 1907. *Recherches sur l'évolution du Moustérien dans le gisement de La Quina (Charente).* Premier Fascicule. Schleicher Freres, Paris.

HENRI-MARTIN, H. 1909. *Recherches sur l'évolution du Moustérien dans le gisement de La Quina (Charente).* Deuxième Fascicule. Schleicher Freres, Paris.

HOFMAN, J. L. 1981. 'The refitting of chipped-stone artefacts as an analytical and interpretive tool.' *Current Anthropology* 22: 691–3.

HOFMAN, J. L. 1986. 'Vertical movements in alluvial and stratified deposits.' *Current Anthropology* 27: 163–71.

ISAAC, G. 1981. 'Stone age visiting cards; approaches to the study of early land use patterns.' *In* I. Hodder, G. Isaac & N. Hammond (eds.), *Pattern of the Past: Studies in Honour of David Clarke*, pp. 131–55. Cambridge University Press, Cambridge.

JELINEK, A. J. 1976. 'Form, function, and style in lithic analysis.' *In* C. E. Cleland (ed.), *Cultural Change and Continuity: Essays in Honor of James Bennett Griffin*, pp. 19–33. Academic Press, New York.

LEROI-GOURHAN, A. 1984. *Pincevent: Campement Magdalenien de Chasseurs de Rennes.* Guides archéologiques de la France.

LEROI-GOURHAN, A. & A. BRÉZILLON 1966. 'L'habitation magdalénienne n. 1 de Pincevent près Montrau.' *Gallia Préhistoire* 9: 263–385.

LINDENBECK, J. 1988. 'Siteplot. Ein programme zur archälogischen Mengenkartierung.' *Bulletin Société Préhistorique Luxembourgoise* 10: 55–63.

LINDENBECK, J. 1990. 'SITEFIT — 3d-refitting with the PC.' *In* Cziesla *et al.* (1990), pp. 73–80.

LUBBOCK, J. 1872. *Prehistoric Times, as illustrated by Ancient remains, and the Manners and Customs of Modern Savages.* Williams and Norgate, London, third edition.

MEIJER, T. 1985. 'The pre-Weichselian nonmarine molluscan fauna from Maastricht-Belvédère (Southern Limburg, the Netherlands).' *Mededelingen Rijks Geologische Dienst* 39(1): 75–103.

OTTE, M. 1995. 'The nature of Levallois.' *In* H. L. Dibble & O. Bar-Yosef (eds.), *The Definition and Interpretation of Levallois Technology*, Monographs in World Archaeology 23, pp. 117–24. Prehistory Press, Madison.

PELEGRIN, J., C. KARLIN & P. BODU 1988. 'Chaînes opératoires: un outil pour le préhistorien.' *In* J. Tixier (ed.), *Technologie Préhistorique*, Notes et Monographies Techniques No 25, pp. 55–62. CNRS, Paris.

PENGELLY, W. 1867. 'The antiquity of man in the South-West of England.' *Transactions of the Devonshire Association for the Advancement of Science, Literature, and Art* 1867: 1–33.

PENGELLY, W. 1874. 'The cavern discovered in 1858 in Windmill Hill, Brixham, South Devon.' *Transactions of the Devonshire Association* 6: 775–856.

PETRIE, W. M. F. 1892. *Medum.* David Nutt, London.

PRESTWICH, J. 1873. 'Report on the exploration of Brixham Cave, conducted by a Committee of the Geological Society, and under the superintendence of Wm. Pengelly, Esq., FRS aided by a local committee; with descriptions of the animal remains by George Busk, Esq., FRS, and of the flint implements by John Evans, Esq., FRS.' *Proceedings of the Royal Society* 163(2): 471–575.

RENSINK, E. 1992. 'Eyserheide: A late Magdalenian site on the fringe of the northern loessbelt (Limburg, The Netherlands).' *Archäologisches Korrespondenzblatt* 22: 315–27.

ROEBROEKS, W. 1980. 'De 'Middenpaleolithische' vindplaats Sint Geertruid (L). Hypothesen voor nader onderzoek.' *Archaeologische Berichten* 8: 7–37.

ROEBROEKS, W. 1988. *From Find Scatters to Early Hominid Benaviour: a study of Middle Palaeolithic riverside settlements at Maastricht-Belvédère (The Netherlands).* Analecta Praehistorica Leidensia 21. University of Leiden, Leiden.

ROEBROEKS, W., N. CONARD & T. VAN KOLFSCHOTEN 1992. 'Dense forests, cold steppes, and the Palaeolithic settlement of Europe.' *Current Anthropology* 33(5): 551–86.

ROEBROEKS, W., D. DE LOECKER, P. HENNEKENS & M. VAN IEPEREN 1993. 'On the archaeology of the Belvédère pit.' *Mededelingen Rijks Geologische Dienst* 47: 69–79.

ROEBROEKS, W. & P. HENNEKENS 1990. 'Transport of lithics in the Middle Palaeolithic: Conjoining evidence from Maastricht-Belvédère (NL).' *In* Cziesla *et al.* (1990), pp. 283–95.

ROEBROEKS, W., J. KOLEN & E. RENSINK 1988. 'Planning depth, anticipation and the organization of Middle Palaeolithic technology: the 'archaic natives' meet Eve's descendants.' *Helinium* 28(1): 17–34.

ROEBROEKS, W., M. VAN POECKE, J. KOLEN & A. VAN GIJN 1997. "Site j': an Early Weichselian (Middle Palaeolithic) find scatter at Maastricht-Belvédère, the netherlands.' *PALEO: Revue d'Archéologie Préhistorique* 9: 143–72.

SCHIFFER, M. B. 1972. 'Archaeological context and systemic context.' *American Antiquity* 37(2): 156–65.

SCHIFFER, M. B. 1987. *Formation Processes of the Archaeological Record.* Albuquerque, University of New Mexico Press.

SCHLANGER, N. 1994. *Flintknapping at the Belvédère: Archaeological, Technological and Psychological Investigations at the early Palaeolithic Site of Maastricht-Belvédère (Limburg, The Netherlands).* Ph.D. thesis, University of Cambridge.

SCHLANGER, N. 1996. 'Understanding levallois: Lithic technology and cognitive archaeology.' *Cambridge Archaeological Journal* 6(2): 231–54.

SCHLANGER, N. & D. DE LOECKER 1992. *Techno- en typologische analyse van Midden-Palaeolithische vuursteencomplexen.* Unpublished typescript, University of Leiden.

SMITH, W. G. 1881. 'Replacing flakes on Palaeolithic implements.' *Nature* 24: 582.

SMITH, W. G. 1883. 'Flint flakes replaced.' *Nature* 28: 490.

SMITH, W. G. 1884. *On a Palaeolithic floor at North-East London.* Harisson and Son, London.

SMITH, W. G. 1892. 'Primaeval man: a Palaeolithic Floor near Dunstable.' *Natural Science* 1(9): 664–70.

SMITH, W. G. 1894. *Man the Primeval Savage, his Haunts and Relics from the Hill-tops of Bedfordshire to Blackwall.* Edward Stranford, London.

SMITH, W. G. 1916. 'Notes on the Palaeolithic floor near Caddington.' *Archaeologia* 67: 49–74.

SPURRELL, F. C. J. 1880a. 'On implements and chips from the floor of a Palaeolithic workshop.' *Archaeological Journal* 37: 294–9.

SPURRELL, F. C. J. 1880b. 'On the discovery of the place where Palaeolithic implements were made at Crayford.' *Quarterly Journal of the Geological Society* 36: 544–9.

STAPERT, D. 1976. 'Some natural surface modifications on flint in the Netherlands.' *Palaeohistoria* 18: 7–41.

STAPERT, D. 1992. *Rings and sectors: intrasite spatial analysis of Stone Age sites.* Ph.D. thesis, University of Groningen.

THIEME, H. 1983. *Der paleolithische fundplatz Rheindahlen.* Ph.D. thesis, Cologne.

THIEME, H. 1990. 'Wohnplatzstrukturen und fundplatzanalysen durch das zusammensetzen von steinartefacten: ergebnisse vom Mittelpäleolithischen fundplatz Rheindahlen B1 (Westwandkomplex).' *In* Cziesla *et al.* (1990), pp. 543–68.

VAN GIJN, A. L. 1988. 'A functional analysis of the Belvédère flints.' *In* Roebroeks (1988), pp. 151–7.

VAN GIJN, A. L. 1989. *The Wear and Tear of Flint. Principles of functional analysis applied to Dutch Neolithic assemblages.* Analecta Praehistorica Leidensia 22. University of Leiden, Leiden.

VAN KOLFSCHOTEN, T. 1993. 'The mammal fauna from the interglacial deposits at Maastricht-Belvédère.' *Mededelingen Rijks Geologische Dienst* 47: 51–60.

VAN KOLFSCHOTEN, T. & W. ROEBROEKS (eds.) 1985. *Maastricht-Belvédère: stratigraphy, palaeoenvironment and archaeology of the Middle and Late Pleistocene deposits.* Analecta Praehistorica Leidensia 18. Leiden University Press, Leiden.

VANDENBERGHE, J., W. ROEBROEKS & T. VAN KOLFSCHOTEN (eds.) 1993. *Maastricht-Belvédère: Stratigraphy, palaeoenvironment and archaeology of the Middle and Late Pleistocene deposits: Part II.* Mededelingen Rijks Geologische Dienst 47. Haarlem.

VYNCKIER, P., P. M. VERMEERSCH & L. BEECKMANS 1988. 'Le Paléolithique Moyen de Vollezele-Congoberg (Brabant, Belgique).' *Revue archéologique de Picardie* 1(2): 133–9.

All enquires regarding this paper should be directed to Wil Roebroeks

Dimitri De Loecker
Jan Kolen
Wil Roebroeks
Dept. of Archaeology
University of Leiden
Faculty of Archaeology
P.O. Box 9515
2300 RA Leiden
The Netherlands
w.roebroeks@arch.leidenuniv.nl

Paul Hennekens
Holstein Bastion 33
6217 LG Maastricht
The Netherlands

11
Size as a Factor in Middle Palaeolithic Assemblage Variation in the Old World: a North American Perspective

Michael J. Shott
Department. of Sociology, Anthropology & Criminology, University of Northern Iowa

Abstract

Conventional views understand assemblage variation in essentialist terms that correspond to ethnographic patterns at various levels. In North American Palaeoindian data, however, some variation is categorical but much is continuous and some is determined by assemblage size, a factor more often controlled than studied. A similar view emerges in the study of 146 Middle Palaeolithic assemblages from Europe and the Near East. There, degree of tool reduction also explains some variation in proportional frequency of tool 'types'. Despite obvious differences of time and place, similar formation processes operated in the cases. Norm and activity contribute to assemblage variation, but only among other factors like formation processes.

11.1 Introduction

Assemblage analysis long has been a cornerstone of archaeology. In the 1960s, it was the setting for a debate about the meaning of Middle Palaeolithic assemblage variation in the Perigord. Everyone knows François Bordes' position in this debate and, at least in the Anglophone world, Lewis Binford's opposing view. Earlier, Bordes had defined an elaborate set of Middle Palaeolithic tool types and used the typology to define assemblage facies that he invested with meaning in terms of identity-conscious social groups. To him, assemblage variation registered the boundaries between cultures. Bordes then charted the complex comings and goings of Middle Palaeolithic cultural variants — facies — in the stratigraphic sequences revealed in Perigordian rockshelters. In contrast, Binford interpreted Middle Palaeolithic assemblage variation in toolkit or 'functional' terms. To him, assemblage variation registered behavioural variation within culture.

Binford's view was justly criticised on methodological grounds, and many archaeologists also opposed it on interpretative ones. At least in Europe and the Levant, time seemed kinder to Bordes' view. Yet it too had critics. Bordes' scenario had a musical-chairs quality that inspired Clarke's (1973, 10) memorable quip: 'to interpret the French Mousterian sequence, of more than 30,000 years duration, in terms of the acrobatic manoeuvrings of five typological tribes is tantamount to an attempt to explain the Vietnam War in terms of electron displacements'.

The Bordes-Binford debate never was resolved owing partly to fundamental ontological differences in views, a condition that persists today in other contexts (Clark 1996). Yet it had the salutary effect of forcing Palaeolithic specialists to attend to how assemblages formed and how best to interpret them. Eventually, some of them reinterpreted the Bordes-Binford debate by altering its terms and rejecting the premises shared by the protagonists (Dibble & Rolland 1992; Freeman 1994). No longer were assemblages viewed as categorical things, whether ethnic calling cards or functional toolkits. Instead, they were merely the result of variable and often complex relationships between the constituent tool types they contained, relationships that cross-cut the boundaries erected between assemblage 'types'. Assemblages were reimagined as entities constantly in a state of becoming, never fixed in form or content as essential things. Emphasis shifted from assemblage form and content to process of assemblage formation.

Ironically, this revised view is not as renowned as the opposing ones of Bordes and Binford that it replaced, at least outside the community of Palaeolithic scholars. North American Palaeoindian specialists, for instance, recognise at least broad technological and cultural similarities between Upper Palaeolithic and Palaeoindian industries (Irwin & Wormington 1970) and still hew primarily to either Bordes' or Binford's view of the meaning of assemblage variation.

This paper summarises a recent study of process in North American Palaeoindian assemblages (Shott 1997a). It then applies the same perspective to a large sample of Old World Middle Palaeolithic assemblage data to identify similarities and differences between the cases. In this way, the classic Middle

Palaeolithic controversy receives fresh treatment from a New World perspective.

11.2 Assemblage properties and measures

Assemblages must be described before they can be analysed. There are many ways to describe them, but size and composition are fundamental properties of all assemblages. Size simply is the number of specimens found in the assemblage. Composition is their distribution across types, the latter defined by any combination of technology, form, function, style, material and other criteria. Size and composition seem straightforward properties, but may not be. Most assemblages contain tool fragments whose presence begs the question of their treatment. Ignoring them or considering them equal to complete tools both seem unsatisfactory. Pottery analysts know that quantification engages fundamental if refractory problems but have developed methods to solve them (Orton & Tyers 1990), a task still confronting lithic analysis (Shott 1998a). Composition is measured with respect to tool types, so raises typological questions. Size and composition simultaneously engage as well the problem of distinguishing tools from other categories of chipped stone. Many specimens once thought to be tools, for instance, may instead be debris (*e.g.*, Barton *et al.* 1996; Crovetto *et al.* 1994). Before we measure size and composition, we must know what specimens are tools and calculate size from them alone. The problematics engaged by measurement of size and composition are acknowledged but otherwise blissfully ignored here.

Size and composition ordinarily are assumed to be independent, yet they can be related in ways helpful to our understanding of formation processes (Shott 1997a). Most measure assemblage composition by richness (number of types present), evenness (dispersion of type proportions, measured there by the standard deviation) (Bobrowsky & Ball 1989, 7) and perhaps heterogeneity. There is great dispute regarding the validity of heterogeneity measures. If any consensus exists, it is to deny their validity because of their frequent size-dependence, the tendency of composition measures to vary with assemblage size. Richness and evenness are championed in comparison. Ironically, those measures are equally size-dependent, and heterogeneity is not always size-dependent so must sometimes measure something besides size (Shott 1997a).

Many archaeologists and statisticians regard size dependence as an unmitigated evil to be rigorously suppressed before interpretation begins. For instance, Kaufman's (1998) valuable jackknife method to calculate assemblage measures and test for significant differences between assemblages removes size dependence not in jackknifing itself but in the critical further step of calculating pseudo values, necessary for parametric tests of differences in means. Like any method, Kaufman's can answer the wrong question, a criticism of its possible use not its intrinsic merit. The problem in this case owes to its assumption that size and composition are independent. Yet assemblage measures can be size-dependent because size and composition are related, not independent. They are related from the complex interactions between the spans over which assemblages accumulate, tool use lives and tool use rates. If size dependence possesses intrinsic analytical value, methods that removes it risk eliding a relationship of interest.

11.3 North American Palaeoindian assemblages

Palaeoindian culture is the earliest major, if not absolute earliest, human occupation of North America and spans a few centuries either side of approximately 11,000 bp. Palaeoindians are famous for their proficiency in fluting bifaces but they also made unfluted bifaces, unifaces and flake tools. There are relatively few Palaeoindian tool types, and most have integrity in the sense that they are not merely resharpened and reduced versions of other types.

In assemblage analysis, typical North American practice resembles Old World practice: it apportions variation in composition among assemblage types. In the classic study, Bordean methods supported Binfordian conclusions, by defining types or facies identified with purposes like hunting and residence (Judge 1973). Later studies essentially were variations of this practice. Such assemblage types were identified with functional poses or task-group activities documented ethnographically in North American and elsewhere. Thus, archaeologists directly reproduced ethnographic pattern in the archaeological record. In the process, they assumed that each assemblage accumulated either in a single brief episode of days, weeks or at most months, or in many episodes that differed insignificantly in the number of tools used and their rates of use. Archaeologists also assumed that assemblage variation is categorical and that it represents types that possess unambiguous meaning. Some variation is permitted within types, but it is insignificant. In contrast, there must be great variation - qualitative differences — between assemblage types. As in biology (Mayr 1988, 15), this is essentialist thinking.

Paleontologists do not ordinarily suppose that the animals whose remains form their assemblages were arrayed around the margins of a waterhole, as in some museum diorama, at the moment of their common demise. They know that most fossil assemblages accumulated over intervals that greatly exceed individual life spans, let alone fleeting moments in tableaux. They also know that assemblage composition does not mirror species proportions in living communities. Instead, it varies systematically with assemblage size

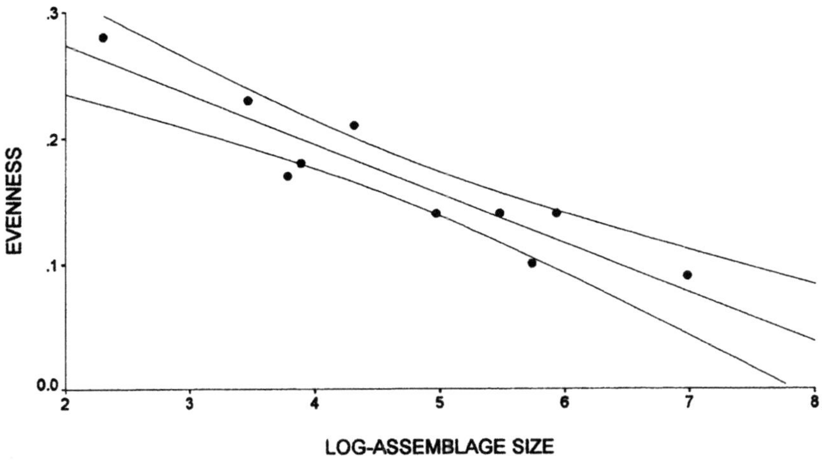

Figure 11.1: Continuous variation in evenness in North American Palaeoindian assemblages. Evenness measured as the standard deviation of assemblage proportions; lower values indicate greater evenness. Source: Shott 1997b, Fig. 2b.

through the interaction of differences in species mean life spans, accumulation span, and changing biogeographic patterns. Paleontologists call this quality the 'time-averaging' of their assemblages.

Great Lakes Palaeoindian assemblages exhibit continuous variation between assemblage 'types' (Shott 1997a) (*e.g.,* Fig. 11.1). Even types defined in Correspondence Analysis are linked to continuous dimensions of variation. However the validity of assemblage types was tested, it was found wanting. Yet types routinely are discovered in assemblage analysis. No doubt this often is because they are there to be found. However, if we assume *a priori* that assemblage types exist, and if we use analytical methods commensurate with that assumption, we always will find types. If they exist, we will find them. If they do not exist, we will create them.

It seems better to study assemblage composition without assuming that types exist. If they do, their presence will emerge in analysis. If they do not, we will not create them (Freeman 1994). In the Palaeoindian study, there seemed to be fewer Great Lakes assemblage types than often supposed, but there were fascinating and sometimes complex relationships among type counts, and between them and size. Evidence, in short, revealed the process of assemblage formation — how assemblages constantly are in a state of becoming — rather than the form of assemblage 'types'. The process owes chiefly to the interdependence of size and composition, governed by the relationship between three quantities: occupation span, tool use rates and tool use lives.

11.4 Time-averaging

In effect, this is time-averaging (Stern 1994), and a hypothetical example illustrates its archaeological expression. Imagine one Palaeoindian who uses five types of tools, one at a time of each. The types are used the same amount or at the same rate, time passes in unit increments, and the types' use lives differ, the first lasting one time unit, the second two time units, and so on. The Palaeoindian occupies a series of places for varying lengths of time, doing exactly the same thing and using the same types of tools in the same frequencies at each. This is another way to say that the tool-using activity carried out at the settlements is identical. There is variation here, both in use life between tool types and in settlement occupation span, but there is none in tool-using activity. Yet the assemblages differ in size and composition purely from a relationship between tool use lives and settlement occupation spans.

Thus, a settlement occupied for one time unit will experience the discard of one specimen of the shortest-lived type. Here, assemblage size and richness are one. A second settlement occupied for two time units experiences the discard of two specimens of the shortest-lived type and one of the next shortest-lived type. Obviously, richness is two and assemblage size three. Tool-using activity is constant, yet assemblage size is three times larger over an occupation span twice as long. Carrying through to a settlement occupied for five time units, richness attains the maximum of five and assemblage size is 10. Plotting richness and evenness against assemblage size will yield a simple version of the size-dependent pattern so common in archaeological data, yet there is no variation whatsoever in the activity involved (Fig. 11.2). All that differs is occupation span and therefore assemblage size. (Similar results would be obtained using evenness or by holding occupation span constant and varying either use life or use rates.) The extreme and simplistic nature of this hypothetical example is too obvious to belabour, but archaeologists must attend to formation's manifest effects in real assemblages. At some point determined by the relationship between occupation span and use lives, assemblage measures stabilise and no longer vary as a function of span. Until then, however, they vary systematically with assemblage size.

In this simple model, assemblage variation is real and informative but does not correspond neatly to types. Thus, the earlier study questioned the validity of types of Palaeoindian assemblages and a very satisfying interpretation of the archaeological record

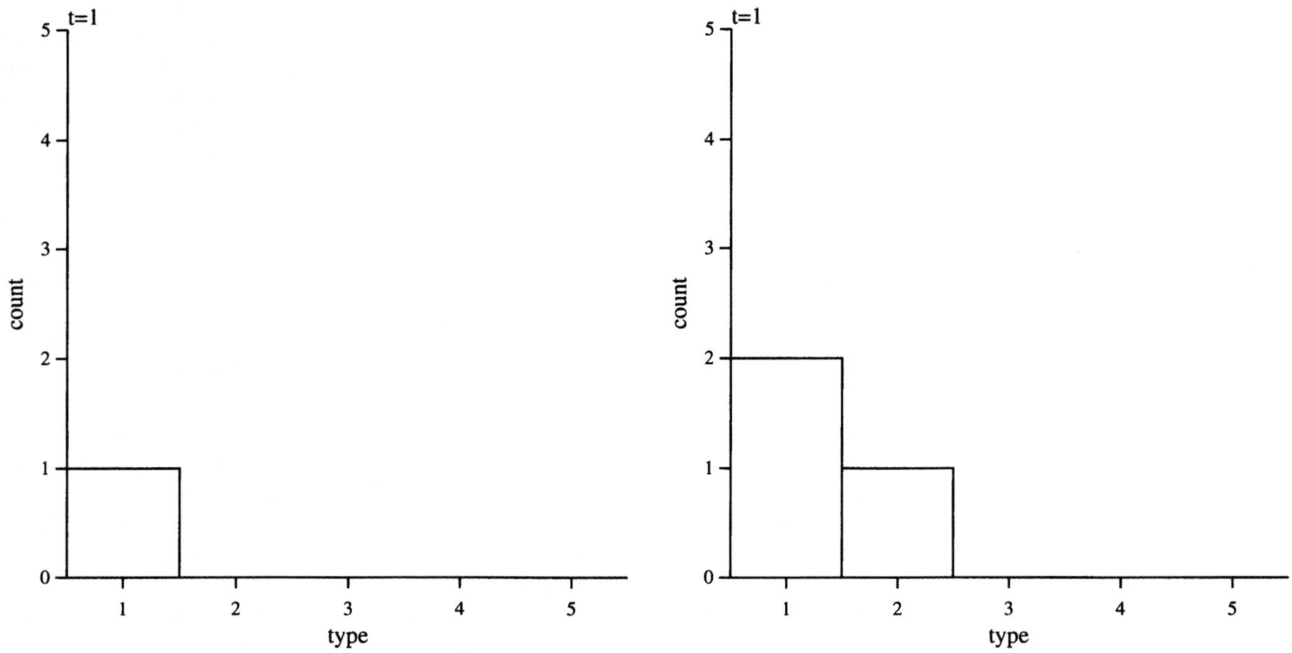

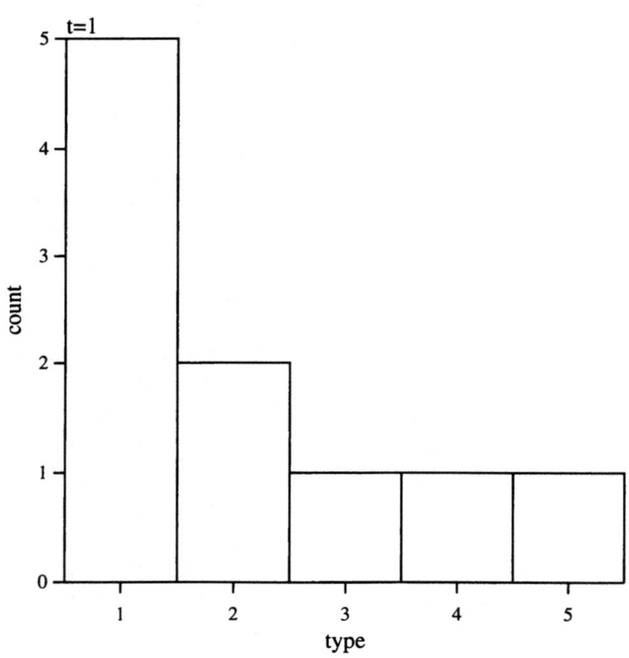

Figure 11.2: Hypothetical example showing the interaction of use life and occupation span to produce size-dependence in assemblage composition — see text for explanation. Source: Shott 1997b, Fig. 5.

in synchronic ethnographic terms. Yet its conclusion is no counsel of despair that the record fails to reveal unambiguous meaning. On the contrary, the study helped identify the various factors that in complex interaction largely determine continuous and categorical variation in assemblage size and composition. A simple story yields to a more complicated one that nevertheless holds great promise for interpreting the archaeological record commensurate with its nature and the time scale over which it accumulated. It is worth assaying a similar approach to Middle Palaeolithic assemblages.

11.5 Middle Palaeolithic assemblages

A great deal separates New World Palaeoindian cultures and assemblages from Old World Middle Palaeolithic ones. Their differences are too obvious to belabour, but a few deserve mention.

1. By comparison to the Old World, the New World record obviously covers much less time and was created only by modern humans. Thus, places have accumulated stone tools for comparatively short periods. Indeed, single Middle Palaeolithic taxa and strata may span more time than the entire New World record does, certainly than the Palaeoindian interval does.

2. Biface technology is much more common and more highly developed in the New World. By comparison, Middle Palaeolithic assemblages seem dominated by flake technology, Upper Palaeolithic ones by blade technology.

3. For whatever reason, Old World archaeologists favour caves and rockshelters more than do New World ones. Trivially, there are caves in the New World; ancient people used them and archaeologists excavate them. But New World archaeologists perhaps recognise better, as did ancient people, that the landscape is much bigger than just caves and have examined open areas more extensively than have their Old World counterparts.

4. Perhaps because of the New World's lesser antiquity and greater emphasis on bifaces, Palaeoindian archaeology enjoys more refined time-space systematics, subdividing a shorter record of human occupation into phases that often are much shorter than Middle Palaeolithic ones.

5. Finally, the New World is closer in time and space to most ethnographic cultures broadly analogous to the forager cultures thought to be represented in much of its archaeological record. Combined with archaeology's development in the New World from anthropology, not history or paleontology, New World archaeologists are much more inclined to equate the ethnographic and archaeological records, which is at once good and bad. Ethnographers of forager cultures describe various types of settlements (*e.g.*, residences, hunting stands) that archaeologists dutifully reproduce in their record.

On balance, New World archaeologists typically study many more but smaller assemblages in comparatively small areas, confined to comparatively short intervals. The New World record is much more concentrated in time, the Old World's in locations. The New World's is much more extensive across the landscape, the Old World's in time.

11.6 The sample

Data are tool counts from 146 assemblages in 19 Middle Palaeolithic deposits. Industry or facies identification were recorded as reported by original authors. Some sources did not identify facies (*e.g.*, Callow & Cornford 1986, for La Cotte de Saint Brelade).

Counts are in Bordes' 63 Middle Palaeolithic uniface types (Bordes 1961; see also Debenath & Dibble 1994). (Bifaces were recorded as a single category and are not analysed.) Bordes' types were defined by combinations of blank attributes, morphology, and especially retouch type, placement and extent. Obviously, typological criteria were not always explicit or replicable (Debenath & Dibble 1994; Kolpakov & Vishnyatsky 1989), a potential problem with all typologies. Here, it is possible only to hope that inconsistent identification has no significant effect. The number of specimens summed across those types is called an assemblage's 'real' count (Debenath & Dibble 1994, 174). But Types 1–3 and 45–50 literally do not count in conventional 'essential' counts and today some analysts also omit pseudo-Levallois 'points' and naturally-backed knives (Types 5 and 38, respectively) (McPherron 1994). Obviously, there are several ways to count Middle Palaeolithic tools. This study uses McPherron's type set, which omits Types 1–3 (unretouched Levallois flakes and 'points'), 39 (raclettes), 45–50 (varieties of retouched flakes) and 62–63 ('miscellaneous' and bifacial foliates, respectively) and leaves 51 uniface types (McPherron 1994, Table 6). 'Type' is used nominally, not to signify an essential revealed category.

A sample of this nature at once possesses great virtues and flaws. Its sheer size and the distribution of assemblages in time and space make it highly diverse and any patterns demonstrated robust. Yet the sample freely combines assemblages that original authors classified as Acheulean or Mousterian facies. Admittedly, the distinction no longer can be maintained on chronological grounds (McPherron 1994, 328; Dibble & Rolland 1992, 2). The sample's distribution is broad but highly uneven, unavoidable considering the history of Middle Palaeolithic research and the vagaries of publication. The Perigord is especially well represented; other French regions, the Levant and the Zagros Mountains are represented by sizeable assemblages. The very large area between France and the Near East is entirely unrepresented.

Within the sample's broad time-space limits, assemblages accumulated under what must have been at least somewhat variable cultural conditions and practices and certainly under environmental conditions of varying nature and frequent change. Raw-material supply and form also must have differed between places, cyclically by landform processes and ground cover and through time as consumption gradually exhausted sources. Moreover, if territorial range varied through time for social and material reasons, as it surely did, raw material effective abundance also varied. Raw-material supply is especially pertinent to the testing of reduction theses (McPherron 1994, 313). Disentangling the effects of norm, activity, assemblage formation and other factors of interest should be difficult, to put it mildly, in the face of so many and so effective complicating factors.

In practice, assemblages are recognised by the joint effects of stratigraphy and excavation sampling. Strata are distinguished by various kinds and degrees of depositional processes that are largely independent of assemblage formation. Their accumulation spans vary greatly and the tempo of their formation probably bears little relationship to the tempo of assemblage formation. Of course, some assemblages also may include erosional remnants left by stratigraphic unconformities. Stratigraphic boundaries are arbitrary with respect to the tempo of culture change registered in the archaeological record, at times capriciously dividing cultural continua and at times just as capriciously combining the products of very different cultural practices and processes. Sampling effects are too many and too obvious to belabour. Clarke (1972, 27) suggested how they might create spuriously the 'interdigitation' often observed in French Middle Palaeolithic sequences.

On balance, assemblages are distinguished by several complicating factors. What unites them as much as anything else is our ignorance of the exact effects in each case and their common reduction to Bordes' Middle Palaeolithic typology. The latter is perhaps not by itself a good thing (Kolpakov & Vishnyatsky 1989), but comparison would be impossible in its absence. The replicability and integrity of Middle Palaeolithic types and the meaning of typological variation between assemblages seems less clear today than they did to Bordes. But criticisms of Bordean typology are at once tributes to it, as is this study. Our ability to recognise the defects of Bordean typology is a measure of its value, because it alone made possible the systematic studies that revealed the defects.

Obviously, there are many ways to analyse 146 assemblages that contain thousands of tools distributed across 51 types. Even if possible, attempting exhaustive analysis would be folly in a short study. Instead, this highly selective treatment seeks patterns between tool types across assemblages, because such patterns reveal the processes that created the assemblages. Only when we understand formation processes can we proceed to culturally meaningful interpretation (Shott 1998b).

11.7 Sample adequacy

Assemblages are samples of larger deposits and processes. All assemblages are not created equal, and their adequacy always should be gauged before analysis. Years ago, Bordes showed due concern for sample adequacy (Debenath & Dibble 1994, 174), using criteria of size and context reasonable at the time. Shott (1996, 204–5) gauged the adequacy of Great Lakes Palaeoindian assemblages as samples for the legitimate versus size-related chance absence of tool types and for the precision of and legitimate differences in type proportions. Assemblage size affects adequacy, one reason why assemblages so often exhibit size-dependence in measures like richness. But size does not exclusively determine adequacy; evenness simultaneously affects the validity of other quantities like richness and also bears on adequacy (Lipo *et al.* 1997, 312).

This large data set cries out for exhaustive evaluation of sample adequacy. As a lame expedient, Lipo *et al.*'s (1997, 312–6) bootstrapping method gauges adequacy of empirical richness values, which simultaneously considers size and evenness. Bootstrapping is a computer-intensive method of sampling with replacement that is best explained in a hypothetical example. If an archaeological assemblage contained 100 tools distributed in any way among any (reasonably small) number of types, bootstrapping begins by random selection of ten tools from the assemblage. Richness is calculated from this small sample. Then the procedure is repeated many times (100 in this case) at the same sample size. This produces 100 richness estimates at a sample size of 10, and the distribution of these estimates can be summarised by its mean and standard deviation. Then 100 samples of 20 tools are drawn and the same estimates compiled and summarised. The process repeats at sample intervals of ten up to the empirical assemblage size. The result is a mean and standard deviation for estimated richness at regular intervals up to the assemblage's actual size.

Following Lipo *et al.*, Middle Palaeolithic assemblages were distinguished by whether they stabilised (*i.e.*, reached an asymptote for richness and the standard deviation of its estimate) before attaining the empirical richness value. Assemblages thus were assigned to one of three types, but these types are not the essential kinds of classic typological thinking. Rather, they are sets of assemblages *qua* samples dis-

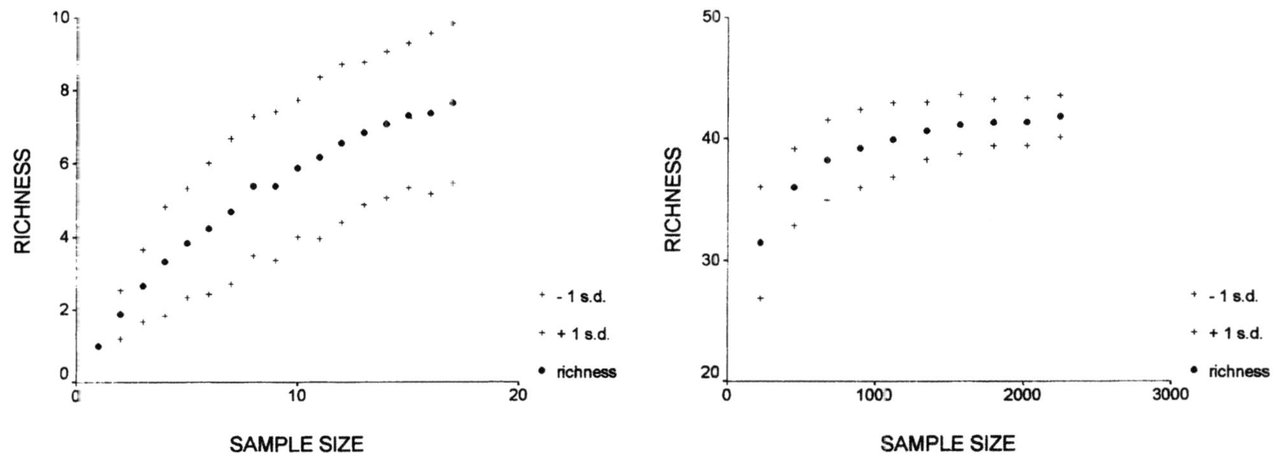

(a) Type 1 assemblage, Vaufrey IV

(b) Type 2 assemblage, La Cotte de Ste. Brelade A

Figure 11.3: Bootstrapped estimated richness and its standard deviation plotted against bootstrap sample size.

tinguished by different degrees — not kinds — of adequacy.

Figure 11.3 shows examples of Type 1 assemblages, in which estimated richness and its standard deviation do not stabilise, and Type 2 assemblages in which they do. (Many assemblages did not fall clearly into either type and were assigned to a third intermediate category omitted in some analyses.) Generally, Type 1 assemblages are not adequate samples of true assemblage richness; Type 2 assemblages are. Bootstrapping is a valid procedure but not entirely satisfactory because it requires subjective judgements of sample plots. A conservative policy, assigning samples to Type 3 when any doubt existed about their character, seemed most prudent, yet this too is subjective.

Comparing the abscissas of Figures 11.3a and 11.3b, demonstrates that size substantially but not entirely distinguishes assemblage types. Mean size of Type 1 assemblages is 126.7 specimens, of Type 2 assemblages 516.5 specimens. But, as above, evenness also affects adequacy. As a result, the types' size ranges overlap considerably (Fig. 11.4), and a simple distinction at a size value *c.* 500–750 would combine some Type 1 and 2 assemblages. For what it is worth, assemblages distribute among adequacy types independent of Bordean facies ($\chi^2 = 3.6, \mathrm{df} = 10, \mathrm{p} = 0.96$) and geographic region ($\chi^2 = 6.4, \mathrm{df} = 4, \mathrm{p} = 0.17$).

11.8 Size and assemblage properties

Not surprisingly, assemblage richness plots clearly as a function of size to an asymptote on size *c.* 750 (Fig. 11.5). Separately, Type 1 and 2 assemblages also regress upon size (Fig. 11.6, a–b) but the former at a considerably higher slope, no surprise considering their instability in richness. Log-transformed size linearises the joint plot (Fig. 11.7). Evenness (here, calculated as the coefficient of variation of type proportions) also plots against size, but less neatly (Fig. 11.8); again, Type 1 assemblages regress at considerably higher slope than Type 2 assemblages. Although the latter's regression is significant (p $=$ 0.054), $r^2 = 0.15$ only. Evenness and size also linearise upon log-transformation, but evenness covaries with size less than it does richness.

The Palaeoindian and Middle Palaeolithic studies differ greatly in number of assemblages and on the contextual grounds noted above. Certainly richness seems equally size-governed in the two cases, but evenness (measured in different ways) more size-governed in the Palaeoindian case. This may owe either to differences in sample sizes or sample diversity in time, space and context. Size-dependence is more robust in richness than in evenness.

Scrapers (Types 12–24) and notched tools (combined notches and denticulates, Types 42, 43 and 52) dominate many Middle Palaeolithic assemblages (Dibble & Rolland 1992; Freeman 1994, 45). Their proportions vary inversely in these data, but not symmetrically. Where scrapers are many, notched tools always are few, but where scrapers are few notched tools may be many or as few as about 20% of the assemblage total. The general inverse pattern is a trivial arithmetic effect of closed arrays even if the types' ratio has meaning; if scrapers and notched tools dominate assemblages, their combined proportions approach unity. When this happens, a rise in one's proportion must lower the other's.

11.9 The reduction thesis

To this point, New World and Old World assemblages exhibit similar if unremarkable patterns of size-

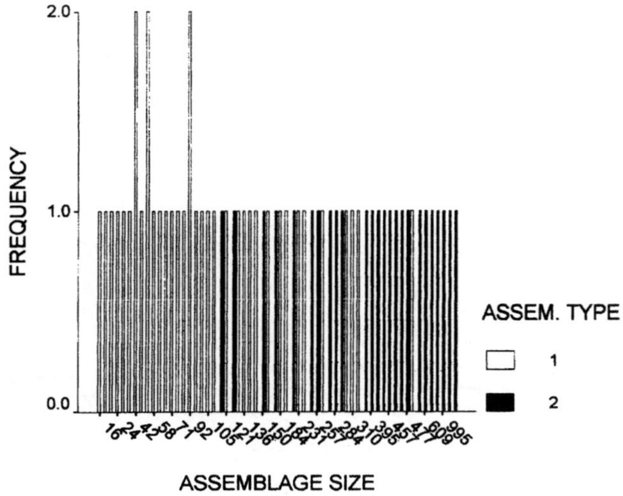

Figure 11.4: Distribution of assemblage size by assemblage type.

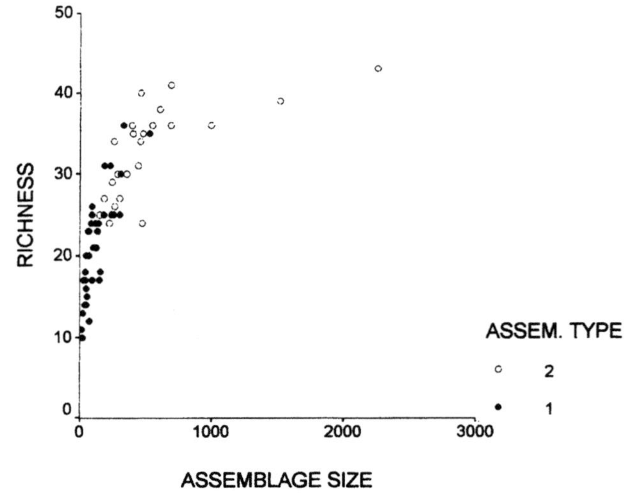

Figure 11.5: Size and richness in Middle Palaeolithic assemblages.

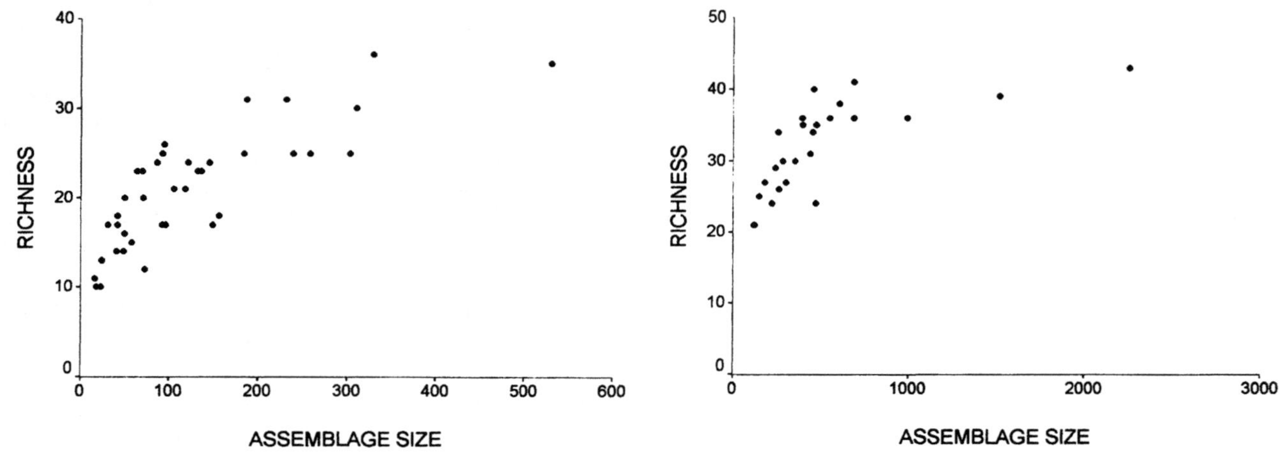

(a) Type 1

(b) Type 2

Figure 11.6: Size and richness by assemblage type in Middle Palaeolithic assemblages.

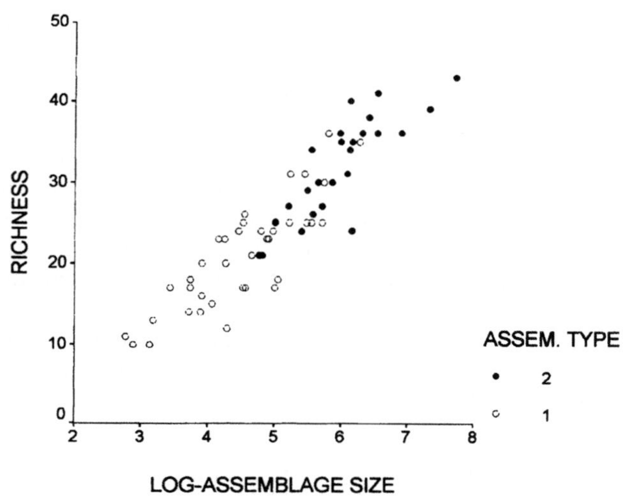

Figure 11.7: Log-size and richness in Middle Palaeolithic assemblages.

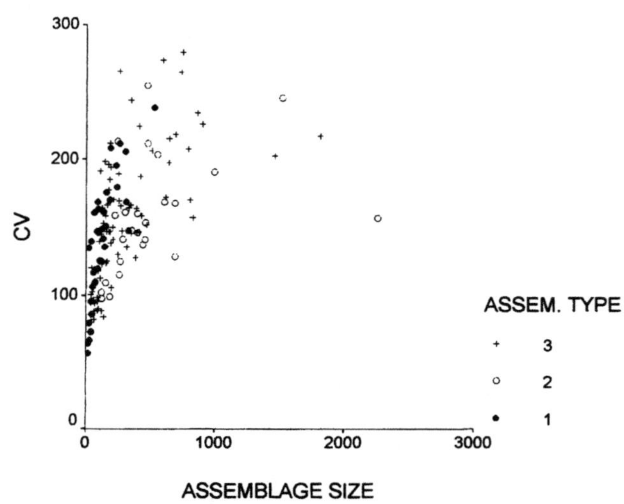

Figure 11.8: Size and evenness in Middle Palaeolithic assemblages.

dependence in composition. But Palaeoindian systematics recognise many fewer tool types, most of which are elaborately retouched bifaces and unifaces. All else equal, fewer types means fewer opportunities for type transitions, the tendency of tools to change from one recognised type to another via size and form change produced by reduction. However, the many Middle Palaeolithic types finely distinguish on, among other grounds, amount and placement of unifacial retouch, so offer much more latitude for typological transitions. Enter the reduction thesis.

By now Dibble's reduction thesis for Middle Palaeolithic scrapers is well known, if not well liked. Briefly, Dibble argued against the integrity of Bordes' various scraper types by linking them in two reduction sequences: single scrapers to double scrapers to convergent scrapers, and single scrapers to transverse scrapers (Dibble 1987, Fig. 4). Single scrapers are unifaces that bear a single (hence the name) retouched edge usually normal to the longitudinal axis of the flake from which the tool was fashioned. Double scrapers bear two opposing, subparallel retouched edges, convergent scrapers two retouched edges oblique to the flake's longitudinal axis that converge, usually at the flake's distal end. Transverse scrapers bear a single retouched edge perpendicular to the longitudinal axis (Debenath & Dibble 1994, 70–81). To use a crude biological syllogism whose size relations are reversed, Dibble argued that single scrapers are to other scrapers as tadpoles are to frogs: different stages of the same taxon, not different taxa.

The reduction thesis was tested in the study of several assemblages (*e.g.,* Close 1991a,b, *cf.* Dibble 1991b; Kuhn 1992, *cf.* Dibble 1995). Significantly, no critic has yet tested it independently using the platform dimensions so critical to its original development. Dibble's thesis is extended to Middle Palaeolithic bifaces (McPherron 1994) and Mousterian 'points' (Holdaway 1989, *cf.* Shea 1997), to Upper Palaeolithic burins (Barton *et al.* 1996) and scrapers (Hiscock 1996) and to Epipalaeolithic blades and burins (Neeley & Barton 1994, *cf.* a cast of thousands). A similar view of the relationship between notched tools and denticulates invoked flake size, not reduction, as the agent of change between Bordean types (Holdaway *et al.* 1996).

The reduction thesis bears a strong but not exclusively American cast (*e.g.,* Verjux 1988). It possesses far-reaching implications for Palaeolithic typology and interpretation, a quality recognised to judge from the *sturm und drang* it has inspired. The reduction thesis is developed substantially in metric and contextual data. Here, assemblage data extend its development. Yet the thesis is contextual to a significant degree, since its operation is held to depend on raw-material supply (Dibble 1991b, 264). McPherron (1994, 313) even suggested that it cannot be tested by comparing assemblages from different places but only by comparing stratigraphic assemblages of the same place.

According to the reduction thesis, the transformation from single to other scrapers occurs, trivially, through reduction in use. Use may be continuous or episodic. Raw-material supply may, depending on circumstances, encourage or discourage extensive use and the reduction that follows. Substantially, such use and reduction mark greater tool curation (*sensu* Shott 1996); as curation rises, scrapers tend to pass from single to more reduced forms. Importantly, they may but need not do so, since specimens may be fully reduced, in Dibble's sense of reaching minimum width suitable for prehension, before changing from single scrapers. Thus, original blank and core size and form and reduction technology — all uncontrolled variables in assemblage data — can produce original blanks of varying width and therefore varying potential reduction.

In tools, reduction is measured in metric attributes (Dibble 1987). In assemblages, increased reduction must be inferred in other ways. Specifically, reduction may covary with what Dibble & Rolland (1992, 11) called 'intensity of occupation'. They calculated this quantity in several ways (1992, 13), including the ratio of scrapers to notched tools. Plotted against this ratio, the proportions of single, double and convergent scrapers yield ambiguous patterns. Recall, though, that only Type 2 assemblages are adequate samples. When these alone are plotted against the ratio, very clear patterns emerge. Single scrapers decline steadily in proportion with the ratio ($r = -0.71, \mathrm{p} < 0.01$), and double ($r = 0.85, \mathrm{p} = 0.00$) and convergent ($r = 0.55, \mathrm{p} = 0.004$) scrapers increase with it (Fig. 11.9). Bisitun E is the conspicuous outlier in these plots; its removal would further strengthen each result.

In regression, the double scraper slope coefficient (0.017; s.e. = ±0.002 units of the ratio) significantly exceeds convergent scrapers' comparable figure (0.007; s.e. = ±0.002 units). (Bisitun E's omission does not significantly alter the differences in regression slopes). The rate of increase in double scrapers is higher than in convergent ones, as suggested by the reduction thesis, because single scrapers must first become double scrapers before they can become convergent ones. Transformation to convergent form is a joint probability; as an arithmetic necessity, its rate of occurrence must be lower than that of double scrapers. Both are transformed single scrapers but double ones are transformed sooner because they form directly from single ones.

Scrapers increase proportionally in Type 2 assemblages as measured by the scraper:notch ratio because the ratio measures increasing scraper use and discard. As these rise, relatively more of the used and discarded specimens are double and convergent scrapers, fewer single scrapers. As with time-averaging, the process of tool reduction systematically alters composition as assemblage size varies. In this respect too, assemblage size and composition are related.

The second reduction sequence is from single to

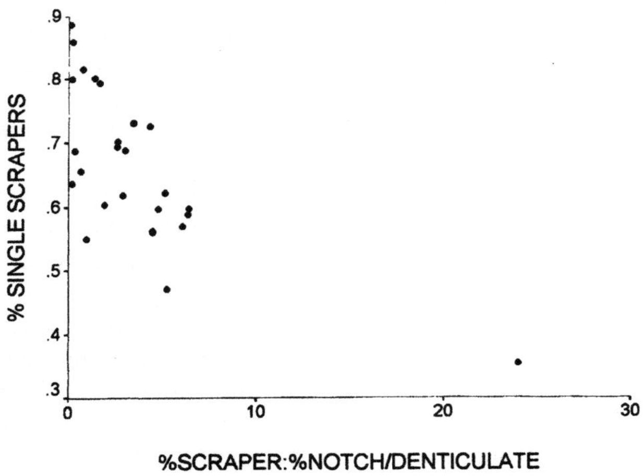

(a) Single scrapers.

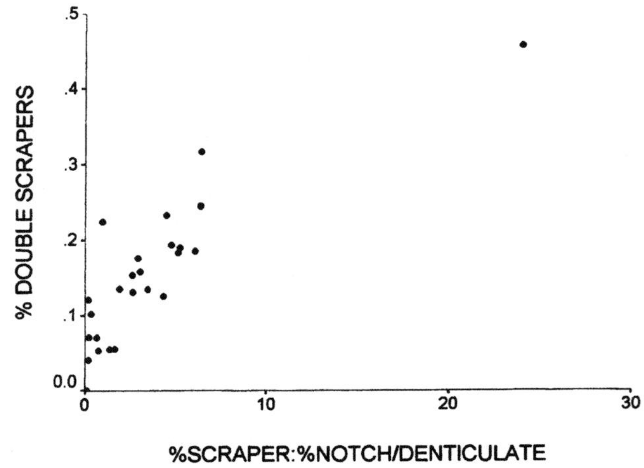

(b) Double scrapers.

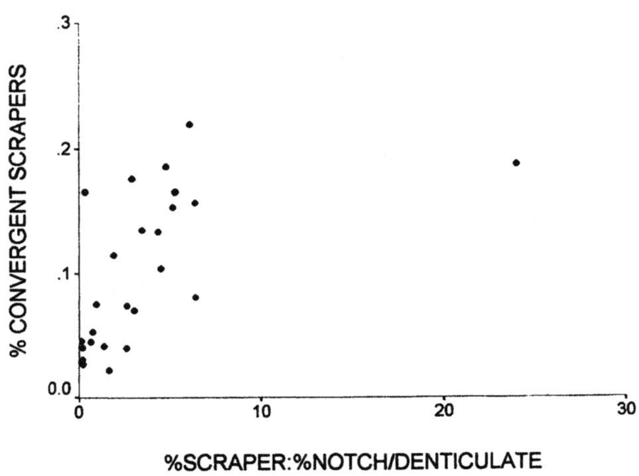

(c) Convergent scrapers.

Figure 11.9: Scraper proportions against reduction intensity.

transverse forms. Even in Type 2 assemblages, transverse scrapers do not conform to the pattern in double and convergent scrapers, declining with the scraper:notch-tool ratio. This aspect of the reduction thesis does not bear out in these data.

11.10 Site assemblages

As above, Bordean typology possesses intrinsic ambiguity. Among other things, ambiguity permits analysts to assign the same specimens to different types (Kolpakov & Vishnyatsky 1989, 110). One way to avoid inconsistency is to confine treatment to assemblages classified by the same analyst. (This controls for inconsistency, but not necessarily error because a hypothetical typologist can be consistently wrong. Certainly most Palaeolithic typologists are competent and conscientious, even if they do not always agree among themselves. Individual treatment of any site assemblages makes no assumption about its analysts' ability.) This treatment has the added virtue of controlling, if crudely, for local raw-material abundance.

Briefly, Type 2 assemblages are underrepresented in the 35 Combe Grenal stratigraphic assemblages ($n = 3$) and other types correspondingly over represented. Collectively, Combe Grenal samples exhibit the same size-dependence in richness and evenness that resides in the larger data set. Assemblage types combined resemble the larger data set in the relationship between proportions of scraper types and the scraper:notch ratio. The relationship is especially strong among the three Type 2 assemblages, but they are too few in which to invest great meaning. Comparable patterning should be sought in other sample assemblages.

11.11 Discussion

Everywhere, archaeological assemblages were constantly in a state of becoming, never arrived at a fixed form. It is no surprise, therefore, that New World

Palaeoindian and Old World Middle Palaeolithic assemblage patterns are similar in some respects. As in the Palaeoindian case, evenness and especially richness are substantially size-governed in Middle Palaeolithic assemblages. One of Dibble's two scraper reduction sequences bears out in this data set, the other does not. Therefore, this is no brief for the 'minimalism' (Henry 1996, 136) charged of reduction advocates. Surely, however, Palaeolithic specialists would not dogmatically deny the possibility that reduction contributes significantly among other factors to assemblage variation. To some extent but not entirely, the reduction thesis is supported.

Varying proportions of scraper types are recognised in traditional assemblage interpretations by terms like Charentian Mousterian. Such labels describe but do not explain variation. Instead, they merely partition a continuum of variation in assemblage composition that is explained more satisfactorily in terms of assemblage formation (Dibble & Rolland 1992; Freeman 1994).

Generally, archaeologists must pay more serious regard to formation processes and the quality of assemblage data, determined only in part by their size and the quality of excavation. Until then, we will continue to regard all absence of tool types or lack of association as inherently meaningful. A quarter-century ago, Clarke (1972, 27) suggested that the distinction between Acheulean and Clactonian industries was more apparent than real, more a sample effect than a legitimate difference. Recent discoveries bear out this suggestion (Ashton *et al.* 1994). In Africa, Developed Oldowan and Acheulean seem distinguished only by the former's lack of a few types (Gowlett 1986) whose absence may owe more to sample effects than legitimate difference. This suggestion is easily tested by common plots of richness in Developed Oldowan and Acheulean assemblages and by bootstrapping and other measures of sample adequacy. If the former lie at the lower left of richness-size plots and do not reach stability in estimated richness below observed assemblage sizes, then the difference may not be real.

11.12 Conclusion

There are important similarities in assemblages as disparate as New World Palaeoindian and Old World Middle Palaeolithic ones. Formation processes, including different degrees of reduction and thus curation, link the assemblages. Through the joint agency of accumulation span, tool use rate and tool use life, assemblage size and composition are not independent but systematically related. The relationship is not a complication or cause for despair but a real process of interest and analytical value in its own right.

Only when the effects of formation processes (themselves culturally determined) are controlled can we attribute variation to cultural norm and other causes. Due respect for formation processes is imperative because they exist, because their influence is manifest, and because they affect the character of archaeological assemblages. We do not need formation theory besides genuine theory. Indeed, formation theory is genuine theory (Shott 1998b). Instead, we need formation theory to limit and therefore properly situate the application of genuine cultural theory.

Thus, the reduction thesis may help explain continuous variation in Middle Palaeolithic assemblage composition. In a broader sense, though, it merely begs a new interpretative question: Why was reduction more 'intense' or extensive in some cases than in others? Why, in effect, was reduction so variable? Raw material abundance certainly is part of the answer (Dibble 1991a) but other, more purely social or cultural, factors also are. We can seek those factors better once we understand the nature and proximate cause of variation in assemblage composition. In this way, formation processes and the reduction thesis help to situate the application of general theory.

Bordes' facies interpretation was a good story — in that word's best sense — that satisfied an understandable desire to order the remote past. The reduction thesis and its assemblage-level counterpart are much less satisfying stories, but they are necessary preludes to the more nuanced, contextualized and better grounded stories that will replace Bordes' facies argument. Linnaeus' biological taxonomy rested on religious premises that biologists today find unnecessary to invoke. But the taxonomy remains perfectly adequate, except to cladists, because the similarities that Linnean taxonomy reveal have an evolutionary explanation. Similarly, Middle Palaeolithic tool types have lost their original meaning but acquired equally significant new ones (Dibble 1987, 116) that recognise systematic relationships between apparently distinct taxa. Mousterian facies as types retain, if not integrity, at least a measure of significance as segments of continua of complexly varying assemblages. The Middle Palaeolithic archaeological record is not as simple as Bordes perceived it, a realisation made ironic by the recognition that Bordes' contribution was critical to achieving it; today Middle Palaeolithic practice has a strong 'schizophrenic' (Freeman 1994, 47) quality. In this sense, the reduction thesis and its assemblage-level counterpart are testaments to Bordes more than criticisms of him.

Acknowledgements

Norah Moloney and the Institute of Archaeology, University College London, kindly invited me to participate in this symposium. Harold Dibble supplied the assemblage data and Carl Lipo the bootstrapping method and some crucial advice in its use. I alone am responsible for the use made of others' kindness.

References

ASHTON, N., J. MCNABB, B. IRVING, S. LEWIS & S. PARFITT 1994. 'Contemporaneity of Clactonian and Acheulean flint industries at Barnham, Suffolk.' *Antiquity* 68: 585–9.

BARTON, C. M., D. I. OLSZEWSKI & N. R. COINMAN 1996. 'Beyond the graver: Reconsidering burin function.' *Journal of Field Archaeology* 23: 111–25.

BOBROWSKY, P. T. & B. F. BALL 1989. 'The theory and mechanics of ecological diversity in archaeology.' *In* R. Leonard & G. Jones (eds.), *Quantifying Diversity in Archaeology*, pp. 4–12. Cambridge University Press, Cambridge.

BORDES, F. 1961. *Typologie du Paléolithique Ancien et Moyen.* CNRS, Paris.

CALLOW, P. & J. M. CORNFORD (eds.) 1986. *La Cotte de Ste. Brélade, Jersey. Excavations by C. M. B. McBurney 1961–1978.* Geo Books, Norwich.

CLARK, G. A. 1996. 'Plus Français que les Français.' *Antiquity* 70: 138–9.

CLARKE, D. L. 1972. 'Models and paradigms in contemporary archaeology.' *In* D. L. Clarke (ed.), *Models in Archaeology*, pp. 1–60. Methuen, London.

CLARKE, D. L. 1973. 'Archaeology: The loss of innocence.' *Antiquity* 47: 6–18.

CLOSE, A. E. 1991a. 'On the validity of middle paleolithic tool types: A test case from the eastern Sahara.' *Journal of Field Archaeology* 18: 256–64.

CLOSE, A. E. 1991b. 'Reply to Dibble.' *Journal of Field Archaeology* 18: 267–9.

CROVETTO, C., M. FERRARI, C. PERETTO, L. LONGO & F. VIANELLO 1994. 'The carinated denticulates from the paleolithic site of Isernia La Pineta (Molise, Central Italy): Tools or flaking waste? The results of the 1993 lithic experiments.' *Human Evolution* 9: 175–207.

DEBENATH, A. & H. A. DIBBLE 1994. *Handbook of Paleolithic Typology. Volume One: Lower and Middle Paleolithic of Europe.* University Museum, University of Pennsylvania, Philadelphia.

DIBBLE, H. L. 1987. 'The interpretation of Middle Paleolithic scraper morphology.' *American Antiquity* 52: 109–17.

DIBBLE, H. L. 1991a. 'Local raw material exploitation and its effects on Lower and Middle Paleolithic assemblage variability.' *In* A. Montet-White & S. Holen (eds.), *Raw Material Economies among Prehistoric Hunter-Gatherers*, Publications in Anthropology No. 19, pp. 33–46. University of Kansas, Lawrence.

DIBBLE, H. L. 1991b. 'Rebuttal to Close.' *Journal of Field Archaeology* 18: 264–7.

DIBBLE, H. L. 1995. 'Middle Paleolithic scraper reduction: background, clarification, and review of the evidence to date.' *Journal of Archaeological Method and Theory* 2: 299–368.

DIBBLE, H. L. & N. ROLLAND 1992. 'On assemblage variability in the Middle Paleolithic of Western Europe.' *In* H. L. Dibble & P. Mellars (eds.), *The Middle Paleolithic: Adaptations, Behaviour and Variability*, pp. 1–28. University Museum of Philadelphia Museum Monograph 78, Philadelphia.

FREEMAN, L. G. 1994. 'Kaleidoscope or tarnished mirror? Thirty years of Mousterian investigations in Cantabria.' *In* J. Lasheras (ed.), *Homenaje al Dr. Joaquin González Echegaray*, pp. 37–54. Museo y Centro de Investigaci6n de Altamira, Monografias No. 17, Santander.

GOWLETT, J. 1986. 'Culture and conceptualisation: the Oldowan–Acheulian gradient.' *In* G. N. Bailey & P. Callow (eds.), *Stone Age Prehistory: studies in memory of Charles McBurney*, pp. 243–60. Cambridge University Press, Cambridge.

HENRY, D. O. 1996. 'Functional minimalism versus ethnicity in explaining lithic patterns in the Levantine Epipaleolithic.' *Antiquity* 70: 135–6.

HISCOCK, P. 1996. 'Transformations of Upper Paleolithic implements in the Dabba industry from Haua Fteah (Libya).' *Antiquity* 70: 657–64.

HOLDAWAY, S. 1989. 'Were there hafted projectile points in the Mousterian?' *Journal of Field Archaeology* 16: 79–85.

HOLDAWAY, S., S. MCPHERRON & B. ROTH 1996. 'Notched tool reuse and raw material availability in French Middle Paleolithic sites.' *American Antiquity* 61: 377–87.

IRWIN, H. T. & H. M. WORMINGTON 1970. 'Paleo-Indian tool types in the Great Plains.' *American Antiquity* 35: 24–34.

JUDGE, W. J. 1973. *Paleo-Indian Occupation of the Central Rio Grande Valley in New Mexico.* University of New Mexico Press, Albuquerque.

KAUFMAN, D. 1998. 'Measuring archaeological diversity: An application of the jackknife technique.' *American Antiquity* 63: 73–85.

KOLPAKOV, E. M. & L. B. VISHNYATSKY 1989. 'The Bordes method?' *Norwegian Archaeological Review* 22: 107–18.

KUHN, S. 1992. 'Blank form and reduction as determinants of Mousterian scraper morphology.' *American Antiquity* 57: 115–28.

LIPO, C. P., M. E. MADSEN, R. C. DUNNELL & T. HUNT 1997. 'Population structure, cultural transmission, and frequency seriation.' *Journal of Anthropological Archaeology* 16: 301–34.

MAYR, E. 1988. *Towards a New Philosophy of Biology: Observations of an Evolutionist.* Belknap Press, Cambridge.

MCPHERRON, S. P. 1994. *A Reduction Model for Variability in Acheulian Biface Morphology.* Ph.D. thesis, University of Pennsylvania.

NEELEY, M. P. & C. M. BARTON 1994. 'A new approach to interpreting Late Pleistocene microlith industries in Southwest Asia.' *Antiquity* 68: 275–88.

ORTON, C. R. & P. A. TYERS 1990. 'Statistical analysis of ceramic assemblages.' *Archeologia e Calcolatori* 1: 81–110.

SHEA, J. J. 1997. 'Middle Paleolithic spear point technology.' *In* H. Knecht (ed.), *Projectile Technology,* pp. 79–106. Plenum Press, New York.

SHOTT, M. J. 1996. 'An exegesis of the curation concept.' *Journal of Anthropological Research* 52: 259–80.

SHOTT, M. J. 1997a. 'Activity and formation as sources of variation in Great Lakes Paleoindian assemblages.' *Midcontinental Journal of Archaeology* 22: 197–236.

SHOTT, M. J. 1997b. 'Transmission theory in the study of stone tools: A Midwestern example.' *In* C. M. Barton & G. A. Clark (eds.), *Rediscovering Darwin: Evolutionary Theory and Archaeological Explanation,* Archaeological Papers of the American Anthropological Association No. 7, pp. 193–206. American Anthropological Association, Arlington.

SHOTT, M. J. 1998a. 'Status and role of Formation Theory in contemporary archaeological practice.' *Journal of Archaeological Research* 10.

SHOTT, M. J. 1998b. 'Status and role of Formation Theory in contemporary archaeological practice.' *Journal of Archaeological Research* 6: 299–329.

STERN, N. 1994. 'The implications of time-averaging for reconstructing the land-use patterns of early tool-using hominids.' *Journal of Human Evolution* 27: 89–105.

VERJUX, C. 1988. 'Les denticulés Moustériens.' *In* M. Otte (ed.), *L'Homme de Néanderthal, 4. La téchnique,* Etudes et Recherches Archéologiques de l'Université de Liège 31, pp. 197–204. Université de Liège, Liège.

Michael J. Shott
Department of Sociology, Anthropology and Criminology
University of Northern Iowa
Cedar Falls
IA 50614–0513
USA
`shott@csbs.uni.edu`

12

A Re-Evaluation of Palaeolithic Stone Tool Cutting Edge Production Rates and Their Implications

Joanne C. Tactikos

Anthropology Department, Rutgers University

Abstract

I use results of experiments and analysis of archaeological lithic assemblages to examine Leroi-Gourhan's (1964) claim that the cutting edge-to-mass ratios (CE/M) of stone tools increased from the Lower, Middle to the Upper Palaeolithic. Focusing on whole flakes I found a general temporal trend towards higher CE/M ratio occurs, but CE/M ratios are not significantly different between the Middle and Upper Palaeolithic. The trajectory of technological change as proposed by Leroi-Gourhan is re-evaluated and morphological diversity is considered as an indication of changing technological organisation through time. Experimental archaeology is proposed as a means to inform behavioural interpretation.

12.1 Introduction

Diachronic change in Palaeolithic artefact assemblages has been used to define culture-chronological sequences under the unwarranted assumption that technological progress is an intrinsic property of lithic technology. Such progressivist models of technological evolution have been constructed from lithic assemblages classified on the basis of established typological criteria. Consequently, cultural entities have been defined by typologically based classifications of artefact distributions. These attempts to draw connections between technology and behaviour are deficient due to their reliance on typology as the exclusive measure of lithic assemblage variation.

Early models of lithic technological evolution assessed progress in terms of increased refinement of tool form. Departing from previous paradigms, André Leroi-Gourhan (1964) proposed that Palaeolithic technological variation could be viewed in terms of attributes that were linked to the functional or behavioural context in which the tools were created, used, discarded and lost. These attributes were length of the usable cutting edge and raw material mass. Assuming that all stone tools except hammerstones and spheroids were intended for cutting, scraping or piercing tasks, Leroi-Gourhan focused on the cutting edge of a stone tool as a measure of efficiency. Together with the existing empirical knowledge of the decrease in average tool size from the Lower to the Upper Palaeolithic, Leroi-Gourhan hypothesised that the process of cutting edge production varied in time proportionately with the ratio between cutting edge and the raw material mass. This technological progression, as he saw it, 'reflected the first material relationship between humans and minerals', and developed 'strictly in parallel' with human evolution increasing significantly over time as a result of technological advances between the Lower and Middle Palaeolithic, the Middle and Upper Palaeolithic, and especially Upper Palaeolithic and Mesolithic. Linking lithic material and biological progress he correlated this technological progression with an increase in hominid cranial capacity (1964, 134).

In the 1964 publication *Le Geste et la Parole*, Leroi-Gourhan explains the progression from chopper to biface to levalloisian point and finally to microlith, citing the quantifiable cutting edge to mass ratio as being diagnostic of technological progress. His conclusions reflect the paradigms of the period in which he worked and, in recent years, continue to be cited in various forms (Pfeiffer 1969; Price & Feinman 1993; Stringer & Gamble 1993). Despite its basis in technological attributes rather than aesthetics, Leroi-Gourhan's model is progressivist in its implication that particular tools are developing techniques which evolve toward more efficient designs than those which existed previously, rather than a particular set of solutions to the ecological constraints of a given time and place.

Although there has been a general acceptance of Leroi-Gourhan's hypothesis, there has been little concerted effort to test his results. Close examination of Leroi-Gourhan's model raises crucial questions about the data upon which it is based. Were only retouched tools measured? Were CE/M rates tabulated for numerous tools of each type or were the CE/M rates projected from highly selective observations? Since only one data point per culture-stage is presented, he does not explore intra-stage variability in CE/M ratios, and what might be the basis of it. By emphasising finished tools and neglecting debitage, it is unlikely that Leroi-Gourhan's results accurately reflect the amount of usable cutting edge retrievable from a given mass of raw material.

Notions of qualitatively defined tool morphology continue to cloud interpretations of lithic technology (Odell 1981). Retouched pieces are considered proper tools while unretouched and unmodified pieces continue to be thought of exclusively as waste by-products of tool manufacture. This distinction is inherent in all typologically based criteria. The dual assumptions of predetermination of tool form and of artefacts being the intentional end result of tool manufacture allows for biases in lithic analysis. Quantification of retouched pieces enables a partial descriptive analysis, but such measurements cannot inform the manufacturing process as thoroughly as a technological assessment of unretouched pieces. Lithic 'waste' may be behaviourally diagnostic in ways that curated tools are not. The lithic debitage and assemblage attributes most commonly examined are size and form (for a good review of this, see Shott 1994). The utility of experimental and archaeological debitage is well demonstrated (Fish 1981; Marks & Volkman 1983; Newcomer 1980; Pelcin 1997a,b,c; Schick 1987a; Shea 1985, 1989; Texier & Francisco-Ortega 1995; Toth 1991). Measuring unretouched pieces reduces the effects of factors such as curation, modification and differential function on the composition of shaped tools and tool assemblages. To some degree, acceptance of Leroi-Gourhan's hypothesis may be a consequence of its correspondence with traditional hypotheses about the Middle-to-Upper Palaeolithic transition. This is especially true in Europe where Neanderthals and anatomically modern humans were associated exclusively with the Middle Palaeolithic and Upper Palaeolithic respectively.

Upper Palaeolithic technology did not arise *de novo*, but instead was based on technological and typological trends which first occurred in the Middle Palaeolithic (Marks 1990; Reynolds 1990; Straus 1990). Research on the processes by which Middle and Upper Palaeolithic tools are made, used, discarded and lost (*i.e., chaînes opératoires*, Boëda *et al.* 1990) have called into question any strict correspondence between typology and technology.

Hypotheses arguing that the species-specific technology of the Neanderthal was replaced by the more advanced technology of anatomically modern humans have been confounded by discoveries in the Middle East (Bar-Yosef & Vandermeersch 1993; Bar-Yosef *et al.* 1986; Harold 1983; Leveque & Vandermeersch 1980). Palaeoanthropologists frequently assume that the Upper Palaeolithic represents a technological advance over the Middle Palaeolithic (Bar-Yosef 1989, 1992; Binford 1989; Davidson & Noble 1993; Gamble 1986; Gowlett 1984; Klein 1990; Mellars 1973, 1989; Otte 1990; Pfeiffer 1982; Stringer & Gamble 1993; Trinkaus 1989b). Given the ongoing debates regarding the origins of modern humans, and the aforementioned methodological and theoretical deficiencies of Leroi-Gourhan's model, there is clearly a need to reanalyse the relationship between cutting edge length and raw material mass in various industrial complexes.

The question I will address here is simply stated, 'How much cutting edge can be produced from a given raw material mass using various core reduction strategies?' One important goal of this study is to compare the CE/M ratios of tools from curated archaeological assemblages, such as those studied by Leroi-Gourhan (1964), with controlled, replicated assemblages that include the entire core reduction sequence. In exploring this issue experimentally, this study also:

1. demonstrates the trend over time of technological change as determined by the analysis of cutting edge production rates (calculated using the cutting edge to mass ratio);
2. discerns what (if any) difference exists between the CE/M ratio of retouched 'tools' and unretouched flakes through the comparison of Leroi-Gourhan's data with my experimental and archaeological data; and
3. evaluates some of the problems of employing archaeological assemblages in characterisations of technological change.

This evaluation is conducted through a comparison of archaeologically sampled assemblages and replicated assemblages in which the full range of a reduction sequence is represented.

12.2 Materials

For both the archaeological and experimental assemblages, I examine only unretouched whole flakes. This contrasts with Leroi-Gourhan's methodology which, '[leaves] ... discards out of account' (1964, 134). Avoiding retouched tools allows for the control of variation caused by reworking, reshaping, or reduction. It also serves to decrease the ambiguities of morphological diversity resulting from curation and different activities. Forty-five assemblages are examined; twenty-two derive from archaeological contexts in Europe and the Middle East, and twenty-three are experimental sets replicating the dominant technology of each industrial complex represented (Table 12.1).

The Lower Palaeolithic industrial complex is represented archaeologically by an Oldowan type assemblage from Swanscombe, one from Tabun (level G) and by two Acheulean assemblages from Tabun (levels E, F). It is represented experimentally by five replicated

Industrial Complex	Site, level/Experimental set	Raw Material	*n*
Lower Palaeolithic			
Oldowan	1 Tabun, G	chert	16†
	1 Swanscombe	flint	53†
	5 replicated sets	feldsite/chert	99W
Acheulean	2 Tabun, F, E	flint	40
	5 replicated sets	obsidian	100
Middle Palaeolithic			
Levantine Mousterian	2 Tabun, D, C	flint	40
	1 el Wad, G	flint	20
	1 Rosh Ein Mor	flint	20
	1 Ksar Akil, 27	chert	20
Mousterian	3 La Quina	flint	60
	2 Le Moustier	flint	35†
	2 replicated sets (laminar technique)	flint/obsidian	40
	3 replicated sets (centripetal technique)	obsidian	60
Upper Palaeolithic			
Levantine Aurignacian	1 el Wad, D	chert	20
	5 Ksar Akil, 5, 7, 21	chert	100
UP blade technology	5 replicated sets (indirect percussion)	flint/chert	95‡
	1 replicated set (direct percussion)	chert	20
Mesolithic			
microblade technology	2 el Wad, B1, B2	chert	40
	2 replicated sets (indirect percussion)	glass/obsidian	40

† entire assemblage was examined
‡ includes one entire assemblage, four randomly sampled assemblages

Table 12.1: Archaeological and experimental assemblages.

sets of Oldowan technology made on feldsite and chert that include a complete range of primary, secondary and tertiary flakes from intensively flaked cores, and five Acheulean replications in obsidian, consisting of debitage and/or thinning flakes resulting from the manufacture of handaxes. The handaxe manufacturing process included both hard and soft hammer percussion techniques. The Middle Palaeolithic is represented archaeologically by four Levantine sites including Tabun (levels D and C), el Wad (level G), Rosh Ein Mor (square T16) and Ksar Akil (level 27), and two European sites, La Quina and Le Moustier. It is represented experimentally by five Levallois assemblages made on obsidian using hard hammer percussion in both laminar and centripetal reduction techniques. Preparation flakes as well as Levallois flakes were included in the analysis. The Upper Palaeolithic is represented by archaeological assemblages from el Wad (level D) and Ksar Akil (levels 5, 7b, 7c, 21e, 21f). It is represented experimentally by six Upper Palaeolithic sets consisting of blades made on obsidian, flint and chert, using soft hammer (antler billet) direct and indirect percussion techniques on prismatic blade cores. The Mesolithic is represented by archaeological assemblages from El Wad (levels B1, B2), and by two experimental sets that were made on chert and glass; these consisted of microblades made by indirect percussion using a punch technique (Table 12.1). The archaeological assemblages were chosen in order to sample both relatively complete, regional sequences, as well as a range of industries from different regions. The sample sizes vary depending upon collection size. All the archaeological assemblages analysed are from collections housed at the Peabody Museum, Harvard University, USA. The experimental sets were made by an experienced flintknapper, Dr John J. Shea, of the State University of New York at Stony Brook.

In the experimental assemblages various raw materials were used in order to approximate the diversity of archaeological assemblages. Initially, five sets each of the Oldowan, Acheulean, Mousterian and Upper Palaeolithic blades were replicated (Table 12.1). The Mesolithic and one other Upper Palaeolithic blade sets were added to accommodate comparability with the curated collection at the Peabody Museum. As several different techniques were replicated, different sample sizes were generated. The replicated assemblages consisting of less than 20 flakes were sampled in their entirety, and of the larger assemblages, 20 randomly selected detached pieces were examined.

12.3 Methods

Each archaeological and experimental specimen was weighed to the nearest 0.1g using a digital balance. In addition, the total length of cutting edge for each piece was measured in two ways: with callipers and by rolling the cutting edge carefully across a sheet of millimetre graph paper. The entire circumference of the cutting edge, including the striking platform was measured in calculations of circumference. Two independent researchers measuring identical samples tested the precision of this method. Inter-analyst differences did not exceed 1mm. A ratio of cutting edge length and raw material mass values was calculated creating a CE/M index. It should be noted that it is unclear whether Leroi-Gourhan included striking platforms in his analysis, so it is possible that the estimates of CE/M in this study will be slightly higher than in his original analysis. None of the archaeological specimens examined measured less than 2cm. Therefore, in order to diminish the effect of sampling bias, only those replications measuring 2cm or more were included in the analysis. It should also be noted that individual flintknapper skill is a variable that may affect lithic assemblage character and that such variability may not be reflected in the replicated assemblages examined in this study. Ratios of cutting edge to mass were calculated for each assemblage and for each industry represented, then subjected to various statistical analyses in order to quantify the degree of variability within and between assemblages and industrial complexes.

Archaeological lithic assemblages are often used to make interpretations of human behaviour (Fish 1981). Associated artefacts are used to interpret manufacturing processes and reduction sequences (Collins 1975; Dibble 1988; Jelinek 1976) The resulting interpretations are used to address issues of human behavioural patterns (Henry 1989) and even cultural identity (Sackett 1982). Even if the stratigraphic associations of artefacts is strong, there is still the danger of time averaging (Stern 1994) or other palimpsestic processes shaping the character of a lithic assemblage (Hull 1987; Isaac 1986; Petraglia & Potts 1994; Schick 1987b; Schiffer 1983). In order to evaluate the confidence with which interpretations can be made using empirical evidence alone, it is important to test the correspondence of archaeological assemblages with replicated ones, which reflect a complete set of tool manufacturing debris.

Regarding how closely measures of cutting edge to mass ratio in the archaeological sets examined in this study resemble the cutting edge to mass ratios of the replicated sets, the two groups were compared. The differences in CE/M ratios between industrial complexes and within industrial complexes were statistically tested by an analysis of variance (ANOVA). All statistical analysis were done at the 0.05% significance level.

12.4 Results

Supporting Leroi-Gourhan's hypothesis, a general increase in the overall mean CE/M ratio from the Lower Palaeolithic through to the Mesolithic is evident in both the archaeological and experimental sets (Fig. 12.1). Maximum CE/M values do not show a steady increase over time. Maximum CE/M values for the archaeological sets show a steady increase throughout culture-stages, from 4.0 in the Oldowan to 39.0 in the Mesolithic. However, the experimental sets show higher maximum CE/M values in the Mousterian at 13.40 than in the Upper Palaeolithic at 11.0 (Fig. 12.2). Both mean and maximum CE/M values indicate a substantial increase between the Upper Palaeolithic and Mesolithic culture-stages. Mean CE/M values of the archaeological sets extend from 0.88 in the Oldowan to 18.6 in the Mesolithic; and in the experimental sets from 1.51 in the Oldowan to 6.76 in the Mesolithic (Table 12.2).

The range of CE/M values varies considerably between the archaeological and experimental sets. The archaeological sets display a general increase in range of values (Fig. 12.3) but the experimental sets show the Upper Palaeolithic as having a narrower range than the Acheulean, Mousterian and the Mesolithic culture stages (Fig. 12.4). Note that Leroi-Gourhan's data was not included in the comparison of range as minimum-maximum CE/M values were not available.

Statistical comparisons (ANOVA) between the archaeological and experimental assemblages show that CE/M values for the archaeological sets are significantly different between all pairs of industrial complexes except the Oldowan-Acheulean, the Acheulean-Mousterian and the Oldowan-Mousterian (Table 12.3). Comparisons of experimental sets between Industrial Complexes show significant differences in CE/M ratios between all Industrial Complexes except the Mousterian-Acheulean, the Mousterian-Upper Palaeolithic, and the Acheulean-Upper Palaeolithic (Table 12.4).

Analysis of variance (Fisher's LSD) was examined between Industrial Complexes of all assemblages and within Industrial Complexes between this study's archaeological and experimental assemblages and Leroi-Gourhan's data (Table 12.5). The results demonstrated no statistically significant differences in mean CE/M values between Leroi-Gourhan's mean CE/M values for shaped tools and the archaeological assemblages of whole flakes sampled in this study, except for the Mesolithic culture-stage. Analysis of variance between Leroi-Gourhan's sample and the experimentally replicated assemblages showed significant variance in the Acheulean culture-stage. Comparison of mean CE/M values between this study's experimental and archaeological assemblages revealed significant variation ($p \leqslant 0.0009$ to $p < 0.0001$) in all culture-stages.

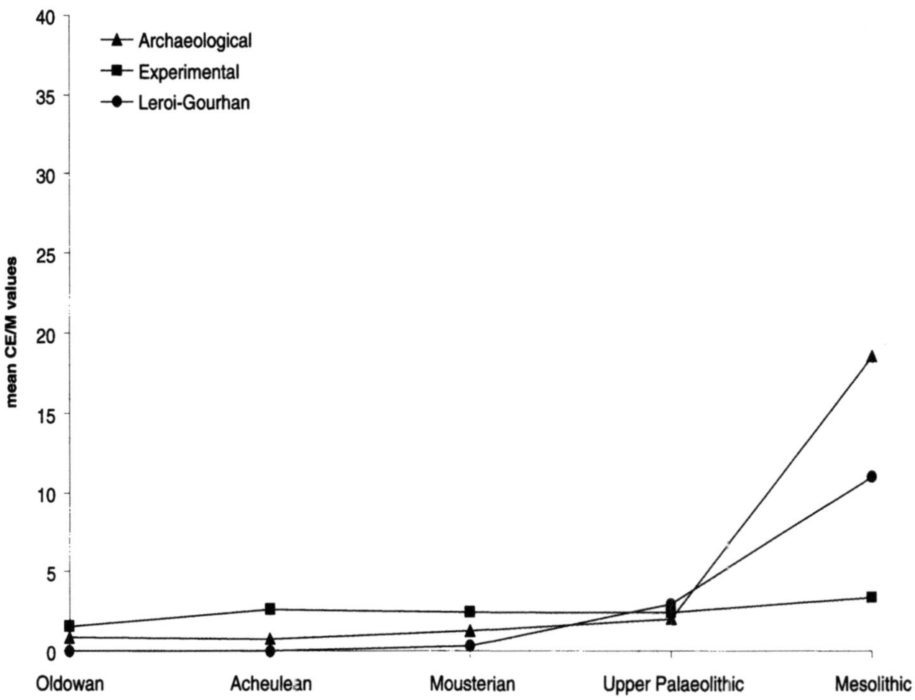

Figure 12.1: Mean CE/M values.

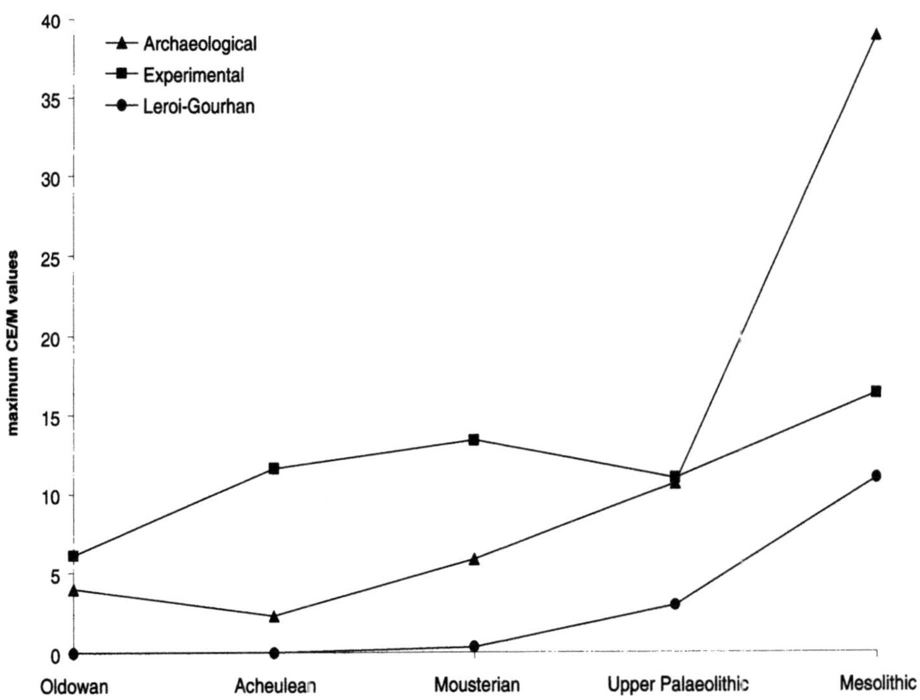

Figure 12.2: Maximum CE/M values.

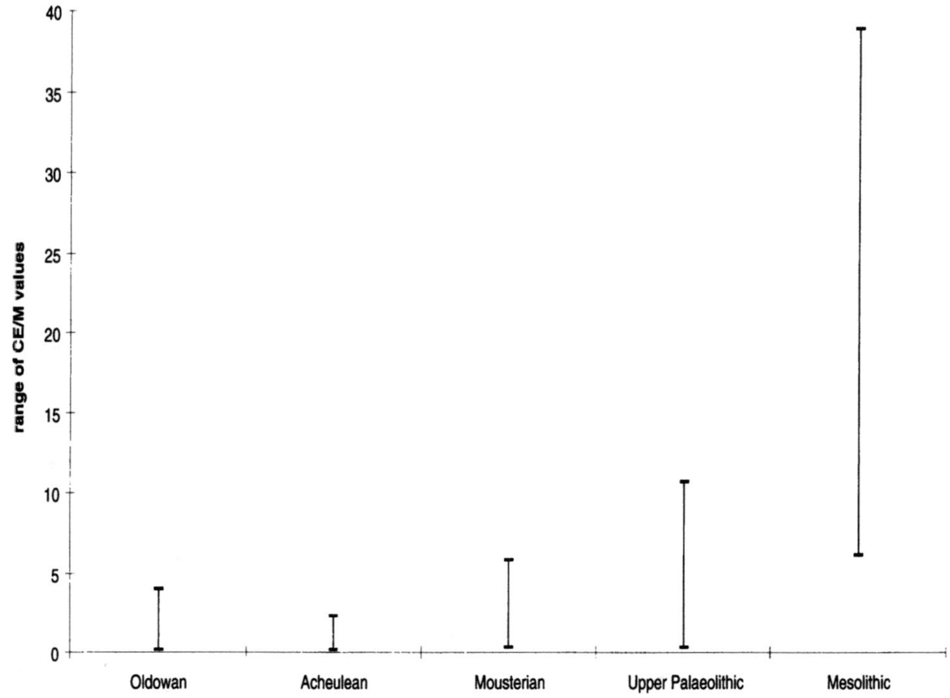

Figure 12.3: Range of CE/M values for archaeological assemblages.

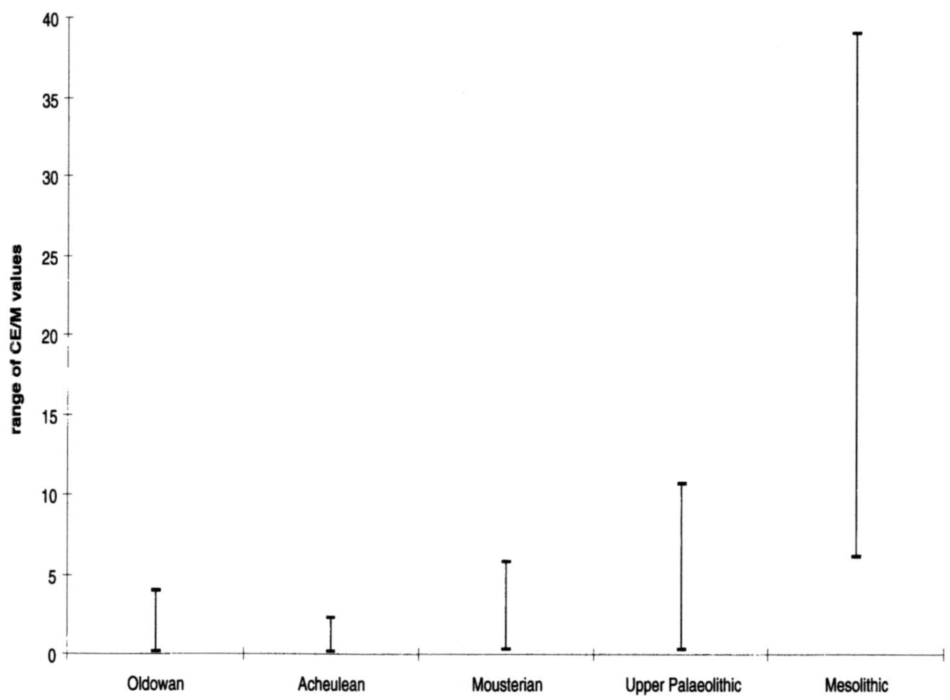

Figure 12.4: Range of CE/M values for experimental assemblages.

		CE/M			
	n	mean	min	max	std. dev.
Archaeological assemblages					
Oldowan	69	0.88	0.15	4.00	0.124
Acheulean	40	0.77	0.19	2.30	0.496
Mousterian	195	1.28	0.29	5.83	0.997
Upper Palaeolithic	120	2.06	0.28	10.71	1.704
Mesolithic	40	18.6	6.18	39.00	9.028
Total	**464**				
Experimental assemblages					
Oldowan	99	1.51	0.14	6.11	1.169
Acheulean	100	2.60	0.31	11.57	2.389
Mousterian	100	2.48	0.19	13.40	2.309
Upper Palaeolithic	115	2.47	0.38	11.00	2.041
Mesolithic	40	3.40	2.91	16.33	3.401
Total	**454**				

Table 12.2: Summary of descriptive statistical data.

	Oldowan	Acheulean	Mousterian	Upper Palaeolithic	Mesolithic
Oldowan		0.8554	0.3240	0.0067*	< 0.0001*
Acheulean			0.3152	0.0146*	< 0.0001*
Mousterian				0.0190*	< 0.0001*
Upper Palaeolithic					< 0.0001*
Mesolithic					

Archaeological Assemblages (P-values, $\alpha = 0.05$)

Table 12.3: ANOVA comparison of mean values for the archaeological assemblages.

12.5 Discussion

The significant difference demonstrated between the CE/M ratios of all the culture-stages confirms the sensitivity and diagnostic potential of CE/M ratios to technological change through time. The comparison of Leroi-Gourhan's (1964) model with the archaeological assemblages examined in this study generally support his empirically based hypothesis of technological change. The difference between Leroi-Gourhan's model and the replicated assemblages is significant for the Acheulean period. This may be attributed to methodological differences as Leroi-Gourhan examined finished tools (handaxes) while this study examined flakes removed during the manufacture of a handaxe. The greatest variation is evident between the experimentally replicated and archaeological assemblages examined in this study.

Lithic Industrial Complexes, or technological 'modes' are often characterised by the frequency of various morphological attributes or by recognisable landmarks on artefacts, both of which can lead to misinterpretations. For instance, it may be true that the classification of tool types in the Mousterian is vague due to the irregularity of the Mousterian blank form. The common Upper Palaeolithic blank form allows for easier identification of landmarks such as laterals, ends, point of percussion, *etc.* (see Reynolds 1990). As demonstrated by the narrow range of CE/M values in the Upper Palaeolithic, a blade is a more standardised form than, for instance, are Mousterian flake tools, hence the number of identifiable reference points increases as does the number of classifiable 'tool types'. Thus the use of a more standardised blank form in the Upper Palaeolithic, suggesting a greater focus on morphological standardisation, will allow for a larger number of different tool types to be classified than for the Mousterian. In Upper Palaeolithic artefact classifications, differentiation between blades and bladelets also notably increases the amount of tool diversity from that of the Mousterian. Assemblage richness (Grayson 1998) and flake size are technologically diagnostic variables considered in lithic analysis (Dibble & Pelcin 1995; Sheppard & Kleindienst 1996; Shott 1994). Bettinger & Eerkens (1997) argue that even in the North American Great Basin, more complex point shapes displayed less metrical variation than simple forms. When classificatory criteria are taken into account, increased standardisation in Upper Palaeolithic technology is a factor affecting a perceived increase in tool diversity. The results of this study show that high cutting edge recovery rates may not substantiate the idea of a watershed technological innovation that ushered in the Middle to Upper Palaeolithic transition.

Human behavioural interpretations based solely on the archaeological record can be problematic. This is evident when the disparity between archaeological and experimental assemblage CE/M values are con-

	Oldowan	Acheulean	Mousterian	Upper Palaeolithic	Mesolithic
Oldowan					
Acheulean	0.0005*				
Mousterian	0.0017*	0.7114			
Upper Palaeolithic	0.0004*	0.6669	0.9619		
Mesolithic	< 0.0001*	< 0.0001*	< 0.0001*	< 0.0001*	

Experimental Assemblages (P-values, $\alpha = 0.05$)

Table 12.4: ANOVA comparison of mean values for the experimental assemblages.

		Fisher's LSD	
P-value (b/twn industries)	F-value	comparison–P-value (w/in industries)	
Oldowan			
0.0025*	6.209	arch–exp	0.0009*
		arch–LG	0.4882
		exp–LG	0.2230
Acheulean			
< 0.0001*	13.523	arch–exp	< 0.0001*
		arch–LG	0.5383
		exp–LG	0.0173*
Mousterian			
< 0.0001*	16.000	arch–exp	< 0.0001*
		arch–LG	0.6677
		exp–LG	0.2245
Upper Palaeolithic			
< 0.0001*	9.800	arch–exp	< 0.0001*
		arch–LG	0.1260
		exp–LG	0.6052
Mesolithic			
< 0.0001*	69.900	arch–exp	< 0.0001*
		arch–LG	< 0.0001*
		exp–LG	0.0212

Table 12.5: ANOVA for CE/M between Palaeolithic culture-stages and intra-stage comparisons of archaeological (arch), experimental (exp), and Leroi-Gourhan's (LG) data (P-values, $\alpha = 0.05$)

sidered. Occupational interpretations must be augmented with analogous behavioural models (as discussed in Gifford-Gonzalez 1991). The value of information derived from sources independent of the archaeological record, such as experimental archaeology is well documented (Dibble 1985; Jones 1981, 1980; Keeley 1980; Newcomer 1980; Sahnouni *et al.* 1997; Schick 1987a; Shea 1993a; Toth 1987, 1991). In his ethnographic study of Inuit hunters, Binford (1977) recorded a great disparity between the amount of material transported and used in the field with the amount actually discarded there. His study demonstrated that artefacts archaeologically sampled might reflect only a small portion of the lithic material involved in actual activities. In order to reach informed conclusions about lithic technological organisation (*cf.* Nelson 1988; Torrence 1989c) ethnographic observation and experimental archaeology are needed. These actualistic models lend meaning to quantified analysis of archaeological assemblages, which also help to identify activity-related differences in lithic assemblages.

General models of technological change may not explicate all technological variability. Examining technological change in the Early and Middle Stone Age at Kalambo Falls, Zambia, Sheppard & Kleindienst (1996) concluded that there is remarkably little change in basic blank production techniques and blank attributes produced from the Early Stone Age (ESA) to the Middle Stone Age (MSA). They found that blank form and raw material type were related but that there was no systematic change in blank size or proportion correlated with the presence of prepared cores. The authors raise questions concerning the validity of transitional industrial complex designations, stating that technological change may be a very complex process proceeding at different rates at different times and places and may be further complicated by differential tool function or task.

Simplistic models of technological evolution do not take human behavioural ecology into consideration. Many studies of lithic technological organisation include examinations of strategies that involve social and economic concerns, and ecological considerations (Nelson 1988; Torrence 1989a,c). Behavioural differences in mobility and hunting techniques are well demonstrated within a period such as the Middle Palaeolithic (Lieberman & Shea 1994; Shea 1989, 1993b, 1994, 1998) but advanced or innovative technology in the form of higher edge-recovery rates are not. Behavioural ecological studies demonstrate the link between human tool using behaviour and the environment in which humans subsist. Correlating environment and technology, Henry (1989) examines the covariation between resource availability (*i.e.*, water, food, and flint sources) and technological evolution, linking selection pressures that favour technological improvements to increased edge production. Variation in lithic morphology has been viewed as more than a consequence of encephalization, but as an adaptive response to environmental variation (Kuhn 1992, 1993, 1994). In his 1995 monograph, Kuhn examines the rôle of toolmaking in hominid subsistence; he states:

> In order to bring technological data to bear on major changes in human adaptations, it is essential to explore how toolmaking was related to hominid subsistence behaviour.
>
> (Kuhn 1995, 16)

12.6 Conclusions

Re-evaluating Leroi-Gourhan's model of edge production rates and their implications brings other questions to the foreground. The results of this study do not support the assumption that Upper Palaeolithic blade technology afforded increased lengths of cutting edges which conferred a selective advantage upon Upper Palaeolithic toolmakers. Contrary to widely held opinions concerning the transition from Middle to Upper Palaeolithic technology, this study's results suggest that relatively high edge-recovery rates were already in evidence during the Middle Palaeolithic, in some cases, even higher than in the Upper Palaeolithic. If an increase in cutting edge production was not strictly linear then the question of whether more sophisticated and cost-effective technology in the form of greater cutting-edge production was a critical advantage of modern human populations remains unanswered. If, indeed, high edge-recovery rates alone are not indicative of an innovative technology, what, if any advantage did Upper Palaeolithic toolmakers have which facilitated their success and possible replacement of existing archaic populations?

Increased efficiency in the Upper Palaeolithic industries has yet to be incontrovertibly identified according to behavioural ecological models without relying on typological classifications. The disparity in CE/M ratios as seen in the comparison of archaeological and replicated assemblages or retouched and non-retouched pieces may be due to variation in research method. Issues that must be resolved before behavioural interpretations can be made include how changing adaptive strategies can be identified technologically, what the diagnostic technological attributes for mobility, transport and design factors are, and how these design factors covary with environmental constraints. These issues can be addressed through experimental studies that are designed to identify behaviourally sensitive lithic attributes. Experimentation and the analysis of replicated assemblages in their entirety are essential to provide objective parameters of tool utility, thereby establishing relevant technological criteria. Archaeological assemblages procured through unbiased sampling techniques should be examined to make more meaningful comparisons and inferences about the nature of blade manufacture, raw material economy, tool shaping and retouch as technological, cultural and/or economic adaptations. No simple relationships have been shown to exist between human behavioural trends and lithic assemblage characteristics. Clearly in order for lithic analysis to reach its full potential as an interpretative tool in comprehending human behaviour, a better understanding of the place of tool use in human subsistence behaviour must be deduced. Striving towards a comprehensive interpretation of lithic assemblages, experimental archaeology is a compelling proposition.

References

BAR-YOSEF, O. 1989. 'Upper Pleistocene human adaptations in Southwest Asia.' *In* E. Trinkaus (ed.), *Patterns and Processes in Later Pleistocene Human Emergence*, pp. 154–179. Cambridge University Press, Cambridge.

BAR-YOSEF, O. 1992. 'Middle Palaeolithic human adaptations in the Mediterranean Levant.' *In* T. Akazawa, K. Aoki & T. Kimura (eds.), *The Evolution and Dispersal of Modern Humans in Asia*, pp. 189–215. Hokusen-Sha, Tokyo.

BAR-YOSEF, O. & B. VANDERMEERSCH 1993. 'Modern humans in the Levant.' *Scientific American* 268: 94–100.

BAR-YOSEF, O., B. VANDERMEERSCH, B. ARENSBURG, P. GOLDBERG, H. LAVILLE, L. MEIGNEN, L. RAK, E. TCHERNOV & A. M. TILLIER 1986. 'New data concerning the origins of modern man in the Levant.' *Current Anthropology* 27: 63–4.

BETTINGER, R. L. & J. EERKENS 1997. 'Evolutionary implications of metrical variation in Great Basin projectile points.' *In* C. M. Barton & G. A. Clark (eds.), *Rediscovering Darwin: Evolutionary Theory and Archaeological Explanation*, Archaeological Papers of the American Anthropological Association No. 7, pp. 177–91. American Anthropological Association, Arlington.

BINFORD, L. R. 1977. 'Forty-seven trips: a case study in the character of some formation processes of the archaeological record.' *In* E. S. Hall (ed.), *Contributions to Anthropology: The Interior Peoples of Northern Alaska*, pp. 299–351. National Museum of Canada, Ottawa.

BINFORD, L. R. 1989. 'Isolating the transition to cultural adaptation: an organizational approach.' *In* Trinkaus (1989a), pp. 18–41.

BOËDA, E., J. M. GENESTE & L. MEIGNEN 1990. 'Identification de Chaînes Opératoires lithiques du Paléolithique Ancien et Moyen.' *Paléo* 2: 43–80.

COLLINS, M. B. 1975. 'Lithic technology as a means of processual inference.' *In* E. Swanson (ed.), *Lithic Technology: Making and Using Stone Tools*, pp. 15–34. Mouton, The Hague.

DAVIDSON, I. & W. NOBLE 1993. 'Tools, language, and cognition in human evolution.' *In* K. Gibson & T. Ingold (eds.), *Tools, Language, and Cognition in Human Evolution*, pp. 363–88. Cambridge University Press, Cambridge.

DIBBLE, H. L. 1985. 'Technological aspects of flake variation: a comparison of experimental and prehistoric flake production.' *American Archeology, Albuquerque* 5(3): 236–40.

DIBBLE, H. L. 1988. 'Typological aspects of reduction and intensity of utilization of lithic resources in the French Mousterian.' *In* H. L. Dibble & A. Montet-White (eds.), *Upper Pleistocene Prehistory of Western Eurasia*, University Museum Monograph 54. University Symposium Series Vol. I, pp. 181–97. The University Museum, Philadelphia.

DIBBLE, H. L. & A. PELCIN 1995. 'The effect of hammer mass on velocity and flake mass.' *Journal of Archaeological Science* 22: 429–39.

FISH, P. R. 1981. 'Beyond tools: Middle Palaeolithic debitage analysis and cultural inference.' *Journal of Anthropological Research* 37: 374–86.

GAMBLE, C. 1986. *The Palaeolithic Settlement of Europe.* Cambridge University Press, Cambridge.

GIFFORD-GONZALEZ, D. 1991. 'Bones are not enough: analogues, knowledge and interpretive strategies in zooarchaeology.' *Journal of Anthropological Archaeology* 10: 215–54.

GOWLETT, J. A. J. 1984. *Ascent To Civilization; the Archaeology of Early Man.* Alfred A. Knopf, New York.

GRAYSON, D. K. 1998. 'Stone tool assemblage richness during the Middle and Early Upper Palaeolithic in France.' *Journal of Archaeological Science* 25: 927–38.

HAROLD, F. 1983. 'The Chatelperronian and cultural and biological change at the Middle/Upper Palaeolithic transition.' *In* Trinkaus (1983), pp. 123–40.

HENRY, D. O. 1989. 'Correlations between reduction strategies and settlement patterns.' *In* D. O. Henry & G. H. Odell (eds.), *Alternative Approaches to Lithic Analysis*, Archeological Papers of the American Anthropological Association 1, pp. 139–55. American Anthropological Association, Washington D.C.

HULL, K. L. 1987. 'Identification of cultural site formation processes through microdebitage analysis.' *American Antiquity* 52(4): 772–83.

ISAAC, G. L. 1986. 'Foundation stones: early artefacts as indicators of activities and abilities.' *In* G. N. Bailey & P. Callow (eds.), *Stone Age Prehistory: studies in memory of Charles McBurney*, pp. 221–41. Cambridge University Press, Cambridge.

JELINEK, A. J. 1976. 'Form, function, and style in lithic analysis.' *In* C. E. Cleland (ed.), *Cultural Change and Continuity: Essays in Honor of James Bennett Griffin*, pp. 19–33. Academic Press, New York.

JONES, P. 1981. 'Experimental implement manufacture and use: a case study from Olduvai Gorge.' *Philosophical Transactions of the Royal Society* (B)292: 189–95.

JONES, P. R. 1980. 'Experimental butchery with modern stone tools and its relevance for Palaeolithic archaeology.' *World Archaeology* 12(2): 153–65.

KEELEY, L. H. 1980. *Experimental Determination of Stone Tool Uses. A Microwear Analysis.* Prehistory, Archaeology and Ecology Series. The University of Chicago Press, Chicago.

KLEIN, S. 1990. 'Human cognitive changes at the Middle to Upper Palaeolithic transition: the evidence of Boker Tachtit.' *In* Mellars (1990), pp. 499–516.

KUHN, S. L. 1992. 'On planning and curated technologies in the Middle Palaeolithic.' *Journal of Anthropological Research* 48(3): 185–214.

KUHN, S. L. 1993. 'Mousterian technology as adaptive response: a case study.' *In* Larsen-Peterkin *et al.* (1993), pp. 25–31.

KUHN, S. L. 1994. 'A formal approach to the design and assembly of mobile toolkits.' *American Antiquity* 59: 426–42.

KUHN, S. L. 1995. *Mousterian Lithic Technology: An Ecological Perspective.* Princeton University Press, Princeton.

LARSEN-PETERKIN, G., H. M. BRICKER & P. A. MELLARS (eds.) 1993. *Hunting and Animal Exploitation in the Later Palaeolithic and Mesolithic of Eurasia.* Archaeological Papers of the American Anthropological Association 4. American Anthropological Association, Arlington.

LEROI-GOURHAN, A. 1964. *Le Geste et la Parole 1. Techniques et langage.* Albin Michel, Paris.

LEVEQUE, F. & B. VANDERMEERSCH 1980. 'Découverte de restes humains dans un niveau castelperronian a Saint-Césaire (Charente-Maritime).' *Comptes-rendu de l'Académie des Sciences de Paris* 291: 187–9.

LIEBERMAN, D. E. & J. J. SHEA 1994. 'Behavioral differences between Archaic and Modern Humans in the Levantine Mousterian.' *American Anthropologist* 96(2): 300–32.

MARKS, A. 1990. 'The Middle and Upper Palaeolithic of the Near East and the Nile Valley: the problem of cultural transformations.' *In* Mellars (1990), pp. 56–80.

MARKS, A. & P. W. VOLKMAN 1983. 'Changing core reduction strategies: a technological shift from the Middle to the Upper Palaeolithic in the Southern Levant.' *In* Trinkaus (1983), pp. 13–34.

MELLARS, P. 1973. 'The character of middle-upper palaeolithic transition in south-west France.' *In* C. Renfrew (ed.), *The Explanation of Culture Change: Models in Prehistory*, pp. 255–76. Duckworth, London.

MELLARS, P. 1989. 'Technological changes across the Middle-Upper Palaeolithic transition: Economic, social and cognitive perspectives.' *In* Mellars & Stringer (1989), pp. 338–65.

MELLARS, P. (ed.) 1990. *The Emergence of Modern Humans: an Archaeological Pespective.* Cornell University Press, Ithaca.

MELLARS, P. & C. STRINGER (eds.) 1989. *The Human Revolution: Behavioral and Biological Perspectives in the Origins of Modern Humans.* Princeton University Press, Princeton.

NELSON, M. C. 1988. 'The study of technological organization.' *Archaeological Method and Theory* 3: 57–100.

NEWCOMER, M. 1980. 'Experimental flake scatter patterns: a new interpretive technique.' *Journal of Field Archaeology* 7(3): 345–52.

ODELL, G. 1981. 'The morphological express at Function Junction.' *Journal of Anthropological Research* 37: 319–42.

OTTE, M. 1990. 'From the Middle to the Upper Palaeolithic: the nature of the transition.' *In* Mellars (1990), pp. 438–56.

PELCIN, A. 1997a. 'The effect of indentor type on flake attributes: evidence from a controlled experiment.' *Journal of Archaeological Science* 24: 613–21.

PELCIN, A. W. 1997b. 'The effect of core surface morphology on flake attributes: evidence from a controlled experiment.' *Journal of Archaeological Science* 24(8): 749–56.

PELCIN, A. W. 1997c. 'Formation of flakes: the role of platform thickness and exterior platform angle in the production of flake initiations and terminations.' *Journal of Archaeological Science* 24(12): 1107–13.

PETRAGLIA, M. D. & R. POTTS 1994. 'Water flow and the formation of early Pleistocene artifact sites in Olduvai Gorge, Tanzania.' *Journal of Anthropological Archaeology* 13: 228–54.

PFEIFFER, J. E. 1969. *The Emergence of Man.* Harper and Row, New York.

PFEIFFER, J. E. 1982. *The Creative Explosion.* Harper and Row, New York.

PRICE, T. D. & G. M. FEINMAN 1993. *Images of the Past.* Mayfield Publishing, London.

REYNOLDS, T. E. G. 1990. 'The Middle-Upper Palaeolithic transition in Southwestern France: interpreting the lithic evidence.' *In* Mellars (1990), pp. 262–75.

SACKETT, J. R. 1982. 'Approaches to style in lithic analysis.' *Journal of Anthropological Archaeology* 1(1): 59–112.

SAHNOUNI, M., K. SCHICK & N. TOTH 1997. 'Experimental investigation into the nature of faceted limestone "spheroids" in the Early Palaeolithic.' *Journal of Archaeological Science* 24(8): 701–13.

SCHICK, K. D. 1987a. 'Experimentally-derived criteria for assessing hydrologic disturbance of archaeological sites.' *In* D. T. Nash & M. D. Petraglia (eds.), *Natural Formation Processes and the Archaeological Record*, pp. 86–107. British Archaeological Reports International Series 352, Oxford.

SCHICK, K. D. 1987b. 'Modeling the formation of Early Stone Age artifact concentrations.' *Journal of Human Evolution* 16: 789–807.

SCHIFFER, M. 1983. 'Toward the identification of formation processes.' *American Antiquity* 48: 675–706.

SHEA, J. J. 1985. 'Experiments in debitage analysis: some testable hypotheses.' *Bulletin — Massachusetts Archaeological Society* 46(2): 44–50.

SHEA, J. J. 1989. 'A functional study of the lithic industries associated with hominid fossils in the Kebara and Qafzeh Caves, Israel.' *In* Mellars & Stringer (1989), pp. 611–25.

SHEA, J. J. 1993a. 'An experimental investigation of the effects of trampling on the results of lithic microwear analysis.' *Journal of Archaeological Science* 20: 175–94.

SHEA, J. J. 1993b. 'Lithic use-wear evidence for hunting by Neanderthals and Early Modern Humans from the Levantine Mousterian.' *In* Larsen-Peterkin *et al.* (1993), pp. 189–98.

SHEA, J. J. 1994. *Spear point technology in the Middle Palaeolithic/Middle Stone Age: an inter-regional perspective.* Paper presented at the SAA symposium *Projectile Technologies: Archaeological, Ethnoarchaeological, and Experimental Perspectives*, Anaheim, California.

SHEA, J. J. 1998. 'Neandertal and Early Modern human behavioral variability: a regional-scale approach to lithic evidence for hunting in the Levantine Mousterian.' *Current Anthropology* 39: 45–78.

SHEPPARD, P. J. & M. R. KLEINDIENST 1996. 'Technological change in the earlier and Middle Stone Age of Kalambo Falls (Zambia).' *African Archaeological Review* 13(3): 171–96.

SHOTT, M. J. 1994. 'Size and form in the analysis of flake debris: review and recent approaches.' *Journal of Archaeological Method and Theory* 1(1): 69–110.

STERN, N. 1994. 'The implications of time-averaging for reconstructing the land-use patterns of early tool-using hominids.' *Journal of Human Evolution* 27: 89–105.

STRAUS, L. G. 1990. 'The Early Upper Palaeolithic of Southwest Europe: Cro-Magnon adaptations in the Iberian Peripheries, 40,000–20,000 BP.' *In* Mellars (1990), pp. 276–302.

STRINGER, C. & C. GAMBLE 1993. *In Search of the Neanderthals; solving the puzzle of human origins.* Thames and Hudson, New York.

TEXIER, P. J. & I. FRANCISCO-ORTEGA 1995. 'Main technological and typological characteristics of the lithic assemblage from Level I at Bérigoule, Murs-Vaucluse, France.' *In* H. L. Dibble & O. Bar-Yosef (eds.), *The Definition and Interpretation of Levallois Technology*, Monographs in World Archaeology 23, pp. 213–26. Prehistory Press, Madison.

TORRENCE, R. 1989a. 'Retooling: towards a behavioral theory of stone tools.' *In* Torrence (1989b), pp. 57–66.

TORRENCE, R. (ed.) 1989b. *Time, Energy, and Stone Tools.* Cambridge University Press, Cambridge.

TORRENCE, R. 1989c. 'Tools as optimal solutions.' *In* Torrence (1989b), pp. 1–6.

TOTH, N. 1987. 'Behavioral inferences from Early Stone artifact assemblages: an experimental model.' *Journal of Human Evolution* 16: 763–87.

TOTH, N. 1991. 'Importance of experimental replicative and functional studies in Palaeolithic archaeology.' *In* J. D. Clark (ed.), *Cultural Beginnings: Approaches to Understanding Early Hominid Lifeways in the African Savanna*, pp. 109–124. Rudolf Habelt Verlag, Bonn.

TRINKAUS, E. (ed.) 1983. *The Mousterian Legacy: Human Biocultural Change in the Upper Pleistocene.* British Archaeological Reports International Series 164, Oxford.

TRINKAUS, E. (ed.) 1989a. *The Emergence of Modern Humans: Biocultural Adaptations in the Later Pleistocene.* Cambridge University Press, Cambridge.

TRINKAUS, E. 1989b. 'The Upper Pleistocene transition.' *In* Trinkaus (1989a), pp. 42–66.

Joanne C. Tactikos
Department of Anthropology
Rutgers University
32 Bishop Street
New Brunswick, NJ 08901
USA
TACTIX@RCI.RUTGERS.EDU

13
Use-Wear in the 1990s in Western Europe: Potential and Limitations of a Method

Juan José Ibáñez[1] and Jesús Emilio González[2]
[1]*Dpto. de Ciencias Históricas, Universidad de Cantabria;* [2]*Dpto. de Antropología Social y Prehistoria, Universidad Autónoma de Barcelona*

Abstract

In this paper, we review the current state of use-wear analysis in Western Europe during the 1990s. The three main topics of methodological debate during the 1980s (high power/low power approach, diagnostic reliability of use-wear polish, and theories of polish formation processes) are revisited in the light of new methodological positions of use-wear analysts during the 1990s. Finally, the types of archaeological problems which have been dealt with by use-wear specialists in recent years are shown.

13.1 Introduction

Twenty-five years after the development of use-wear studies in the West, this technique has still made a relatively minor contribution to our knowledge of prehistoric ways of life. The structural impact it has had on archaeological research is limited as the number of projects involving specialists devoted to this kind of work are proportionately few. It is difficult for use-wear analysis to make headway in the face of either the scepticism or indifference of many of our colleagues. Reasons behind the sceptical response are to be found in their uncertainties about the reliability of analytical methods. The attitude of indifference can be explained by the fact that, among many archaeologists, use-wear information is regarded as of little value as a means of reconstructing past life forms. This situation was reflected in the conference that led to the writing of this volume. Indeed the section on use-wear analysis was called: *Microwear, Is There a Future?* In discussion sessions some interventions were illustrative of how little many archaeologists are impressed by this method. Their remarks went roughly like this: ***Use-wear analysis can tell us what a tool is for ... but of what use is this information? ... it is just static knowledge that offers nothing to a dynamic understanding of past societies ...*** In this article we attempt to identify the reasons for this state of affairs. We assess the possibilities and limitations of this method of analysis from the more sanguine perspective of the 1990s now that we have left behind extremes, both of optimism and of disproportionately negative criticism, that were to be found during the 1980s.

In the atmosphere of innocent optimism that marked the beginning of the 1980s, nobody could have predicted the difficulties in developing the method that would be encountered in later years. The method of functional analysis, starting with the pioneering work of Semenov (1964) had aroused great expectations, particularly in light of methodological contributions of Tringham *et al.* (1974) and Keeley (1980). Initial expectations gave way to outright optimism after the blind test that was carried out by Keeley (Keeley & Newcomer 1977).

In the years that followed, a series of difficulties put a stop to that initial surge of activity. Firstly, the methodological debate occurring between opposing groups of supporters of high power and low power techniques, often expressed from extreme positions, led to researchers wasting energy on an issue that was probably badly posed. Secondly, there were criticisms of the value of polish as a trace that allows worked material to be identified (Grace *et al.* 1985; Newcomer *et al.* 1986) and, finally, there was the absence of a clear overall theory to explain the origins and formation of use-wear polish.

13.2 High/low power: from confrontation to integration

Keeley (1980, 83), who pioneered the high power method, advocated the joint study of all types of traces. His successors, however, for a variety of reasons, tended to base their identifications almost exclusively on polish analysis. This situation was encouraged, firstly, by Vaughan's critique of the diagnostic reliability of microscarring, whereby he affirmed that there was no consistent relationship between microscarring characteristics, tool motion and material worked (Vaughan 1985). Secondly, the results of Keeley and Newcomer's blind test led to enormous faith being invested in the diagnostic capacities of polish. All of this created a polarisation of positions between high and low power methodological approaches which, at that time, appeared to be irreconcilable.

From the end of the 1980s and throughout the 1990s there was a softening of those positions, bringing into play, in the case of most European researchers, a more pragmatic and integrated approach (Grace 1990; Ibáñez & González 1996; Pawlik 1995; Unger-Hamilton 1989; van Gijn 1989, 147). A number of factors have contributed to this change: from the middle of the 1980s, specialists began to stress the limitations of polish, as we shall see, and consequently supporters of the high power approach, who for some years in Europe seemed to be winning the battle, began to nuance their position. Furthermore, it was observed that Vaughan's declarations about the diagnostic worthlessness of microscarring had to be reconsidered; different kinds of experimental work, followed by quantification of microscarring characteristics, have demonstrated the relevance of this type of trace for the reconstruction of tool motion and the relative hardness of the worked material (González & Ibáñez 1994; van den Dries & van Gijn 1997).

Since the 1990s, high and low power analysis is no longer discussed in Europe (van Gijn 1989, 147). Study of tool function encompasses analysis of the context in which the tool is found, its potential function, use-wear traces (striations, microscarring, edge rounding and polish) and presence of residues of worked material. Actual analysis of traces begins with a preliminary observation using a binocular microscope and continues with an analysis using an incident light microscope. Although we may talk of this process as the most commonly used technique, analysis actually depends on the state of preservation of the tool and on the archaeological problem one is attempting to resolve. In many archaeological cases, patination has destroyed use-wear polish. In these cases, the state of preservation determines the type of analytical techniques to be used, which in the case of microscarring must be restricted to study using a binocular microscope. If the aim is to identify specific tool uses in relation to characteristic microscarring patterns, such as those created when a tool is used as a projectile or for percussive action on hard materials, exclusive study with a binocular microscope of a wide range of collections of tools can be sufficient to establish whether one particular tool shape was used as a projectile or another tool type was used as an axe, a hoe or a ploughshare. Nevertheless, if the archaeological problem involves the need to identify the worked material precisely, use-wear polishes must be analysed via an incident light microscope. Where preservation is good, residue analysis may involve the use of SEM microscopes equipped with microsound (Anderson 1994b; Christensen *et al.* 1991). In present use-wear analysis the question is not so much to establish a standard technique, as to adapt the technique to the problem that must be resolved whilst, of course, having a precise knowledge of the limitations and possibilities of the technique employed (Grace 1989).

13.3 Is polish diagnostic?

The possibility of identifying worked material using polish was called into question in the mid 1980s (Grace *et al.* 1985; Newcomer *et al.* 1986). Their critique was based on the poor results of a blind test carried out at the Institute of Archaeology, London, and on the impossibility of suitably grouping different types of polish which had been characterised according to image analysis. It must be pointed out, with reference to the blind test, that a good many of the identification errors were due to the draconian conditions under which the test was conducted (Bamforth 1988; Hurcombe 1988; Moss 1987).

Where image analysis was concerned (Grace 1989), some aspects of the method applied were hardly appropriate. Surfaces examined included both polished and unpolished areas (Hurcombe 1988; Moss 1987; van Gijn 1989, 6). Moreover, polish was treated as a static phenomenon, without taking into account the state of polish development as a fundamental variable for defining the appearance of this trace. Significantly, in one analysis the system was unable to group together polishes created on both sides of a tool employed to cut wood; that is to say, it was incapable of identifying that these elements were the same: the same worked material, the same movement and the same working time (Moss 1987). We consider a general criticism of Grace's use of image analysis to be that he failed to reproduce the visual criteria of polish identification which are put into practice by use-wear analysts who evaluate, in a co-ordinated fashion, the pattern, topography and degree of polish development.

Different attempts at quantification which have developed since then have revealed that it is possible to group different kinds of polish in a coherent fashion, confirming empirical observations of dozens of researchers working independently of each other (Bietti 1996; González & Ibáñez in press; Kimball *et al.* 1995; Knutsson 1988; Vila & Gallart 1993).

Grace (1989) affirms that polish characteristics depend on worked material and time of use. As time of use cannot be known for archaeological tools, worked material cannot be inferred. We agree with the first statement, but not the second; in fact, polish is a continuous variable, which depends mainly on characteristics of the worked material and on its degree of development. However, degree of polish development can be known and isolated by considering the proportion of polished surface on the analysed area (González & Ibáñez in press). Polish features (pattern, microtopography, distribution along the edge and invasiveness) are evaluated with respect to the state of polish development in order to identify the worked material. Indeed, use-wear polish analysis is only one part of use-wear analysis. It is, therefore, the coherence of all traces (microscarring, striations, edge rounding and polish) as well as consideration of data offered by the archaeological context and the functional potential of the tool which together guarantee the reliability of any inferences made.

Despite criticisms that could be levelled against the conclusions of the researchers from the Institute of Archaeology, it is undeniable that their work served as a salutary shock which helped the progress of methods of analysis. In the first place, faith placed in polish had to be put into perspective, and this enabled researchers to move on to more realistic positions. In the second place, the path opened up by the application of image analysis revealed the need to advance further in the objective characterisation of the appearance of polished surfaces.

13.4 Polish origins

A principal point in the debate about use-wear analysis methodology centred on the origins of use-wear. One hypothesis suggests that abrasion produces polish, whilst a second approach points to a more complex cause, suggesting that a gel of amorphous silica is created, caused by friction, and is then redeposited on the changed surface (Anderson-Gerfaud 1981; Mansur-Franchomme 1983). The importance of this question is to be found in the need to possess an overall theory to explain polish variability and in the reasonable expectations that precise knowledge of how polish is produced could make it possible to advance in developing methods of polish analysis. These methods could cease to be of an eminently empirical nature and take on analytical form.

There have been few new contributions to this debate since the 1990s and, to a great degree, the arguments upheld in the previous decade have been repeated (Hurcombe 1997; Jardón 1996; Levi-Sala 1995; Mansur 1997). The debate is therefore still open.

Whichever theory is right, we believe either that of abrasion or of silica gel could act on its own as an overall theory to explain the relationship between polish appearance and nature of the worked material. This relationship has already been widely borne out in the case of silica gel. In reference to the abrasive origins of use-wear polish, characteristics of hardness, flexibility, presence of abrasive components and degree of humidity which vary from one material to another, might justify the relationship between polish appearance and worked material.

13.5 The state of methodology: between science and craftsmanship

Since the 1990s use-wear analysis in Europe has attempted to integrate all types of use-wear traces, so that a coherence between them might permit reliable identification. After the first general experimental programmes, whose aims were to observe differences in traces generated by work on different materials, other more specific programmes have emerged, focusing on specific activities and contexts, such as stone bowl manufacture (Astruc in press), wild or domestic cereal harvesting (Anderson 1992a; González *et al.* 1994; Juel Jensen 1994), Solutrean projectiles (Geneste & Plisson 1993) *etc.* This has increased the degree of detail with which archaeological reality is reproduced and, correspondingly, reliability of the method.

We can say that the method has gradually been refined. However, limitations of analysis remain. Implements located in sedimentary contexts that have undergone considerable change cannot be studied making it virtually impossible to study the oldest tools. Work on very soft material or tasks carried out for very short time periods do not generate identifiable traces. Therefore, one has to assume that the kinds of work identified do not represent all the work that was developed in a concrete archaeological context. Plus, absence of traces cannot be directly associated with an absence of use.

Today a consensus exists concerning the diagnostic possibilities and limitations of different types of traces. Experimentation has established a series of qualitative criteria for functional interpretation. Thus we know, for example, that working hard materials produces larger chippings than working soft ones, or that polish produced by scraping bone is, at a similar degree of development, smoother than that produced by scraping wood. However, these criteria are merely qualitative and may be influenced by subjectivity on the part of a particular researcher.

Identification of tool use implies considering many criteria in a co-ordinated way. Criteria must be integrated coherently within a system of analysis in which the argument is clearly presented and leads the analyst to understand the use of a prehistoric tool. Therefore, the functional inference requires a complex system of decisions; with a few exceptions (Grace *et al.* 1985; Ibáñez & González 1996), this system of inference has not been explained.

Therefore use-wear analysts have a set of qualitative criteria for functional identification which are not expressed in an explicit system. At this point of methodological development it is possible to contribute interesting information on prehistoric tool use. However, the method has various limitations. As they are not expressed in quantitative form, the criteria present a certain level of subjectivity which has an effect on the degree of reliability of interpretations. The interpretative ability of analysts depend upon their personal experience, their visual memory and capacity for logical argument. Thus, credibility of results is based more on an analyst's personal prestige than on critical knowledge of analytical methods. The qualitative and non-standardised character of the criteria makes development of the discipline more difficult. Training new researchers implies new experimental programmes which for the most part repeat those already carried out, and also makes the gradual refinement and perfecting of the system of interpretation very difficult.

As such, use-wear is more a question of expertise, based on empirical data, than a real scientific method. This criticism should be put into context as, in fact, it could also be applied to most archaeological techniques of analysis. It does not mean that we should be happy with this state of affairs; new efforts must be made to increase the objectivity of the technique to improve reliability of present analysis and on our capacity to transmit criteria of interpretation.

13.6 Results obtained since the 1990s

At the beginning of this article we pointed to the lack of interest felt by many archaeologists towards use-wear analysis. One cause for the limited state of development of use-wear analysis might be that many archaeologists have not grasped the potential of this field of study. Alternatively, perhaps we use-wear analysts have made very little effort to register our research within an anthropological framework. It is true that with use-wear analysis an attempt has been made to throw light on substantial historical problems, as we shall see below, but there is a noticeable lack of theoretical reflection about how knowledge of a tool can contribute important information concerning the person who used it.

Nevertheless, taken as a whole, use-wear studies have progressed considerably, and the framework of interpretation has gradually widened although, in general, this interpretative framework remains more implicit than explicit. Early studies arose within the context of typological conceptions of tools, so research was principally aimed at identifying use of different tool types (see Jardón 1990; Juel Jensen 1988). The scope of these studies later expanded and is reflected in themes of successive use-wear trace conferences: the 1985 conference in Antibes (France) set out to combine tool function with tool manufacture under the heading *Technologie et Fonction* (Beyries 1988); the 1987 Uppsala conference was devoted to studying the relevance of understanding tool use in order to tackle the question of how settlements functioned (Gräslund *et al.* 1990); under the rubric *Traces et Fonction, les Gestes Retrouvés*, the Liège conference in 1989 introduced concepts of technical motion and technical process, integrating the tool into a work process (Anderson *et al.* 1993). The 1990 Jalès conference on palaeoagriculture (Anderson 1992b) represented a certain break, by focusing its attention on an historical issue drawing together, alongside use-wear analysts, specialists from different fields (archaeobotanists, ethnographers *etc.*). It is significant that, since Liège, no general conference has been held on the subject of use-wear traces. Contributions of use-wear analysts have been presented to conferences dealing with archaeological problems of a diverse nature. In our opinion, it would be mistaken to consider the absence of specific meetings in the 1990s concerning use-wear traces as a sign of decline. On the contrary, this emphasises, together with the fact that communication concerning functional analysis has become normal procedure at all types of conferences, that the cohesive force of functional analysts united in their execution of a pioneering technique, has given way to analysts becoming integrated in forums where archaeological issues, in which they are deeply involved, are under discussion. Analytical techniques are becoming accepted and it is not so much the technique now that brings us together as much as historical problems, for which functional analysis has a new perspective to offer which complements other approaches.

Use-wear studies, like any other technique of analysis, present us with static data (tool motion and worked material). These data acquire a dynamic character when they are contextualised, that is to say, placed in relation to the other data gathered in an archaeological survey (*i.e.,* Beugnier 1997) and integrated into a theoretical framework. In the search for such a theoretical framework, a key aspect has been ethnographic sources. During the 1980s, ethnographic sources were mainly used to define experimental programmes which were to serve as an interpretative base (Knutsson 1988; van Gijn 1989). Since the 1990s, ethnography has been more widely addressed and such information has been used to interpret results (Gassin 1996; Ibáñez & González 1996). An example of such new links between the functional studies of archaeological tools and ethnographic information is the project carried out by the CSIC team from Barcelona in Tierra del Fuego. They have tried to undertake a comparative study using ethnographic information on the Ona and the Yámana as well as data gathered from excavation work in villages from these communities (Clemente Conte 1997; Estévez & Vila 1996). While there are extensive accounts of these cultures and their way of life, the amount of technological information on lithic tools is very slight since this data appeared to be of only limited relevance to 18th century ethnographers. This problem is shared by all ethnographic sources which tend to be very sparing when it comes to recording details about working techniques. Functional analysts are attempting to solve this situation by resorting to the direct study of work processes. Such an approach can be found in ethnographic research on how hide is worked in the Canary Islands (Rodriguez in press) and among indigenous communities in Canada (Beyries 1997). Along the same lines, we have recently begun a project investigating hide working and agricultural techniques associated with the cultivation of *T. monoccocum*, in the Moroccan Jbala (Northern Morocco).

At a first level of analysis use-wear traces offer information about relationships between tool shape and use; recent studies confirm work done in the 1980s. During the Middle Palaeolithic different morphological types show a broad functional diversity, as has been observed in the case of Levallois points in the Near East or side-scrapers from different European sites (Lemorini 1997; Lhomme *et al.* 1998; Plisson & Beyries 1998). In the Upper Palaeolithic there is a closer relationship between shape and use; end-scrapers are associated with the scraping of skins, burins with work on bone material and projectiles with hunting. In addition to these retouched instruments there is a wide collection of non-retouched tools used for a rich variety of tasks (Ibáñez & González 1998; Jardón 1996; Kimball 1989).

The amalgamation of use-wear studies with those of knapping and raw materials enables identification of management strategies for lithic raw materials and blanks. These concepts (Binder & Perlès 1990; Binford 1979; Odell 1996; Perlès 1992) have been widely used in the 1990s by use-wear analysts (Beugnier 1997; Gassin 1996; Ibáñez & González 1996). As a rule, Middle Palaeolithic collections reveal expedient tool use strategies where local raw materials are used, knapped and then abandoned after use. Nevertheless, there are cases where the procedure is more complex. On the Combette site most lithic remains consist of side-scrapers knapped from allocthonous flint which were brought into the rock-shelter as finished tools (Texier *et al.* 1998); side-scrapers were used for working hides, wood and animal butchery (Lemorini 1997). The existence of tools for expedient use along with others for curated use has already been extensively noted in the Upper Palaeolithic; curated tools show more evidence of retouching, hafting and greater use, although they were also used for the same activities and on the same worked materials as were expedient tools (Audouze *et al.* 1981; Ibáñez & González 1996; van Noten *et al.* 1978). During the Neolithic, tool management appears more complex. For example, in the burnt Pre-Pottery Neolithic A (PPNA) house from Tell Mureybit (Syria) nearly 10% of retouched tools reveal work carried out on more than one different material (Cauvin *et al.* in press). In this very house a tool store was found and throughout the site sequence a growing tendency is observed towards a greater recycling of tools, an aspect which has also been noted at the Pre-Pottery Neolithic B (PPNB) site at Tell Halula (Molist *et al.* in press). This suggests that the tools were stored after use with a view to being used again. However, from the Late Neolithic onwards a change can be seen in the region towards more expedient strategies; knapping of bipolar blade cores is replaced by unipolar knapping of flake cores. This tendency is also reflected in use as less intense applications appear and the degree of tool recycling drops (Ibáñez & González in press). Greater complexity in tool management during the Neolithic is also to be seen in the Chasséen site at l'Eglise (Gassin 1996). Here, flakes, percussion-knapped blades and heat-treated, pressure-flaked blades show signs of being used differently; unretouched flakes were used for scraping and thrown away after being used once; percussion-knapped blades were used for many different activities, while pressure-knapped blades were selected for their sharp edges, suitable for cutting, and were frequently hafted and re-used.

In the 1990s, functional concepts coming from the field of anthropology (Lemonnier 1976, 1983; Leroi-Gourhan 1964) were integrated into use-wear studies, such as technical process (*processus technique*) and technical sequences (*chaâines opèratoires*) which have filtered down to use-wear analysts through their use by specialists in lithic manufacturing techniques (Geneste 1991; Karlin *et al.* 1991; Pelegrin *et al.* 1988).

When integrated within this theoretical framework, use-wear analysis can also contribute data about work processes in which tools were involved. For example, on the Neolithic site of Maastricht-Klinkers (Schreurs 1992) wood was worked with unretouched, unhafted tools which only rarely show traces of repeated or intense use, from which it was interpreted that the wood was not worked intensely nor in a specialised way. Reconstruction of technical processes can be of a particularly detailed nature where manufactured products and manufactured remains are preserved, as happens with objects made of bone or stone. At the Neolithic site of Khirokitia, a comprehensive study of stone bowls which included debris left over from their manufacture and tools used to work the material enabled a detailed reconstruction of the technical process (Astruc in press) to be made. Recently, agricultural work processes have been particularly subjected to study; in this field we have mainly been able to identify harvesting techniques (Anderson 1992a; Juel Jensen 1994), cereal processing techniques (Anderson 1994a; Clemente & Gibaja 1998; Skakun 1994; Yerkes & Kardulias 1994) and tools used in working the land (Ibáñez *et al.* 1998; Skakun 1994).

As it allows identification of some of the technical activities in progress on a site, use-wear analysis has been used to define function of habitat sites (Binford 1980). The existence of various technical activities, and the fact that jobs are carried out requiring considerable technical effort and time, is associated with stable settlements and prolonged habitat (Jochim 1976). Such is the case of the Epipaleolithic settlement at Berniollo (Ibáñez & González 1996) where blade and bladelet manufacture is amply represented. Moreover, costly activities were also carried out, such as the softening of hides and creation of antler, bone and wood tools. Tools reveal intense use but at the same time deposits of unworked blades and retouched tools made out of better quality flint have been found. In contrast, sites where only one part of the technical system is represented are interpreted as specialised and of short duration as at the Solutrean site of Combe Saunière (Geneste & Plisson 1993) where fragmented Solutrean points appear although they were not manufactured *in situ*. Distal fragments entered the site in the body of the hunted animal, whilst proximal fragments are the result of broken points being thrown away. The site has been interpreted as a temporary settlement used during a hunting period as a resting place or a specialised area of activity.

Another line of research recently undertaken using use-wear analysis relates to social organisation of work. Gibaja's research on the use of tool offerings recovered from Middle Neolithic graves in Catalunya (Spain) has revealed a certain relationship between work represented by tools and age and sex of the person buried (Gibaja & Clemente 1996). Though cereal harvesting is represented in the graves of both men and women, men tend to be accompanied by polished axes, blade cores and projectile elements. Furthermore, children are found with tool offerings similar to those found in women's graves. A further source of investigation which will give access to information of a social nature is the study of differences in distribution of activities within domestic units and between different domestic units in the Neolithic; in the PPNA house at Tell Mureybit (Syria) (Cauvin *et al.* in press) tools revealing a wide range of activities appear (harvesting, projectiles, wood, hides, stone and bone working). Astruc's work on the later site of Kirokoitia (Cyprus) shows the existence of differential distributions of activities between different houses related to the making of stone bowls (Astruc 1999). The complementary nature in technique between different domestic units has also been observed on some Final Neolithic sites in the Jura region, France (Beugnier 1997). As this work progresses it will be possible to recover data on the beginnings of craft specialisation.

13.7 Conclusion

In the 1980s use-wear analysis went through a preliminary stage of excessive optimism regarding both reliability of the method of analysis and interpretative possibilities of information on tool function. Critiques of the method, which began to appear in the second half of the decade, stimulated a process of development towards more realistic, comprehensive positions which continued during the 1990s. This progression from optimism to realism, including a period of crisis, probably goes hand in hand with the introduction of any new technique of analysis.

In recent years an attempt has been made to widen the supporting base for functional interpretation. Inferences are based on the integration of different types of traces and also, when possible, of the remains of worked material and mastic used in hafting. In combination with this, the study of tool characteristics and the archaeological context is essential too. Analyses today form part of interdisciplinary research, complementing information contributed by analyses from other specialised areas (lithic technology, location of raw material supply sources, archaeobotany ...). It has also become usual to carry out experimental programmes adapted to increasingly more specific problems where archaeological reality is reproduced in greater detail. In the interpretative sense use-wear studies have integrated diverse concepts concerning the relationship between implement and economic and social structure of prehistoric groups which has allowed the tackling of substantial questions in archaeological research.

It would appear, then, that use-wear studies have begun to enter a phase of maturity. The method of functional analysis is of an eminently empirical nature and its reliability is founded on the consensus of researchers, that is to say, on a great number of independent and coincidental experiences. Furthermore, trace quantification studies have confirmed, in general terms, the observations of analysts. Nevertheless, this aspect of the discipline, based as it is on the researcher's experience rather than on a set of quantified, explicit criteria, makes it difficult to progress. Another aspect that still requires work is on the processes of use-wear polish formation. These limitations may partially justify the limited participation of use-wear analyses in archaeological research. However, it is our belief also, that many archaeologists still remember the polemics that took place in the 1980s but which have been left behind by present day research.

References

ANDERSON, P. C. 1992a. 'Experimental cultivation, harvest and threshing of wild cereals and their relevance for interpreting the use of Epipalaeolithic and Neolithic artefacts.' *In* Anderson (1992b), pp. 179–209.

ANDERSON, P. C. (ed.) 1992b. *Préhistoire de l'Agriculture*. Monographie du CRA no. 6. CNRS, Paris.

ANDERSON, P. C. 1994a. 'Interpreting traces of Near Eastern neolithic craft activities: an ancestor of the threshing sledge for processing domestic crops?' *Helinium* 34(2): 306–21.

ANDERSON, P. C. 1994b. 'Reflections on the significance of two PPN typological classes in light of experimentation and use-wear analysis: flint 'sickles' and obsidian 'Çayönü tools'.' *In* H. G. Gebel & S. K. Kozlowski (eds.), *Neolithic Chipped Stone Industries of the Fertile Crescent: Proceedings of the First Workshop on PPN Chipped Lithic Industries, Free University of Berlin, 29th March-2nd April, 1993*, Studies in Early Near Eastern Production, Subsistence, and Environment 1, pp. 101–13. Ex Oriente Free University of Berlin, Berlin.

ANDERSON, P. C., S. BEYRIES, M. OTTE & H. PLISSON (eds.) 1993. *Traces et fonction: les gestes retrouvés. Actes du colloque international de Liège 8–9–10 décembre 1990.* ERAUL 50. Université de Liège, Liège.

ANDERSON-GERFAUD, P. 1981. *Contribution méthodologique à l'analyse des microtraces d'utilisation sur les outils préhistoriques.* Ph.D. thesis, Université de Bordeaux I, Talance.

ASTRUC, L. 1999. *Analyse fonctionnelle et spatiale de l'industrie lithique de Khirokitia (Néolithique précéramique, Chypre).* Ph.D. thesis, Université de Paris X.

ASTRUC, L. in press. *Lithic tools involved in the manufacture of stone ornaments and utilitarian products in Khirokitia (Aceramic Neolithic, Cyprus).* Workshop on PPN lithic tools. Venice, Nov. 1998.

AUDOUZE, F., D. CAHEN, L. H. KEELEY & B. SCHMIDER 1981. 'Le site magdalénien du Buisson Campin à Verberie (Oise).' *Gallia Préhistoire* 24(1): 99–143.

BAMFORTH, D. B. 1988. 'Investigating microwear polishes with blind test: The Institute results in context.' *Journal of Archaeological Science* 15: 11–23.

BEUGNIER, V. 1997. *L'usage du silex dans l'acquisition et le traitement des matières animales dans le Néolithique de Chalain et Clairvaux.* Ph.D. thesis, Université de Paris X.

BEYRIES, S. (ed.) 1988. *Industries lithiques. Tracéologie et Technologie.* British Archaeological Reports International Series 411, Oxford.

BEYRIES, S. 1997. 'Ethnoarchéologie: un mode d'expérimentation.' *Préhistoire Anthropologie Méditerranées* 6: 185–96.

BIETTI, A. 1996. 'Image processing in microwear studies of flint artefacts.' *Archeologia e Calcolatori* 7: 387–96.

BINDER, D. & S. C. PERLÈS 1990. 'Stratégies de gestion des outillages lithiques au Néolithique.' *Paléo* 2: 257–84.

BINFORD, L. R. 1979. 'Organization and formation processes: Looking at curated technologies.' *Journal of Anthropological Research* 35: 255–73.

BINFORD, L. R. 1980. 'Willow smoke and dog's tails: hunter-gatherer systems and archaeological site formation.' *American Antiquity* 45: 4–22.

CAUVIN, M. C., F. ABBES, J. E. GONZÁLEZ & J. J. IBÁÑEZ in press. *L'outillage lithique de la structure XLVII de Mureybet (9.200 cal. av JC) en Syrie du nord.* Workshop on PPN lithic industries, Venice, November 1998.

CHRISTENSEN, M., P. WALTER & M. MENU 1991. 'Physico-chimie en tracéologie. Le cas des couteaux égyptiens.' *Nuclear Instruments and Methods in Physic Research* B64: 488–93.

CLEMENTE, I. & J. F. GIBAJA 1998. 'Working processes on cereals: an approach through microwear analysis.' *Journal of Archaeological Science* 25(5): 457–64.

CLEMENTE CONTE, I. 1997. *Los instrumentos líticos de Túnel VII: una aproximación etnoarqueológica.* Treballs d'Etnoarqueología 2. Consejo Superior de Investigaciones Científicas, Madrid.

ESTÉVEZ, J. & A. VILA (eds.) 1996. *Encuentros en los concheros fueguinos.* Treballs d'Etnoarqueología 1. Universidad Autónoma de Barcelona–Consejo Superior de Investigaciones Científicas, Bellaterra, Barcelona.

GASSIN, B. 1996. *Evolution socio-économique dans le Chasséen de la grotte de l'Eglise supérieure (Var).* Monographies du CRA 17. CNRS, Paris.

GENESTE, J. M. 1991. 'Systèmes techniques de production lithique: variations techno-économiques dans les processus de realisation des outillages paléolithiques.' *Téchniques et Culture* 17–18: 1–35.

GENESTE, J. M. & H. PLISSON 1993. 'Hunting technologies and human behaviour: Lithic analysis of solutrean shouldered points.' *In* H. Knecht, A. Pike-Tay & R. White (eds.), *Before Lascaux. The Complex Record of the Early Upper Paleolithic*, pp. 117–35. CRC Press, London.

GIBAJA, J. F. & I. CLEMENTE 1996. 'Análisis funcional del material lítico en las sepulturas de la Bòbila Madurell (Sant Quirze del Vallès, Barcelona).' In *1er Congrès del Neolitic a la Península Ibèrica*, pp. 183–9. Gavà-Bellaterra. Rubricatum 1(1).

GONZÁLEZ, J. E. & J. J. IBÁÑEZ 1994. *Metodología de análisis funcional de instrumentos tallados en sílex.* Cuadernos de Arqueología de Deusto 14. Universidad de Deusto, Deusto.

GONZÁLEZ, J. E. & J. J. IBÁÑEZ in press. 'The quantification of microwear polish using image analysis.' *Journal of Archaeological Science* .

González, J. E., J. J. Ibáñez, L. Peña, B. Gavilan & J. C. Vera 1994. 'Cereal harvesting during the Neolithic of the Murcielagos site in Zuheros (Cordoba, Spain).' *Helinium* 34(2): 322–41.

Grace, R. 1989. *Interpreting the Function of Stone Tools.* British Archaeological Reports International Series 474, Oxford.

Grace, R. 1990. 'The limitations and applications of use-wear analysis.' *In* Gräslund *et al.* (1990), pp. 9–14.

Grace, R., I. D. G. Graham & M. H. Newcomer 1985. 'The quantification of microwear polishes.' *World Archaeology* 17(1): 112–20.

Gräslund, B., H. Knutsson, K. Knutsson & J. Taffinder (eds.) 1990. *The Interpretative Possibilities of Microwear Studies. Proceedings of the International Conference on lithic use-wear analysis, 15th–17th February 1989 in Uppsala, Sweden.* Aun 14. Societas Archaeologica Upsaliensis, Uppsala.

Hurcombe, L. 1988. 'Some criticisms and suggestions in response to Newcomer *et al.* (1986).' *Journal of Archaeological Science* 15: 1–10.

Hurcombe, L. 1997. 'The contribution of obsidian use-wear analysis to understanding the formation and alteration of wear.' *In* Ramos-Millán & Bustillo (1997), pp. 487–98.

Ibáñez, J. J. & J. E. González 1996. *From tools use to site function. Use-wear analysis in some Upper Palaeolithic sites in the Basque country.* British Archaeological Reports International Series 658, Oxford.

Ibáñez, J. J. & J. E. González 1998. 'The production and use of lithic tools at the end of the Upper Palaeolithic in the Basque country.' *In* S. Milliken (ed.), *The Organization of Lithic Technology in Late Glacial and Early Postglacial Europe*, pp. 17–37. British Archaeological Reports International Series 700, Oxford.

Ibáñez, J. J. & J. E. González in press. 'La función de los útiles en el yacimiento de Tell Halula (8,700–7,500 BP), valle del Eufrates (Siria).' *In* M. Molist (ed.), *El yacimiento neolótico de Tell Halula (Valle del Eufrates, Siria).* Universidad Autónoma de Barcelona, Barcelona.

Ibáñez, J. J., J. E. González, A. Palomo & A. Ferrer 1998. 'Pre-Pottery Neolithic A and Pre-Pottery Neolithic B lithic agricultural tools in the Middle Euphrates: the sites of Tell Mureybit and Tell Halula (Syria).' *In* A. B. Damania, J. Valkoun, G. Willcox & C. O. Qualset (eds.), *The Origins of Agriculture and Crop Domestication*, pp. 132–44. ICARDA, Aleppo.

Jardón, P. 1990. 'La metodología de análisis traceológico y su aplicación a conjuntos líticos prehistóricos.' *Saguntum* 23: 9–37.

Jardón, P. 1996. *La funcionalidad de las industrias líticas del Paleolítico Superior. Tipología, tecnología y función de los raspadores.* Ph.D. thesis, Universitat de Valencia.

Jochim, E. 1976. *Hunter-Gatherer Subsistence and Settlement. A Predictive Model.* Academic Press, New York.

Juel Jensen, H. 1988. 'Functional analysis of prehistoric flint tools by high-power microscopy: a review of West European research.' *Journal of World Prehistory* 2: 53–88.

Juel Jensen, H. 1994. *Flint tools and plant working. Hidden traces of Stone Age technology: a use wear study of some Danish Mesolithic and TRB implements.* Aarhus University Press, Aarhus.

Karlin, C., P. Bodu & J. Pelegrin 1991. 'Processus techniques et chaînes opératoires. Comment les prehistoriens s'approprient un concept élaboré par les ethnologues.' *In* H. Balfet (ed.), *Observer l'action technique. Des chaînes opératoires, pour quoi faire?*, pp. 101–17. CNRS, Paris.

Keeley, L. H. 1980. *Experimental Determination of Stone Tool Uses. A Microwear Analysis.* Prehistory, Archaeology and Ecology Series. The University of Chicago Press, Chicago.

Keeley, L. H. & M. H. Newcomer 1977. 'Microwear analysis of experimental flint tools: a test case.' *Journal of Archaeological Science* 4: 29–62.

Kimball, L. R. 1989. *Planning and Functional Variability in the Upper Palaeolithic: Microwear Analysis of Upper Perigordian Tools from Le Flageolet I (Dordogne).* Ph.D. thesis, Northwestern University.

Kimball, L. R., J. F. Kimball & P. E. Allen 1995. 'Microwear polishes as viewed through the atomic force microscope.' *Lithic Technology* 20(1): 6–28.

Knutsson, K. 1988. *Making and Using Stone Tools. The analysis of the lithic assemblages from the Middle Neolithic sites with flint in Västerbotten, northern Sweden.* Aun 11. Societas Archaeologica Upsaliensis, Uppsala.

Lemonnier, P. 1976. 'La description des chaînes opératoires: contribution à l'analyse des systèmes techniques.' *Techniques et Culture* 1: 100–51.

Lemonnier, P. 1983. 'L'étude des systèmes techniques, une urgence en technologie culturelle.' *Techniques et Culture* 2: 11–26.

LEMORINI, C. 1997. *L'organisation du geste des néanderthaliens. Analyse fonctionnelle des industries lithiques de Grotta Breuil (Latium, Italie) et de La Combette (Boonnieux, Vaucluse, France).* Ph.D. thesis, University of Leiden.

LEROI-GOURHAN, A. 1964. *Le Geste et la Parole 1. Techniques et langage.* Albin Michel, Paris.

LEVI-SALA, I. 1995. *A Study of Microscopic Polish on Flint Implements.* British Archaeological Reports International Series 629, Oxford.

LHOMME, V., C. BEMILLI, S. BEYRIES, M. CHRISTENSEN & N. CONNET 1998. 'Soucy 1 (Yonne): interprétations et réflexions sur un site du Pléistocène moyen en contexte alluvial.' In *Economie Préhistorique: les comportements de subsistence au Paléolithique. XVIIIe Rencontres Internationales d'Archéologie et d'Histoire d'Antibes*, pp. 260–72. Association pour la Promotion et la Diffusion des Connaissances Archéologiques, Sophia Antipolis.

MANSUR, M. E. 1997. 'Functional analysis of polished stone-tools: considerations about the nature of polishing.' *In* Ramos-Millán & Bustillo (1997), pp. 465–88.

MANSUR-FRANCHOMME, M. E. 1983. *Traces d'utilisation et technologie lithique: exemples de la Patagonie.* Ph.D. thesis, Université de Bordeaux I.

MOLIST, M., A. FERRER, J. E. GONZÁLEZ, T. PALOMO & J. J. IBÁÑEZ in press. *L'industrie lithique du néolithique des IXème et VIIIème millénaires BP à Tell Halula (Vallée de l'Euphrate). Etat de la recherche.* Workshop on PPN lithic industries, Venice, Nov. 1998.

MOSS, E. H. 1987. 'A review of "Investigating microwear polishes with blind tests".' *Journal of Archaeological Science* 14: 473–481.

NEWCOMER, M. H., R. GRACE & R. UNGER-HAMILTON 1986. 'Investigating microwear polishes with blind tests.' *Journal of Archaeological Science* 13: 203–18.

ODELL, G. H. (ed.) 1996. *Stone Tools: Theoretical Insights into Human Prehistory.* Plenum Press, New York.

PAWLIK, A. 1995. *Die mikroskopische Analyse von Steingeräten: Experimente, Auswertungsmethoden, Artefaktanalyse.* Urgeschichtliche Materialhefte 10. Verlag Archaeologica Venatoria, Tübingen.

PELEGRIN, J., C. KARLIN & P. BODU 1988. 'Chaînes opératoires: un outil pour le préhistorien.' *In* J. Tixier (ed.), *Technologie Préhistorique*, Notes et Monographies Techniques No 25, pp. 55–62. CNRS, Paris.

PERLÈS, S. C. 1992. 'In search of lithic strategies. A cognitive approach to prehistoric chipped stone assemblages.' *In* J. C. Gardin & C. Peebles (eds.), *Representation in Archaeology*, pp. 223–47. Indiana University Press, Bloomington.

PLISSON, H. & S. BEYRIES 1998. 'Pointes ou outils triangulaires? Données fonctionnelles dans le Moustérien levantin. Commentaires de J. Shea, A. Marks and J. M. Geneste et réponse des auteurs.' *Paléorient* 34(1): 5–24.

RAMOS-MILLÁN, A. & M. A. BUSTILLO (eds.) 1997. *Siliceous rocks and culture.* Universidad de Granada, Granada.

RODRIGUEZ, A. in press. 'The reconstruction of ancient leather technology or how to mix methodological approaches. An example from Canary Island prehistory.' *In* L. Owen (ed.), *Ethnoarchaeology and Ethnoanalogy.* University of Tübingen, Tübingen.

SCHREURS, J. 1992. 'The Michelsberg site Maastricht-Klinkers: a functional interpretation.' *Analecta Praehistorica Leidensia* 25: 129–71.

SEMENOV, S. A. 1964. *Prehistoric technology. An experimental study of the oldest tools and artifacts from traces of manufacture and wear.* Cory, Adams and Mackay, London.

SKAKUN, N. 1994. 'Agricultural implements and the problem of spreading of agriculture in southwestern europe.' *Helinium* 34(2): 294–305.

TEXIER, P. J., J. P. BRUGEL, C. LEMORINI & L. WILSON 1998. 'Fonction d'un site du Paléolithique Moyen en marge d'un territoire: l'Abri de la Combette (Bonnenieux, Vaucluse).' *In* J. P. Brugel, L. Meignen & M. Patou-Mathis (eds.), *Economie préhistorique: les comportements de subsistence au Paléolithique*, pp. 325–48. APDCA, Sophia Antipolis.

TRINGHAM, R., G. COOPER, G. H. ODELL, B. VOYTEC & A. WHITMAN 1974. 'Experimentation in the formation of edge-damage: a new approach to lithic analysis.' *Journal of Field Archaeology* 1: 171–96.

UNGER-HAMILTON, R. 1989. 'Analyse expérimentale des microtraces d'usure: quelques controverses actuelles.' *L'Anthropologie* 93(3): 659–72.

VAN DEN DRIES, M. & A. VAN GIJN 1997. 'The representativity of experimental usewear traces.' *In* Ramos-Millán & Bustillo (1997), pp. 499–514.

VAN GIJN, A. L. 1989. *The Wear and Tear of Flint. Principles of functional analysis applied to Dutch Neolithic assemblages.* Analecta Praehistorica Leidensia 22. University of Leiden, Leiden.

VAN NOTEN, F., D. CAHEN, L. H. KEELEY & J. MOEYERSONS 1978. *Les chasseurs de Meer.* Disertationes Archaeologicae Gandenses XVIII. De Tempel, Brussels.

VAUGHAN, P. C. 1985. *Use-wear Analysis of Flaked Stone Tools.* University of Arizona Press, Tucson.

VILA, A. & F. GALLART 1993. 'Caracterización de los micropulidos de uso: ejemplo de aplicación del análisis de imágenes digitalizadas.' *In* Anderson *et al.* (1993), pp. 459–66.

YERKES, R. W. & P. N. KARDULIAS 1994. 'Microwear analysis of threshing sledge flints from Cyprus and Greece: implications for the study of ancient agriculture.' *Helinium* 34(2): 281–93.

Juan José Ibáñez
Dpto. de Ciencias Históricas
Universidad de Cantabria
Spain

Jesús Emilio González
Dpto. de Antropología Social y Prehistoria
Universidad Autónoma de Barcelona
Spain

14
A Method for Quantifying Size and Shape Dimensions of North American Hafted Bifaces

Andrew P. Bradbury[1] and Philip J. Carr[2]

Lexington, U.S.A.[1]*; Dept. of Sociology & Anthropology, University of South Alabama*[2]

Abstract

A method is developed that allows for recording non-traditional size variables for hafted bifaces using computer technology. Application of the method is demonstrated using a sample of hafted bifaces with bifurcated bases which were recovered from several sites excavated as part of the Tellico Archaeological Project. These quantitative measures are used to define a series of shape related variables. This combination of size and shape data are used in the examination of both typological and morphological aspects of the sample. Our analysis suggests that the traditional subdivisions of bifurcate-base hafted biface types are not as distinctive as previously suggested. The specimens examined appear to lie along a continuum of morphological change. An examination of blade symmetry also revealed differences in the types which suggests some were used more often as knives than projectile points. This investigation of morphology, technology, and resharpening would have been difficult, if not impossible, without the use of computer technology.

14.1 Introduction

The definition of cultural historical types has a long history in Americanist archaeology. Throughout the early part of the century, stone tools took a back seat to ceramics in defining cultures (Johnson 1993; Trigger 1989). This is especially true of the Southeast where it was not until the 1960s, with the work of Coe (1964), that stone tools were used for such purposes to any great extent. At this time, an appreciation of the manner that hafted bifaces functioned in the cultural system was not well developed. Soon after, it was recognised that such implements have a life cycle and that successive stages of resharpening can modify overall tool morphology (*e.g.*, Goodyear 1974, see also Flenniken 1985; Flenniken & Raymond 1986; Hoffman 1985; Titmus & Woods 1986; Towner & Warburton 1990). The resulting definition of type clusters was one solution to this recognition of the variability contained within a single hafted-biface type (*e.g.*, Faulkner & McCollough 1973; Justice 1987).

The use of hafted bifaces for cultural-historical purposes has been, and continues to be, an important endeavour in lithic analysis (*e.g.*, Gunn & Prewitt 1975; Hall 1992; Henton & Durand 1991; Thomas 1981). These space-time bound types often form the foundation, or provide a temporal context, for subsequent analyses. As such, once defined and used, these types become entrenched in the literature and are only rarely re-examined. This is unfortunate for at least two reasons. First, new data are more often placed into the original types without consideration of how that data may provide additional information on those types. Second, new technologies and methods of analysis may provide different insights into the original types or provide a classification system that better addresses the purpose for which it was created. Additionally, many of the common types in use today were developed as a means to answer cultural historical questions and may be ill suited for addressing technological, functional, or other areas of research.

Despite the importance of typology, it has not received the attention it deserves in contemporary archaeological literature of method and theory. Dunnell, however, has consistently examined archaeological classification and provided insight into the ramification of different typological approaches (*e.g.*, Dunnell 1971, 1986; also see Thoms 1993; Tomka & Prewitt 1993 for reference to lithic analysis). Critical issues in Americanist typology remain and include: unclear type definitions (Beck & Jones 1989; Sullivan & Rozen 1985), the use of a type defined for one purpose for another purpose (Dunnell 1986), analyst bias in artefact classification (Beck & Jones 1989; Fish 1978; Odell 1996; Whittaker *et al.* 1998), and the failure to re-examine a type once defined even after new data become available.

With a consideration of the above issues, a method is introduced that uses computer technology to examine several traditional hafted bifaces types from the

Southeastern United States. Despite some criticisms of the use of computers in building artefact typologies (*e.g.,* Adams 1988), it is argued here that the use of computer hardware/software and statistics can overcome a number of substantial obstacles to constructing meaningful types. The method presented provides several advantages, including: defining variables in an objective manner, making accurate measurements, and allowing replicability. While the method is demonstrated using hafted bifaces recovered from the Southeastern United States, it is applicable to collections of materials from any part of the world and to other artefact classes.

14.2 Method

There are an infinite number of attributes that could be used to examine variation within any sample of hafted bifaces. For the present study, we needed a way to quantitatively define both size and shape variation. In addition, the method would have to allow for accurate measurements, operationalisation of the variables, and replicability. While a number of authors have used metric measurements in defining types (*e.g.,* Gunn & Prewitt 1975; Hall 1992; Henton & Durand 1991; Hoffman 1985; Thomas 1981), a common problem is the subjective determination of measurement points. Any method that uses linear or angular measurements must attempt to lessen this subjectivity and increase replicability.

In an attempt to address these issues, Hoffman (1985) used a polar grid coordinate system for recording linear and angular measurements in his examination of projectile-point typology through factor and canonical correlation analysis. In his study, aimed at examining the effects of resharpening on hafted biface typology, specimens were placed on a polar coordinate grid and distances from the centre of the grid to the edge of the specimen were taken at 20 degree increments. While the polar grid allowed for all measurements to be operationalised, one still had to orient each specimen on the polar grid which added a degree of subjectivity.

An alternate means of obtaining measurements that adds to the precision of determining reference points is to use computer technology. Such a method was employed by Henton & Durand (1991) to capture digital images of hafted bifaces to record a number of traditional and non-traditional attributes. Based on these attributes, a computer program was written and used to place hafted bifaces into previously established cultural-historical types and a means was developed to directly compare artefact shape. The authors note that the orientation of each specimen, especially with asymmetrical specimens, could be subjective and was a shortcoming of the method (Henton & Durand 1991, 78). Additionally, the equipment used (video camera, analogue monitor, video analogue to digital converter board) is not readily available to many lithic analysts. While the method does produce accurate results, application to other samples is hampered by the somewhat specialised equipment needs.

An alternate method using computer technology was developed by Hall (1992; also see Gunn & Prewitt 1975 for a similar non-computer oriented method) as a way to remove the subjective element in hafted biface classification. Photocopies of hafted bifaces collected during the Columbia Reservoir project in middle Tennessee were digitised using AutoCAD (Hall 1992, 86–87) to aid in establishing reference points and recording various linear and angular attributes. These attributes were then used to quantitatively define a number of hafted biface 'types' for further analysis. A high correct rate of classification (92.9% in the cross validation) was achieved in separating five different Middle to Late Archaic hafted biface types (Hall 1992, 97). The advantage to using AutoCAD was that the measurement precision was greatly enhanced in relation to traditional methods (Hall 1992, 86).

As a means to building on previous approaches, the methods employed by Hoffman (1985) and Hall (1992) are combined here in order to define and accurately record a number of size variables of a hafted biface sample. These variables are then used to compute additional size related variables and define a series of shape variables. Specifically, in this study, AutoCAD running on a desktop PC was employed to digitise the hafted bifaces. The use of AutoCAD allows the examination of a number of variables that would be difficult, or impossible to record accurately using conventional methods. Reference lines and points can be accurately defined that facilitate the definition and replication of all measurements taken. While there remains a degree of subjectivity in determining reference lines, this was reduced considerably in the current study by greatly enlarging the haft area so that it generally filled the entire 17 inch monitor screen during the placement of all reference lines. All measurements were obtained using standard AutoCAD commands.

In this study, a template file was created that had several layers and the polar coordinate grid already established. Different data (*i.e.,* grid, reference lines, specimen, *etc.*) were stored on separate layers. Each layer was also assigned a different colour to aid in the data recording process. The polar grid was created as a series of lines at 20° angles from each other, starting with a horizontal line (Fig. 14.1). A circle was placed in the centre of the grid. Two lines were then placed across the circle, one horizontal, the other vertical. These lines intersected at the centre of the polar grid and served as the reference point for the grid. The grid lines were labelled so that each had a unique designation for each side of the specimen. In all cases, lines A through E represent measurements of the blade while lines F through I represent measurements of the haft area. Eighteen variables can be recorded for each specimen using the polar grid. The recorded measurements represent the specimen in two dimensions, therefore, it is not possible to record thickness using this technique. If such a measurement is deemed important for a particular question being asked of the data (*e.g.,* Hughes 1998), thickness could be included by use of callipers. As thickness is often highly correlated with width, and a number of width measurements are being recorded, the exclusion of thickness should not have an adverse effect on the analysis presented here.

Once the polar grid and reference point was established, a photocopy of the specimen was placed on the digitising tablet. The photocopy was obtained from plates in a series of reports (Chapman 1975, 1977, 1978, 1979). While photocopies of report plates were used here, photocopies of actual specimens could also serve the same purpose (see Hall 1992). The tablet was calibrated to the scale in the photograph and used to digitise the specimen with the outline represented as a polyline. The inclusion of the scale allowed for compensation of any distortion that might have been introduced during the photographing and photocopying of the specimen. The area of the specimen corresponding to the haft was enlarged and three reference lines added (Fig. 14.2). Line 1 was established at the top of the haft area, at the intersection of the blade and the haft. The specimen was then rotated so that this reference line was horizontal. Line 2 was parallel to Line 1. This line started at the widest part of the haft on one side of the specimen to the widest part of the haft on the other side. Line 3 was then drawn perpendicular to these lines and it ran from the centre of Line 2 to the point where it intersected Line 1. The intersection of Lines 1 and 3 represents the centre point at the top of the haft and was used as the reference point for the specimen. The top of the haft was used as the reference point as it is the portion that is least likely to be affected by resharpening. As the reference lines were added while the specimen was greatly enlarged, precise determinations of the haft/blade juncture were possible, even on asymmetrical specimens. The specimen was overlaid on the polar grid so that its reference lines exactly overlaid the reference lines for the centre of the grid. Each specimen was assigned a unique artefact number which was included in the drawing file along with information concerning 'type name', site, and reference. All type names used here were those assigned in the original reports. Each specimen was saved as a separate drawing file (generally 9–58k in size). Representative examples of these files are shown in Figure 14.3.

The recorded measurements were taken from the centre of the polar grid to the point where the edge of the specimen intersected the grid line. In cases where a deep serration occurred at the measurement point, a line was draw from the tip of the two adjacent serrations and the measurement was made to this line. In addition to the polar grid measurements, two other measurements were added. Maximum blade extension was defined as the most distant point of the blade (the hafted biface tip) from the centre of the polar grid. Minimum basal depth was defined as the shallowest point on the base (within the bifurcation) from the centre of the polar grid. A few specimens that were used in the analysis were fragmentary, usually missing a part of the tip or base. In these cases, measurements corresponding to the broken portion were not recorded.

Given that there is the potential for some error as a result of the digitising process, it was necessary to determine the level of precision that was justifiable. To accomplish this, a random sample of ten specimens was re-digitised two additional times and saved as separate drawing files. Measurements were then taken on all three files and the results compared. In all but one case, the same measurements were obtained at two decimal places and often to three decimal places. In the one exception, two of the measurements were correct at one decimal place. The remainder of the measurements on this specimen were correct at two or three decimal places. Based on these results, measurements were taken for all specimens to two decimal places (to the nearest 0.1mm).

The polar grid method described here allows for the direct measurement of 20 (18 polar grid plus maximum blade and minimum base) variables. Additional variables can also be computed from these original variables. For example, the distance between any two polar grid points can be calculated using basic trigonometry as the distances form two sides of a triangle and the angle between these lines is known (law of cosines). Employing the polar grid as discussed here, a total of 210 possible linear measurements can be defined for each specimen. Shape variables can also be calculated from these variables (see below). While we focus on one suite of measurements in this paper, a number of other linear and angular measurements are also possible. These could easily be recorded for artefacts other than hafted bifaces. Angular measurements, in particular, are difficult to record accurately and reliably with traditional methods. These measurements are easily and accurately recorded using standard AutoCAD commands.

14.3 Sample for analysis

In this analysis, we applied the polar grid method described above to a sample of bifurcate-base hafted bifaces. The 'types' under consideration consist of MacCorkle, St. Albans, LeCroy, Kanawha, and Stanly. Examination of types and varieties of bifurcate hafted bifaces has a long history in the Southeastern U.S. (*e.g.,* Kneberg 1956). This is despite the fact that cultural historians in the mid 20th century who were working in the region tended not to work with stone tools for chronology building (Johnson 1993). The

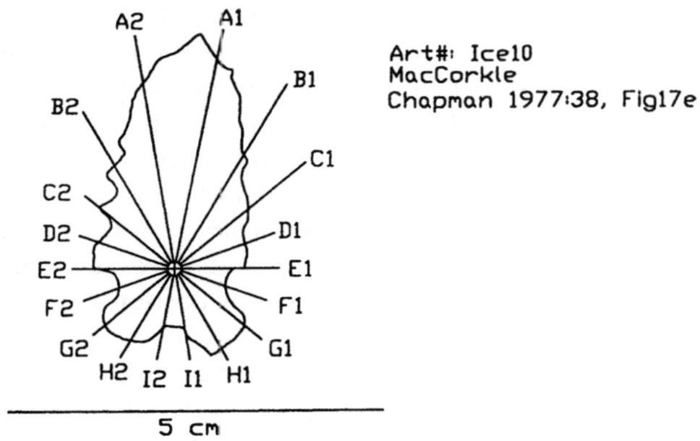

Figure 14.1: Polar grid with hafted biface overlaid.

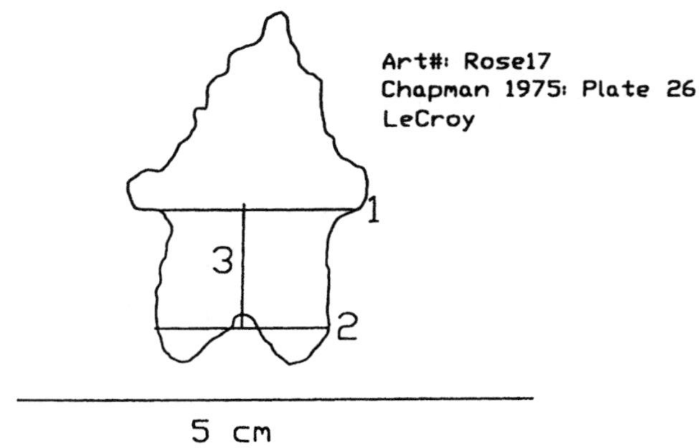

Figure 14.2: Reference lines for hafted biface.

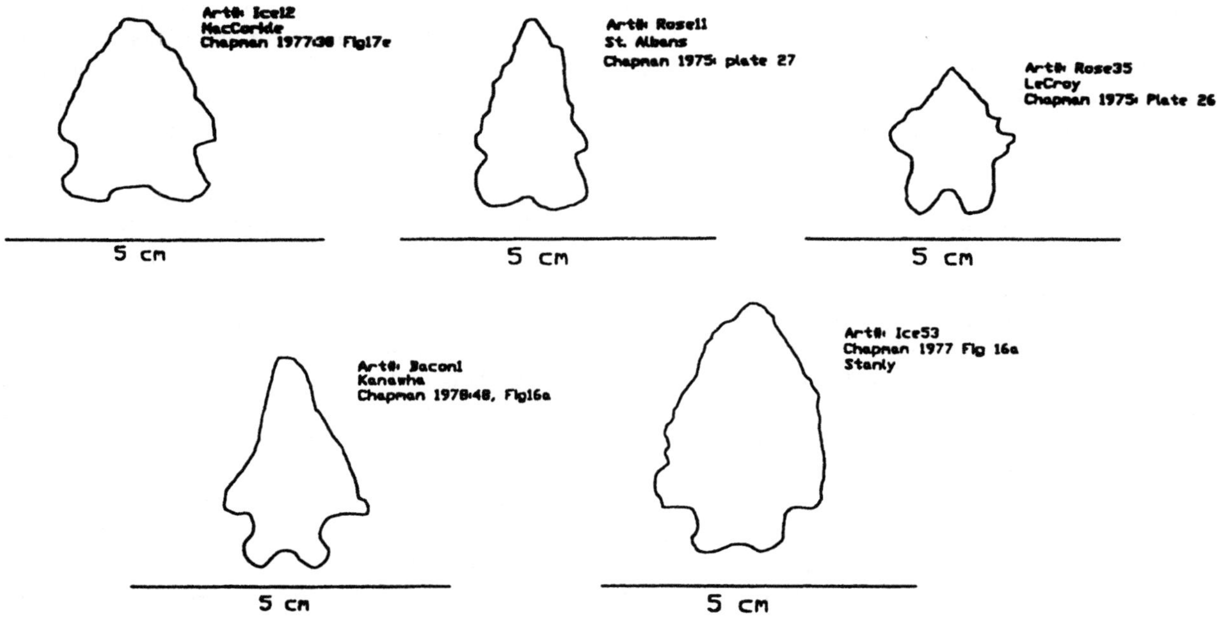

Figure 14.3: Representative sample of digitised hafted bifaces.

morphological types and definitions for the bifurcate hafted bifaces were revised and changed as new assemblages were recovered, especially those in stratigraphic context such as from the St. Albans site and the Tellico sites (Icehouse Bottom, Rose Island, Bacon Farm, Patrick, Calloway Island). Broyles (1966, 1971) developed a sequence of Early Archaic bifurcate forms from MacCorkle to St. Albans to LeCroy to Kanawha based on excavations at the St. Albans site. The Tellico data generally supports this bifurcate chronology developed by Broyles, although Kimball (1996, 159) has suggested that a MacCorkle phase with a distinct hafted-biface type is not evident in the Tellico assemblages. For those who recognise a MacCorkle phase and type (Broyles 1971; Chapman 1975), the date ranges assigned to it (9000–8500bp) do overlap with those for the St. Albans type (8900–8500bp). This suggests that these two types deserve further consideration.

In an overview of hafted bifaces from the Eastern U.S., Justice (1987) employs a type cluster approach to aid in grouping hafted bifaces from disparate geographic ranges with comparable type definitions. In his scheme, MacCorkle and St. Albans types are grouped within the Rice Lobed Cluster (Justice 1987, 86–90). The LeCroy (8500–7800 bp) and Kanawha types (8200–7800 bp) are part of the LeCroy Cluster. Radiocarbon dates indicate that these types overlap temporally. Though not part of either of these clusters, the Stanly type, as sole member of the Stanly Stemmed Cluster (7800–7500 bp), deserves consideration as a bifurcate which post dates the LeCroy and Kanawha types. Other researchers have suggested an ancestral relationship between Kanawha and Stanly forms (*e.g.*, Coe 1964, 35, 54).

All of the bifurcate forms discussed share the common characteristics of a basal notch (bifurcation) and are often resharpened in a generally similar manner, and exhibit comparable geographic ranges. These types also form something of a temporal sequence (MacCorkle, St. Albans, LeCroy, Kanawha, and Stanly from earliest to latest). The MacCorkle and St. Albans types share a deep basal notch or bifurcation with rounded basal ears which gives the types a 'lobed appearance' (Justice 1987). These types differ in that the St. Albans type exhibits slight side notching. The LeCroy type also has a relatively deep bifurcation, but the basal ears are more pointed than lobed. The Kanawha type differs in that it has a much more shallow bifurcation as does the Stanly type. The Stanly type is distinct in having a narrow squared stem. These types also show some internal and external variability in blade shape and basal grinding.

The distinctiveness of the MacCorkle, St. Albans, LeCroy, Kanawha, and Stanly types has been questioned. It is unclear whether some are contemporaneous based on associated radiocarbon dates. This raises the possibility that two 'types' coexisted. This suggests that one type may have served as a knife and the other as a projectile point. Investigation of the distinctiveness of these types and their function can provide important insights into the prehistory of the Southeastern United States.

The Early Archaic lithic assemblage recovered during the Tellico Archaeological Project in east Tennessee is arguably one of the best available for restudy (Fig. 14.4). Lithic artefacts were recovered from stratified deposits in substantial quantities and included a large sample of hafted bifaces. The sites investigated and the materials recovered during the Tellico Project are well documented in detailed published reports (*e.g.,* Chapman 1973, 1975, 1977, 1978, 1979). The five sites included in this restudy; Icehouse Bottom, Rose Island, Calloway Island, Patrick, and Bacon Farm sites all produced substantial Early Archaic assemblages from deeply buried, stratified contexts (Chapman 1973, 1975, 1977, 1978, 1979). These characteristics make these assemblages ideal for this particular restudy which involves an examination of morphological typology, technology, and use. In addition, Knox chert was the predominate raw material type used for production of the bifurcate hafted bifaces recovered during the Tellico Project. This reduces variability in artefact shape and size caused by raw material size and quality. The analysis presented here used measurements of 83 bifurcate-cluster hafted bifaces (Table 14.1).

As this reanalysis is concerned with bifurcate-base forms, our largest sample is from the Rose Island site. Additional specimens were added from other sites excavated as part of the Tellico Archaeological Project to increase the sample size. A number of Stanly forms manufactured from quartz could have been included from the Icehouse Bottom site. However, it was decided to include only the chert specimens to minimise the effect of raw material variability.

14.4 Examination of typology

One of the main areas of interest for the use of the database was examining the relationship between the various types identified by the original analysts through a comparison of the size and shape of each specimen. It is assumed that size and shape of the specimen are determined by several factors:

1. the technology used to produce the implement;
2. the intended use of the implement; and,
3. resharpening/reworking of the implement.

The last factor is particularly important as a number of studies have demonstrated that morphologically similar hafted bifaces may represent the resharpening or reworking of one general form or that implements may change 'type' due to resharpening of either the blade (Goodyear 1974; Hoffman 1985) or the haft area (Flenniken 1985; Flenniken & Raymond 1986; Titmus & Woods 1986; Towner & Warburton 1990). The polar grid method used here provides for obtaining a number of size measurements and from these shape can also be quantified.

For the analysis presented here, we were interested in defining variables that measure the size and shape of the blade and haft areas of each specimen. Variables that relate to size can be defined by taking the linear distance between the pairs of polar grid points (*i.e.,* A1–A2, B1–B2, *etc.*) as representing various width measurements and two other linear measurements (maximum blade, minimum base depth). The width variables can be computed from the polar grid measurements using the law of cosines. Variables A-E define blade widths while variables F to I define haft widths. To define a series of shape variables for each specimen, the method developed by Darroch & Mosimann (1985) is used to convert the size measurements into dimensionless, shape related variables (log-shape variables). In short, the shape variables can be defined by 'taking the log of each variable and subtracting the arithmetic mean of the logged variables from each logged variable' (Kidder *et al.* 1992, 159). Six shape variables relating to the blade area and five shape variables relating to the haft area were so defined. The combination of size and shape variables gives a total of 22 variables for further analysis.

The five bifurcate base types (MacCorkle, St. Albans, LeCroy, Kanawha, and Stanly) were defined based on the recovery of specimens from stratified deposits, therefore, it is of interest to determine how well these types can be defined using the quantitative variables (size and shape) defined above. As the blade often undergoes resharpening and reworking to a greater extent than the haft (*e.g.,* Thomas 1981), only haft related variables are considered in this particular analysis. Other authors have also noted that haft related variables provide the best discriminators (Gunn

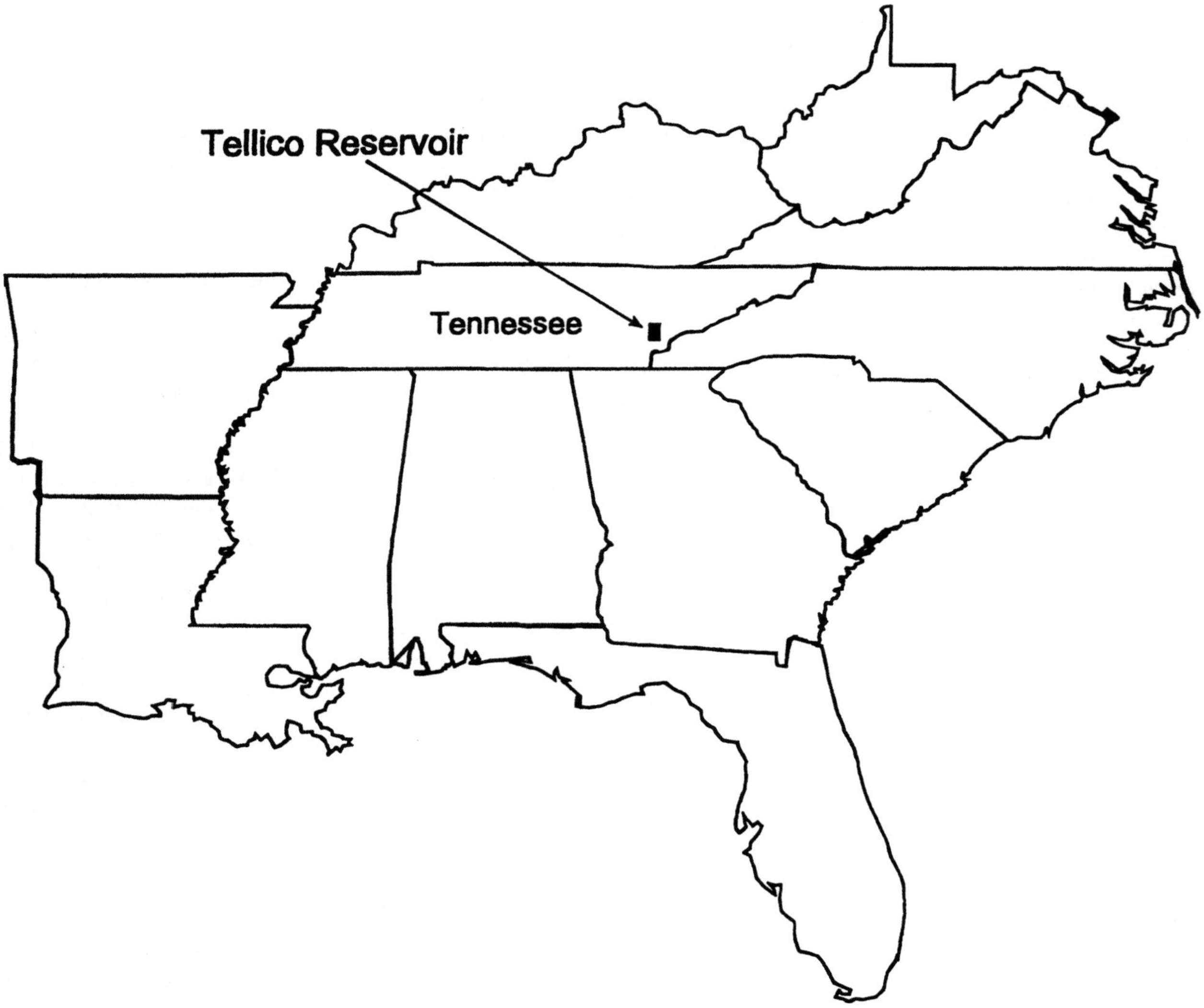

Figure 14.4: Location of study area.

Site	Types					Totals
	MacCorkle	St. Albans	LeCroy	Kanawha	Stanly	
Bacon Farm	1	3	4	4	0	12
Calloway Island	0	0	0	2	2	4
Icehouse Bottom	5	1	2	2	11	21
Patrick	1	0	0	2	2	5
Rose Island	2	17	22	2	0	43
Totals	9	21	28	12	15	83

Table 14.1: Sites and types used in analysis.

& Prewitt 1975; Hall 1992). The haft related variables were used in a K-means cluster analysis (Baxter 1994, 147-9; Johnson & Wichern 1992, 596–602). If the types are 'real entities', then each type should dominate a single cluster. Five clusters, to represent the five types, were requested in the SPSS K-means cluster analysis. As is evident in Table 14.2, each cluster is not dominated by a single type. Instead, several types are represented in each cluster. To test Kimball's (1996) hypothesis that MacCorkle is not represented in the Tellico data, the analysis was re-run, this time with four clusters being requested (Table 14.3). Again, no single cluster is dominated solely by a single type. Of note is that MacCorkle and St. Albans often cluster together as do St. Albans and LeCroy as well as Kanawha and Stanly. In other words, temporally close specimens are grouping together suggesting that there is some overlap in the types.

Given that the above analysis failed to define the five types, the data were further investigated using a K-means cluster analysis as a means of assessing how best to group the specimens. The above analysis suggested that haft variables alone could not account for enough of the variation between the types to successfully group them. Therefore, the inclusion of blade variables, in addition to the haft variable, could provide the necessary criteria. However, simply including all the variables in the analysis is not a viable option given that there is some degree of correlation between them. The use of highly correlated variables can have an adverse effect in defining clusters (see Baxter 1994, 167–70 for a more in depth discussion). As means of decreasing the number of variables, while retaining most of the information contained in them, a principal component analysis (without rotation) was conducted of the size and shape variables. Five components were retained and account for 88% of the variance of the original variables. These components represent a mix of both size and shape of the blade and haft. As an initial examination, plots of the various components were made, however, no obvious groupings were discernible in any of these plots (Fig. 14.5).

A K-means cluster analysis was then run using Keith Kintigh's *Tools for Quantitative Archaeology* computer package. The component scores were used in place of the original variables. The program allows for the creation of a number of randomised data sets. Both the original and randomised data sets are analysed using the K-means algorithm. Plots can then be requested showing the sum squared error (SSE) versus cluster level for both the original and randomised data sets. Inflections in the SSE curve of the original data that are not seen in the randomised data indicates the optimum clustering level(s). If there is no good clustering in the data, the SSE curve of the original data can not be distinguished from the randomised data. Such is the case with the current data set (Fig. 14.6). This is consistent with the principal component plots and suggests that the cluster analysis is simply partitioning the data, but not finding any 'real' groups. While no real groups were defined, an examination of the cluster compositions at various levels revealed that clusters defined by the K-means analysis exhibited a temporal component. That is, temporally closer specimens were being grouped together in a cluster while temporally segregated specimens were assigned to separate clusters. Kanawha and Stanly were most often grouped together as were MacCorkle and St. Albans as well as St. Albans and LeCroy. This suggests that there is significant overlap in the morphology of the bifurcate-base types. While 'classic' examples of each type can be found, the amount of overlap between the types is blurring the distinctions. As Dunnell (1992, 213) has suggested 'the empirical world is always in a state of becoming. Kinds are the consequence of observation, not inherent in the phenomenological world itself.' This certainly appears to be the case with the bifurcate-base forms examined here. We also note that in the original site reports, there are examples of several types occurring within the same excavation level (see for example Chapman 1975, Tables 11–13) which also supports a lack of distinction between the types. A suggestion for future research in this area would be to use stratigraphic levels to group specimens to redefine types. As we currently do not have access to the stratigraphic data, we explored other avenues of inquiry.

Some morphological differences should exist that help to differentiate the originally defined types. As a preliminary examination, several discriminant analyses were run using different subsets of the variables:

1. inclusion of all variables
2. only include size related variables
3. only include shape variables
4. include only blade related variables (size and shape)
5. include only haft related variables (size and shape)

All specimens were separated into five 'types' based on the original descriptions of these implements. It is expected that if the 'types' are real entities (*i.e.*, they represent distinct hafted biface forms), there should be correspondence in the classification results and there should be no pattern to miss-classifications. In addition, as the blade area of the implements undergoes modification due to use more so than the haft area, it is expected that haft related variables will be more useful for classification purposes (Gunn & Prewitt 1975; Hall 1992). In all cases, stepwise discriminant methods using Mahalanobis D2 were used to help identify the best discriminating variables. We note several potential problems with stepwise methods such as:

1. not identifying the best variables for the analysis;

Type	Clusters					Total
	1	2	3	4	5	
MacCorkle	5	2	0	0	2	9
St. Albans	4	9	0	1	7	21
LeCroy	3	5	6	8	6	28
Kanawha	0	0	1	9	0	10
Stanly	0	0	0	12	3	15
Total	12	16	7	30	18	83

Table 14.2: Types by cluster, five cluster solution.

Type	Clusters				Total
	1	2	3	4	
MacCorkle	5	0	4	0	9
St. Albans	4	1	16	0	21
LeCroy	3	2	14	9	28
Kanawha	0	9	0	1	10
Stanly	0	8	4	3	15
Total	12	20	38	13	83

Table 14.3: Types by cluster, four cluster solution.

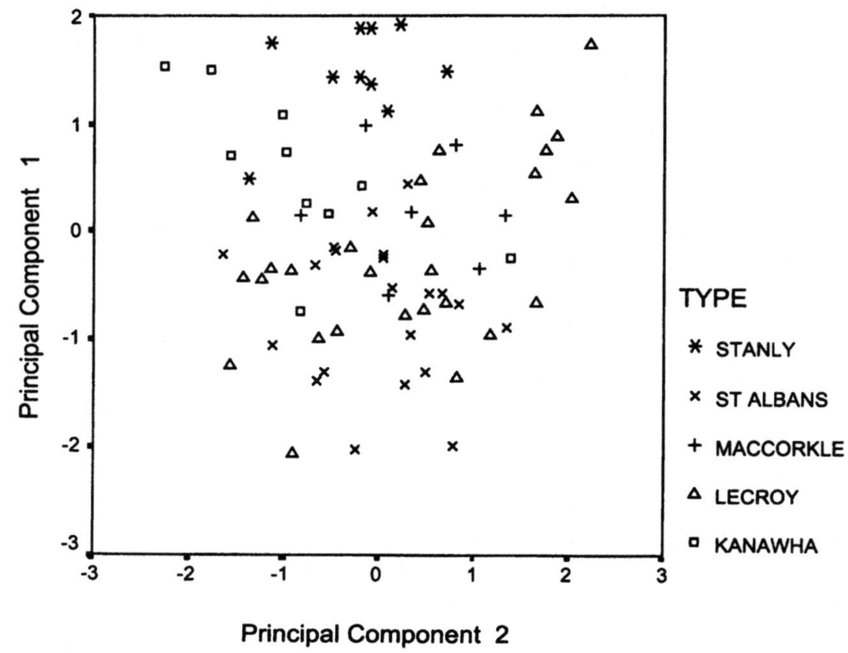

Figure 14.5: Plot of scores on Components 1 and 2.

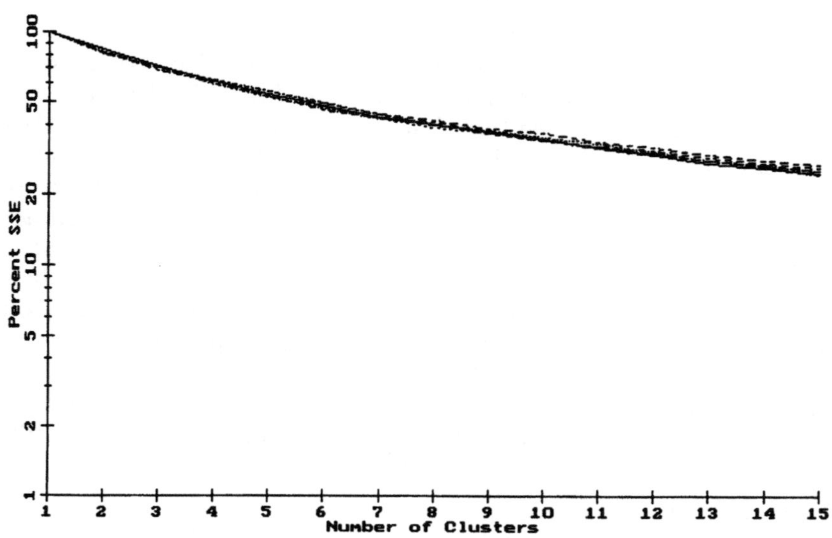

Figure 14.6: Log SSE plot, original data plus five random runs.

2. identifying variables that perform best only for a given data set;

3. relationships between variables not already in the model are not taken into account;

4. with large numbers of variables, several different subsets may perform equally well

More in-depth discussions of these problems are provided by Baxter (1994, 206–9), Fatti *et al.* (1982, 27–31), Johnson & Wichern (1992, 554-5) and Klecka (1980, 52–60). However, these problems are not a major concern for the present analysis as the main goal is to determine if the types, as originally defined, can be separated with the quantitative variables and not as a means of defining classification functions to classify other specimens (Baxter 1994, 208; Johnson & Wichern 1992, 554–5). As a means of assessing the success of the classifications, cross validation was used because it gives a less biased result as each case is classified by the functions derived from all cases excluding that case. As with the K-means results, these analyses indicated that both blade and haft variables are needed to give the best classification results. Additionally, there also appeared to be a temporal component to the miss classifications. That is, specimens are miss-classified as the type occurring prior to, and just after the actual type.

Given that the preliminary analyses indicated that both haft and blade variables were needed to produce good classification results, we were interested in identifying which variables were the best predictors of group membership. To accomplish this, a series of stepwise discriminant function analyses were conducted with SPSS using the size and shape variables. A random sample of the original data set was selected (approximately 85% of the original cases) and a stepwise discriminant analysis was conducted. The variables chosen by the analysis as 'best' were recorded. Another subset was then selected and the analysis rerun. Ten such subsets were subjected to the analysis in this way. Based on the number of times a variable was selected, it was possible to have an understanding of the variable's worth in discrimination. A final analysis was then conducted where by the variables that appeared most often in the above analyses were employed in the model. A combination of size and shape variables (both blade and haft) were used. The cross validation rate (Table 14.4) indicates a 79.5% correct classification. A temporal component to the miss-classifications is apparent with specimens being miss-classified as the type that occurred prior to, and just after the misclassified type. This is graphically depicted in Figure 14.7. The earliest form, MacCorkle, appears in the upper right hand corner of the plot. St. Albans below this followed by LeCroy at the lowest part of the plot. Kanawha is seen slightly above LeCroy followed by Stanly (the latest form) in the upper left hand corner. Of note is that, in addition to the haft related variables, two blade widths are included. This suggests that there are differences associated with how the different types are being resharpened. This also has implications for how these specimens were being used.

14.5 Examination of the differences

Several authors (*e.g.*, Chapman 1977; Coe 1964; Justice 1987; Kimball 1996) have noted that there is a gradual change from the earlier corner notched forms (*i.e.*, Kirk Corner Notched) through the bifurcate tradition. The above analyses also suggest a gradual change through time. The observed differences between the types do not appear to be a function of size, but a combination of size and shape of both the haft and the blade. It is of use then to examine these differences as a means of discerning possible causes of this variation. In the following analyses, we continue to use the type designations given by the original authors. While we recognise that some of these types are probably not real entities, or could be combined with others as a single type cluster, the specimens belonging to these types were generally recovered together from stratified deposits. Therefore, using the type names allows for the examination of changes through time.

Through use, resharpening, and the rejuvenation of broken specimens, changes are expected in both the size and shape of the blade. Further, tools used for different purposes, such as projectiles versus knives, will likely exhibit differential resharpening. These differences in resharpening should have an effect on the symmetry of the specimen as different portions of the blade will become dulled through use, exhibit fractures, and undergo subsequent resharpening to allow for continued use of the specimen. Resharpening will continue until the specimen reaches a point that it can no longer be used for its intended purpose. Based on expectations from design criteria, implements used as projectiles should have corresponding differences in morphological attributes in comparison to those used as knives (see discussions in Bradbury 1997; Christenson 1986; Cotterell & Kamminga 1990; Friis-Hansen 1990; McEwen *et al.* 1991; Shott 1993, for projectile design considerations). Implements used as projectiles must be designed with the primary requirements of 'accuracy, killing power and range' (Christenson 1986, 115). In relation to the actual projectile point, the point must have a sharp tip and be able to penetrate the prey and cause a wound needed to dispatch the animal (Christenson 1986, 117; Friis-Hansen 1990, 499). In addition, as the point must articulate with other parts of the whole projectile system (*i.e.*, method of delivery, shaft, *etc.*), there is a limit to the size range that can be tolerated for the point. Specimens outside this range will not function as intended. It is also

Actual	Classified as				
	MacCorkle	St. Albans	LeCroy	Kanawha	Stanly
MacCorkle	9	0	0	0	0
St. Albans	2	17	2	0	0
LeCroy	0	4	23	0	1
Kanawha	0	0	1	7	2
Stanly	2	0	1	2	10

Table 14.4: Cross validation results of discriminant function analysis.

expected that specimens used primarily as projectiles will exhibit a more symmetrical blade. This is due to the way such implements would be resharpened. The resulting morphology of any projectile is, therefore, a compromise between a number of different factors.

For cutting tools, a sharp edge and a stable haft area are the main consideration. A greater range in the variation of size is allowable than with those implements used as projectile points. As long as the edge can be maintained and the implement is stable in the haft, the implement can be used for its intended purpose. An increase in the size of the blade, as long as this does not make it more susceptible to breakage during use, may produce an increase in the efficiency of the tool for its required purpose and increases the tool's uselife. In addition, a greater degree of asymmetry would be allowable than with a projectile. These differences in design criteria should have an effect on both the size and shape of the original and discarded specimen.

Blade and haft area were calculated directly from the drawing files and used to calculate a blade to haft ratio for each specimen. This is graphically depicted in Figure 14.8. While there is some overlap in the earlier types (*i.e.*, MacCorkle, St. Albans and LeCroy) there is an increase in blade area and a decrease in haft area with the later types (*i.e.*, Kanawha and Stanly) exhibiting the largest blade areas and smallest haft areas. A one way ANOVA indicates significant differences in the blade to haft ratio between the five types (F: 40.74, Df: 4,70, $p < 0.0001$). Fisher's LSD was then used for pairwise comparisons to determine where these differences were occurring (Table 14.5). It was determined that Kanawha and Stanly were both significantly different than all other types and LeCroy and MacCorkle were significantly different. St. Albans and MacCorkle were not significantly different.

As the measurements recorded for this sample are in pairs, direct measurements of blade asymmetry are possible. A measure of the total asymmetry for a specimen is computed by summing the absolute difference between each pair of observations. This produces an index of asymmetry (IA) for each hafted biface. Separate IA values are determined for the blade and the haft area. The IA value for a symmetrical blade is close to zero, while highly asymmetrical blades will have values much greater than zero. It is expected that hafted bifaces used as projectiles will be much more symmetrical than those used for other tasks. Modifications to the haft area may occur to facilitate hafting (*e.g.*, Flenniken & Raymond 1986; Keeley 1982; Towner & Warburton 1990), but are unlikely to occur after the implement has been hafted. Therefore, it is expected that the haft area will be quite symmetrical. A higher IA is, therefore, expected for the blade than for the haft.

There is some disagreement in the literature concerning damage to projectiles caused by use. Odell & Cowan (1986, 204) found that a high frequency of points in their experimental collection exhibited tip damage after use. Damage to the base occurred less frequently. Other researchers (*e.g.*, Flenniken 1985; Flenniken & Raymond 1986; Titmus & Woods 1986; Towner & Warburton 1990) report higher instances of basal damage in experimental projectile points. It is possible that the use of chert by Odell & Cowan (1986) and obsidian by the other researchers is the cause of this difference. Also, Thomas (1986) suggests that the choice of point type employed in an experiment can affect whether damage to the haft area will occur. Nonetheless, there is the potential for damage and subsequent reworking of the haft area which could have an effect on its symmetry.

An initial examination of each type used in the study showed similar results. A greater range and higher average was seen for blade IA than for haft IA. To determine if these differences were significant, a paired t-test was run. As expected, the blade exhibits a significantly higher IA than the haft ($t = 10.102$, DF: 80, $p < 0.0001$).

In the next step of the analysis, blade and haft IA were compared across types using a one-way ANOVA. Blade IA showed significant differences between the types ($F = 4.019$, DF: 4, 70; $p = 0.005$). Fishers LSD was again used for pairwise comparisons. Kanawha was significantly different than all others at $p < 0.03$ (Table 14.6), while the remaining types were not significantly different than each other. In testing haft IA, no significant differences ($F = 1.074$, DF: 4, 78; $p = 0.375$) were found between the types.

An additional examination can be accomplished through the use of the coefficient of variation (CV) which is defined as the standard deviation divided by the mean (Shennan 1997, 43–4). This produces a standardised measure of dispersion and removes the effect of a correlation between the mean and standard deviation. The greater the CV, the greater the variation within the data. A lower CV is expected for imple-

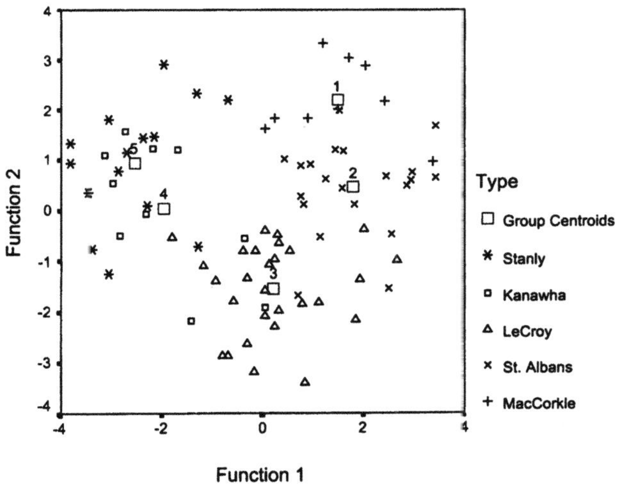

Figure 14.7: Function 1 versus function 2.

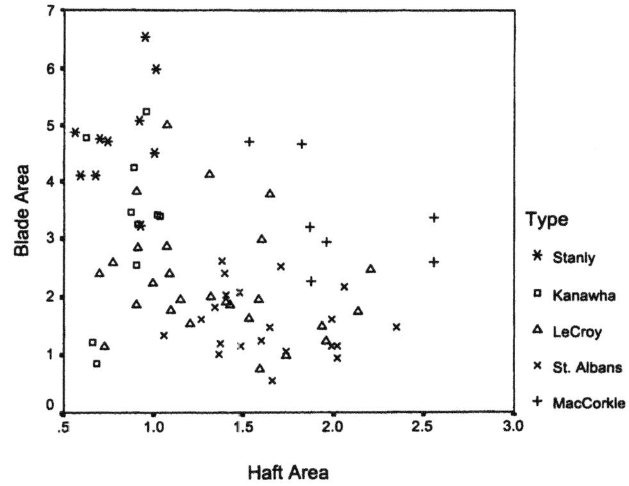

Figure 14.8: Blade versus haft area.

(I) Type	(J) Type	Mean Difference (I–J)	Std. Error	Sig.	95% Confidence Interval	
					Lower Bound	Upper Bound
MacCorkle	St. Albans	0.7660	0.489	0.122	−0.2101	1.7421
	LeCroy	−0.1762	0.476	0.712	−1.1248	0.7725
	Kanawha	−2.0206†	0.553	0.000	−3.1228	−0.9184
	Stanly	−4.3266†	0.553	0.000	−5.4288	−3.2244
St. Albans	MacCorkle	−0.7660	0.489	0.122	−1.7421	0.2101
	LeCroy	−0.9421†	0.326	0.005	−1.5929	−0.2914
	Kanawha	−2.7866†	0.431	0.000	−3.6459	−1.9273
	Stanly	−5.0926†	0.431	0.000	−5.9519	−4.2333
LeCroy	MacCorkle	0.1762	0.476	0.712	−0.7725	1.1248
	St. Albans	0.9421†	0.326	0.005	0.2914	1.5929
	Kanawha	−1.8445†	0.415	0.000	−2.6724	−1.0165
	Stanly	−4.1505†	0.415	0.000	−4.9784	−3.3225
Kanawha	MacCorkle	2.0206†	0.553	0.000	0.9184	3.1228
	St. Albans	2.7866†	0.431	0.000	1.9273	3.6459
	LeCroy	1.8445†	0.415	0.000	1.0165	2.6724
	Stanly	−2.3060†	0.502	0.000	−3.3062	−1.3058
Stanly	MacCorkle	4.3266†	0.553	0.000	3.2244	5.4288
	St. Albans	5.0926†	0.431	0.000	4.2333	5.9519
	LeCroy	4.1505†	0.415	0.000	3.3225	4.9784
	Kanawha	2.3060†	0.502	0.000	1.3058	3.3062

†The mean difference is significant at the 0.05 level or less.

Table 14.5: Pairwise comparisons for blade/haft ratio.

ments used for a single purpose, while a higher CV would be expected for implements used for multiple purposes (see Bettinger & Eerkens 1997; Shott 1997, for more in depth discussions). As can be seen in Table 14.7, there is a decrease in the CV from St. Albans through Stanly forms.

The above data suggests that there are differences in how the blade is being modified, but not the haft. Also, this analysis supports the above suggestion that the types may not be real and that some could be considered as variants that are part of a single type cluster. Further, the data suggest that there are differences in how these types are being resharpened, and therefore, used. Based on these analyses, and expectations from design criteria, it is suggested that MacCorkle, St Albans and LeCroy were used as projectiles in addition to secondary use as knives. In contrast, Kanawha and Stanly were primarily used as knives. The use of bifurcate forms as knives was suggested by Chapman (1975, 268), who saw the bifurcation as a way to increase lateral stability of the hafted biface. Such would be advantageous for non-projectile activities. Chapman further suggests that specimens reworked to a high degree are examples of projectile points that were broken then reworked and used for other tasks. This is certainly possible given evidence for the conservation of raw materials and an increase in bipolar reduction through time (Carr 1995). At the Main site in eastern Kentucky, based on high magnification microwear analysis, Kimball (1994) suggests that bifurcate forms were used primarily as projectiles and secondarily as knives. An understanding of hafted biface function would be increased through better stratigraphic control of the specimens.

14.6 Summary and conclusions

A method was developed that allows in improving precision for the recording of a number of linear and angular measurements of hafted bifaces. The method removes much of the subjectivity inherent in many traditional approaches and has application for the use of existing collections to examine both old and new questions such as typology, tool use, and resharpening. This method allows for the direct recording of a number of size variables. These variables can be used to define a number of additional variables, both size and shape related. The method reduces much of the subjectivity prevalent in conventional methods. Additionally, as AutoCAD is commonly used for drafting purposes, the software is readily available to most archaeologists. Utility of the method was demonstrated through application to a sample of hafted bifaces though it is certainly applicable to other artefact classes.

Several conclusions were reached based on the analysis presented here. In the examination of bifurcate typology, it was determined that the 'types' are not quite as distinct as originally defined. While the five types form one temporal sequence, the morphological distinctions between each successive type are minimal. In effect, these types can be seen as points along a continuum of morphological change, and because of this miss-classifications (in terms of temporal assignment) could result if such specimens are recovered from unstratified contexts. It is suggested that MacCorkle and St Albans can be combined into a single hafted biface cluster as could Kanawha and Stanly. While the latter two may not overlap temporally, morphologically they are very similar. It was further determined that morphological changes in hafted bifaces were occurring gradually during the period that these forms were in use. At least some of the differences between the types are likely due to resharpening and reworking of the blade. Similar results were obtained in an earlier analysis of this data using the polar grid measurements instead of the size and shape variables defined here (Bradbury & Carr 1997). In addition, this earlier study suggested that only minor modifications were occurring between the earlier corner notched forms (*i.e.,* Upper Kirk) and MacCorkle/St Albans. Other researchers have also suggested that MacCorkle and St Albans are not separate types or that they can be considered the same cluster (Justice 1987; Kimball 1996). Additionally, it is suggested that there was a change through time in how these forms were being used. These changes are thought to represent a decreased use of these forms as multifunctional tools (projectile and knife) and an increased use for single tasks (knife). Other than modifications to the hafted bifaces, the remainder of the lithic assemblage from these components are very similar from MacCorkle through Kanawha. Several changes in the lithic assemblage occur with the appearance of Stanly forms and include the addition of several artefact classes such as atlatl weights and an increased use of quartz and quartzite (Chapman 1977; Coe 1964). It is also noted, based on radiocarbon dates, that LeCroy and Kanawha forms overlap temporally. While the radiocarbon dates overlap, these forms have been recovered in stratigraphic sequence suggesting some temporal separation.

The above conclusions are preliminary and are viewed as hypotheses to be tested with the collection of additional data. An increased sample size and a close examination of provenance data, should facilitate a better understanding of the cultural-historical typology and morphological changes. The addition of microwear analysis of a larger sample is needed to test hypotheses concerning tool use. The method presented here is just one of the many ways that existing collections could be fruitfully re-examined. Such analyses take data out of boxes and bring it into the open where it can receive greater consideration and use. Employing computer technology in typological analyses overcomes difficulties with conventional methods and opens new areas of research.

(I) Type	(J) Type	Mean Difference (I–J)	Std. Error	Sig.	95% Confidence Interval	
					Lower Bound	Upper Bound
MacCorkle	St. Albans	0.1578	0.2175	0.4704	−0.2759	0.5915
	LeCroy	0.1599	0.2113	0.4519	−0.2616	0.5813
	Kanawha	−0.5346†	0.2455	0.0328	−1.0243	−0.0449
	Stanly	0.0398	0.2455	0.8716	−0.4499	0.5295
St. Albans	MacCorkle	−0.1578	0.2175	0.4704	−0.5915	0.2759
	LeCroy	0.0020	0.1450	0.9889	−0.2871	0.2912
	Kanawha	−0.6925†	0.1914	0.0006	−1.0743	−0.3107
	Stanly	−0.1180	0.1914	0.5396	−0.4998	0.2638
LeCroy	MacCorkle	−0.1599	0.2113	0.4519	−0.5813	0.2616
	St. Albans	−0.0020	0.1450	0.9889	−0.2912	0.2871
	Kanawha	−0.6945†	0.1844	0.0003	−1.0623	−0.3266
	Stanly	−0.1200	0.1844	0.5174	−0.4879	0.2478
Kanawha	MacCorkle	0.5346	0.2455	0.0328	0.0449	1.0243
	St. Albans	0.6925†	0.1914	0.0006	0.3107	1.0743
	LeCroy	0.6945†	0.1844	0.0003	0.3266	1.0623
	Stanly	0.5745	0.2228	0.0120	0.1301	1.0189
Stanly	MacCorkle	−0.0398	0.2455	0.8716	−0.5295	0.4499
	St. Albans	0.1180	0.1914	0.5396	−0.2638	0.4998
	LeCroy	0.1200	0.1844	0.5174	−0.2478	0.4879
	Kanawha	−0.5745†	0.2228	0.0120	−1.0189	−0.1301

†The mean difference is significant at the 0.05 level or less.

Table 14.6: Pairwise comparisons for blade IA.

Type	Mean	St. Deviation	CV
MacCorkle	0.8978	0.4921	0.5481
St. Albans	0.74	0.5122	0.6922
LeCroy	0.738	0.4077	0.5524
Kanawha	1.4325	0.7298	0.5095
Stanly	0.858	0.4167	0.4857

Table 14.7: Coefficient of variation for blade symmetry by type.

Acknowledgements

An earlier version of this paper was presented at the 54th annual Southeastern Archaeological Conference Meetings in Baton Rouge, Louisiana, 1997. The authors would like to thank Norah Moloney and Michael Shott for their invitation to contribute to the volume. Michael Shott provided a number of insightful comments on a draft of this paper that helped us to clarify a number of points. The authors would also like to acknowledge Jefferson Chapman at the McClung Museum in Knoxville, Tennessee for his support and encouragement over the years and his willingness to allow access to existing collections. The authors take full responsibility for any errors or shortcomings.

References

ADAMS, W. Y. 1988. 'Archaeological classification: Theory versus practice.' *Antiquity* 62(234): 40–56.

BARTON, C. M. & G. A. CLARK (eds.) 1997. *Rediscovering Darwin: Evolutionary Theory and Archaeological Explanation.* Archaeological Papers of the American Anthropological Association No. 7. American Anthropological Association, Arlington.

BAXTER, M. J. 1994. *Exploratory Multivariate Analysis in Archaeology.* Edinburgh University Press, Edinburgh.

BECK, C. & G. T. JONES 1989. 'Bias and archaeological classification.' *American Antiquity* 54: 224–62.

BETTINGER, R. L. & J. EERKENS 1997. 'Evolutionary implications of metrical variation in Great Basin projectile points.' *In* Barton & Clark (1997), pp. 177–91.

BRADBURY, A. P. 1997. 'The bow and arrow in the Eastern Woodlands: Evidence for an Archaic origin.' *North American Archaeologist* 18(3): 207–34.

BRADBURY, A. P. & P. J. CARR 1997. *Towards an Explanation of Variation in Hafted Bifaces: An Early Holocene Example.* Paper presented at the 54th annual Southeastern Archaeological Conference, Baton Rouge, Louisiana.

BROYLES, B. J. 1966. 'Preliminary report: The St Albans site (46KA27), Kanawha County, West Virginia.' *The West Virginia Archaeologist* 19: 1–43.

BROYLES, B. J. 1971. *Second Preliminary Report: The St Albans Site, Kanawha County, West Virginia, 1964–1968.* Report of Archeological Investigations No. West Virginia Geological and Economic Survey, Morgantown.

CARR, P. J. 1995. *Hunter-Gatherers, Mobility, and Technological Organization: The Early Archaic of East Tennessee.* Ph.D. thesis, Department of Anthropology, The University of Tennessee, Knoxville.

CHAPMAN, J. 1973. *The Icehouse Bottom Site, 40MR23.* Report of Investigations 13. University of Tennessee, Department of Anthropology, Knoxville.

CHAPMAN, J. 1975. *The Rose Island Site and The Bifurcate Point Tradition.* Report of Investigations 1. University of Tennessee, Department of Anthropology, Knoxville.

CHAPMAN, J. 1977. *Archaic Period Research in the Lower Little Tennessee River Valley, 1975: Icehouse Bottom, Harrison Branch, Thirty Acre Island, Calloway Island.* Report of Investigations 18. University of Tennessee, Department of Anthropology, Knoxville.

CHAPMAN, J. 1978. *The Bacon Farm Site and a Buried Site Reconnaissance.* Report of Investigations 23 and Tennessee Valley Authority Publications in Anthropology No. 21. The University of Tennessee, Department of Anthropology, Knoxville.

CHAPMAN, J. 1979. *The Howard and Calloway Island Sites.* Report of Investigations 27 and Tennessee Valley Authority, Publications in Anthropology No. 23. The University of Tennessee, Department of Anthropology, Knoxville.

CHRISTENSON, A. L. 1986. 'Projectile point size and projectile aerodynamics: An exploratory study.' *Plains Anthropologist* 31: 109–28.

COE, J. L. 1964. 'The formative cultures of the Carolina Piedmont.' *Transactions of the American Philosophical Society* 54(5): 1–130.

COTTERELL, B. & J. KAMMINGA 1990. *Mechanics of Pre-Industrial Technology.* Cambridge University Press, Cambridge.

DARROCH, J. & J. MOSIMANN 1985. 'Canonical and Principal Components of shape.' *Biometrika* 72(2): 241–52.

DUNNELL, R. C. 1971. *Systematics in Prehistory.* Free Press, New York.

DUNNELL, R. C. 1986. 'Methodological issues in Americanist artifact classification.' *Advances in Archaeological Method and Theory* 9: 149–207.

DUNNELL, R. C. 1992. 'Archaeology and evolutionary science.' *In* L. Wandsnider (ed.), *Quandaries and Quests: Visions of Archaeology's Future*, Occasional Paper No. 20, pp. 209–24. Center for Archaeological Investigations, Southern Illinois University, Carbondale.

FATTI, L. P., D. M. HAWKINS & E. L. RAATH 1982. 'Discriminant Analysis.' *In* D. M. Hawkins (ed.), *Topics in Applied Multivariate Analysis*, pp. 1–71. Cambridge University Press, Cambridge.

FAULKNER, C. H. & M. C. R. MCCOLLOUGH 1973. *Introductory Report of the Normandy Reservoir Salvage Project: Environmental Setting, Typology and Survey.* Tennessee Valley Authority Publications in Anthropology No. 4, and Normandy Archaeological Project Vol. 1, Report of Investigations No. 11. Tennessee Valley Authority and Department of Anthropology, University of Tennessee, Chattanooga and Knoxville.

FISH, P. R. 1978. 'Consistency in archaeological measurement and classification: A pilot study.' *American Antiquity* 43: 86–9.

FLENNIKEN, J. J. 1985. 'Stone tool reduction techniques as cultural markers.' *In* M. G. Plew, J. C. Woods & M. G. Pavesic (eds.), *Stone Tool Analysis: Essays in Honor of Don E. Crabtree*, pp. 265–76. University of New Mexico Press, Albuquerque.

FLENNIKEN, J. J. & A. W. RAYMOND 1986. 'Morphological projectile point typology: Replication, experimentation, and technological analysis.' *American Antiquity* 51: 603–14.

FRIIS-HANSEN, J. 1990. 'Mesolithic cutting arrows: Functional analysis of arrows used in the hunting of large game.' *Antiquity* 64: 494–504.

GOODYEAR, A. C. 1974. *The Brand Site: A Techno-Functional Study of a Dalton Site in Northeast Arkansas.* Publications on Archaeology, Research Series No. Arkansas Archaeological Survey.

GUNN, J. & E. PREWITT 1975. 'Automatic classification: Projectile points from West Texas.' *Plains Anthropologist* 20: 139–50.

HALL, C. L. 1992. *Exploring Archaic Settlement in the Midsouth: The Surface Archaeology of Cannon and Cheek Bends of the Duck River, Tennessee.* Ph.D. thesis, Department of Anthropology, The University of Tennessee, Knoxville.

HENTON, G. H. & S. R. DURAND 1991. 'Projectile point measurement and classification using digital image processing.' *Journal of Quantitative Anthropology* 3: 53–82.

HOFFMAN, C. H. 1985. 'Projectile point maintenance and projectile point typology: Assessment with factor analysis and canonical correlation.' *In* C. Carr (ed.), *For Concordance in Archaeological Analysis*, pp. 566–612. Westport Publishers, Inc. University of Arkansas, Fayetteville.

HUGHES, S. S. 1998. 'Getting to the point: Evolutionary change in prehistoric weaponry.' *Journal of Archaeological Method and Theory* 5(4): 345–408.

JOHNSON, J. K. 1993. 'Lithics.' *In* J. K. Johnson (ed.), *The Development of Southeastern Archaeology*, pp. 36–52. The University of Alabama Press, Tuscaloosa, Alabama.

JOHNSON, R. A. & D. W. WICHERN 1992. *Applied Multivariate Statistical Analysis.* Prentice Hall, New Jersey, third edition.

JUSTICE, N. D. 1987. *Stone Age Spear and Arrow Points of the Midcontinental and Eastern United States: A Modern Survey and Reference.* Indiana University Press, Bloomington and Indianapolis.

KEELEY, L. H. 1982. 'Hafting and retooling: Effects on the archaeological record.' *American Antiquity* 47(4): 798–809.

KIDDER, J. H., R. L. JANTZ & F. H. SMITH 1992. 'Defining modern humans: A multivariate approach.' *In* G. Brauer & F. H. Smith (eds.), *Continuity or Replacement? Controversies in* Homo sapiens *Evolution*, pp. 157–77. A. A. Balkema, Rotterdam.

KIMBALL, L. R. 1994. *Upper Cumberland Archaic and Woodland Period Archeology at the Main Site (15BL35), Bell County, Kentucky*, volume 2, chapter Appendix F: Microwear Analysis of Archaic and Early Woodland Tools from the Main Site (15BL35), pp. 94–6. Cultural Resource Analysts, Publication Series, Lexington, Kentucky.

KIMBALL, L. R. 1996. 'Early Archaic settlement and technology: Lessons from Tellico.' *In* D. G. Anderson & K. E. Sassaman (eds.), *The Paleoindian and Early Archaic Southeast*, pp. 149–86. The University of Alabama Press, Tuscaloosa, Alabama.

KLECKA, W. R. 1980. *Discriminant Analysis.* Quantitative Applications in the Social Sciences, Sage University Paper, 19. Sage.

KNEBERG, M. D. 1956. 'Some important projectile point types found in the Tennessee area.' *Tennessee Archaeologist* 12(1): 25.

MCEWEN, E., R. C. MILLER & C. A. BERGMAN 1991. 'Early bow design and construction.' *Scientific American* 265(6): 76–82.

ODELL, G. H. 1996. 'Economizing behavior and the concept of "curation".' *In* G. H. Odell (ed.), *Stone Tools: Theoretical Insights into Human Prehistory*, pp. 51–80. Plenum Press, New York.

ODELL, G. H. & F. COWAN 1986. 'Experiments with spears and arrows on animal targets.' *Journal of Field Archaeology* 13: 195–212.

SHENNAN, S. J. 1997. *Quantifying Archaeology.* Edinburgh University Press, Edinburgh, second edition.

SHOTT, M. J. 1993. 'Spears, darts, and arrows: Late Woodland hunting techniques in the Upper Ohio Valley.' *American Antiquity* 58(3): 425–43.

SHOTT, M. J. 1997. 'Transmission theory in the study of stone tools: A Midwestern example.' *In* Barton & Clark (1997), pp. 193–206.

SULLIVAN, A. P. & K. C. ROZEN 1985. 'Debitage analysis and archaeological interpretation.' *American Antiquity* 50: 755–79.

THOMAS, D. H. 1981. 'How to classify the projectile points from Monitor Valley, Nevada.' *Journal of California and Great Basin Archaeology* 3(1): 7–43.

THOMAS, D. H. 1986. 'Points on points: A reply to Flenniken and Raymond.' *American Antiquity* 51(3): 619–27.

THOMS, A. V. 1993. 'Knocking sense from old rocks: Typologies and the narrow perspective of the Angostura point type.' *Lithic Technology* 18(1–2): 16–27.

TITMUS, G. L. & J. C. WOODS 1986. 'An experimental study of projectile point fracture patterns.' *Journal of California and Great Basin Anthropology* 8(1): 37–49.

TOMKA, S. A. & E. R. PREWITT 1993. 'Projectile point types and archaeological interpretations from Texas.' *Lithic Technology* 18(1–2): 49–58.

TOWNER, R. H. & M. WARBURTON 1990. 'Projectile point rejuvenation: A technological analysis.' *Journal of Field Archeology* 17: 311–21.

TRIGGER, B. C. 1989. *A History of Archaeological Thought.* Cambridge University Press, Cambridge.

WHITTAKER, J. C., D. CAULKINS & K. A. KAMP 1998. 'Evaluating consistency in typology and classification.' *Journal of Archaeological Method and Theory* 5(2): 129–64.

Andrew P. Bradbury
Cultural Resource Analysts Inc.
143 Walton Avenue.
Lexington
Kentucky 40508
USA
apbradbury@crai-ky.com

Philip J. Carr
Dept. of Sociology and Anthropology
University of South Alabama Mobile
HUMB 34
Alabama 36688-0002
USA
pcarr@jaguar1.usouthal.edu

15

Dating the Manufacture of Projectile Points using Thermoluminescence

Kris H. Wilhelmsen[1] and James K. Feathers[2]

[1]*InterNAP Network Services, Seattle;* [2]*Department of Anthropology, University of Washington, Seattle*

Abstract

Current projectile point studies address function, manufacturing technology, performance efficiency, hunting strategies, and social boundaries. Central to all these efforts and to the larger goal of explaining the evolution of projectile technology is the accurate ordering of points in time and the documentation of patterns in temporal variation. The use of existing projectile chronologies to simultaneously order points and examine temporal variation, however, is circular and many problems inherent in projectile point chronologies argue against the use of projectile point types. An examination of the prevalent assumptions involved in dating projectile points by association and by typological cross-dating indicates that current techniques are flawed but that an alternative strategy, using thermoluminescence (TL), may be profitably applied to this problem. In this strategy, TL is applied directly to individual points that were catastrophically heated during prehistoric manufacture. Although there are a number of technical problems with using TL to date projectile points, the results of this technique suggest that it is possible to build interval-level age associations for projectile points. Significantly, this provides an accurate way to document temporal variation and this is critical for building an evolutionary explanation of projectile technology.

15.1 Introduction

Projectile points are widely distributed over North and South America in contexts as early as 11,500bp, and persisting until after European contact (Justice 1987). During this time projectile point morphology varied with changes in manufacturing technologies, engineering requirements, raw material properties and distribution, and stylistic trends. The lengthy history of projectile manufacture, coupled with the characteristic abundance and durability of the lithic components, has secured a prominent role for projectile points in archaeological research over the last century. Second only to ceramics, projectile points have long been useful for developing cultural chronologies. Relative to ceramics, however, projectile point change tends to occur over longer spans of time, at least given the use of projectile point 'types' as units with which to measure change. Although recent studies of time-dependent metric changes in projectile point attributes (rather than types) suggest that the resolution of projectile point chronologies can be increased (*e.g.*, Fawcett 1998; Litfin *et al.* 1993; Shott 1996), temporal control must still be based on indirect dating methods having an inestimable magnitude of error.

Current research on projectile technology addresses, among other things, projectile point function, manufacturing technology, performance efficiency, hunting strategies, and social boundaries (*e.g.*, see papers in Knecht 1997). To understand projectile use as an evolving technology, it is necessary to evaluate how each of these issues contribute to morphological variation in projectile points, and attempt to explain why this variation changes over time. Yet, given the inherent problems of extant projectile point chronologies (*e.g.*, Flenniken & Raymond 1986; Thomas 1986) it can be argued that a better method of chronologically ordering projectile points should be a matter of first priority. In support of this explanatory goal, the present paper examines the underlying assumptions involved in dating projectile points by association and by typological cross-dating. Revealing the problematic nature of these assumptions is an important starting point for developing an alternative dating strategy. Within this framework we introduce a new methodology for building projectile point chronologies using thermoluminescence dating applied directly to projectile points, and preliminary results of this work are presented. Finally, the technical problems associated with using thermoluminescence to date projectile points are examined, and these are used to suggest directions for future research in projectile point dating.

Prior to radiocarbon, projectile point types from superposed strata provided a means for assigning relative orders to assemblages or sites. Consequently, projectile points played an important role in developing the chronological framework of culture history (*e.g.,* Broyles 1966; Chapman 1947, 1948a,b; Coe 1964; Justice 1987). After the advent of radiocarbon dating, projectile points associated with carbon-dated materials were used to build interval-scale chronologies in which temporal units such as phases were typified by particular point types. Archaeological deposits or assemblages of unknown age were assigned to phases based upon their inventory of morphologically-distinct projectile point types, a procedure known as typological cross-dating (Lyman *et al.* 1997). Typological cross-dating is facilitated by the development of local or regional chronologies in which at least some of the projectile points are found in association with carbon-dated materials. Even today, typological cross-dating is an important method for assigning non-ceramic deposits to archaeological phases, particularly in regions where most archaeological sites have poor conditions for organic preservation. In the Great Basin of North America, for example, pottery use was not widespread until after about 1,000 years bp (Grayson 1993). Thus, the majority of surface assemblages dating to between 11,000bp to 1,000bp are temporally ordered on the basis of projectile point chronologies established from point sequences in stratified rockshelter deposits (Beck 1994; Grayson 1993).

For over a century researchers have recognized that inferring age from projectile point morphology can be problematic. In 1876, while organizing Smithsonian collections for display at the Centennial Celebration in Philadelphia, Charles Rau stated that:

> A classification of the arrow-heads with regard to their chronological development is not attempted, and hardly deemed necessary. North American Indians of the same tribe (as, for instance, the Pai-Utes of Southern Utah) arm their arrows with stone points of different forms, the shape of the arrow-head being with them merely a matter of individual taste or of convenience. (Rau 1876, 8–9)

Rau *identified individual* taste and *convenience* as two possible sources of projectile point variation within a single group or lineage that could make chronological ordering on the basis of shape problematic. By using the term 'individual taste' Rau was alluding to the role of idiosyncratic behaviour, or style, as a source of morphological variation. In retrospect, style is not problematic and stylistic attributes have actually become the keystone for fleshing out time/space systematics (Lyman *et al.* 1997). Convenience, on the other hand, results from a wide variety of causal processes that influence morphological variation irrespective of time. The disproportionate frequency of lithic types having different fracture properties in a given raw material source, or the distances travelled to acquire raw materials, often make some materials more convenient to use than others (Wiant & Hassan 1985). The morphological, structural, and mechanical properties of lithic raw materials also influence morphological variation during manufacture and affect the range of possible functions for which projectile points and other lithic artefacts can be used (Ahler 1971). Furthermore, the non-lithic materials necessary for projectile technology (*e.g.,* wood, bone, antler, fletching, sinew, and mastic), as well as the environmental context of projectile use, influence the engineering design of projectile points (Bergman & McEwen 1997; Bleed 1986; Christenson 1997; Finlayson & Mithen 1997; Knecht 1997). Thus, a broad range of mechanisms that have little or no directional influence in terms of time can affect variation in projectile point morphology.

15.2 Dating by association and typological cross-dating

As noted above, the two common methods of dating projectile points are by:

1. association with carbon-dated materials and by
2. typological cross-dating.

Both of these methods require a bridging argument to demonstrate that the dated events and target events are historically related (Dean 1978). Dated events are the events actually dated by any chronometric technique whereas the target event is the archaeological event of interest. Dating by association relies most often on radiocarbon dates, and requires an argument relating the death of an organism to a target event such as tool manufacture or use. For example, a bridging argument may cite 'evidence' supporting the notion that organic remains were charred in a fire (and therefore isolated from the ^{14}C reservoir) at the same time a spatially associated projectile point was deposited. The assumption underlying such a bridging argument is that spatial proximity and co-occurrence within 'undisturbed' deposits is a measure of historical relatedness. In short, these arguments assume that the shorter the spatial distance between two objects within excavated or surface deposits the greater the probability that the same individual, or individuals of the same lineage, used and/or deposited the objects during the same period of occupation.

Typological cross-dating relies on the additional assumption that morphological similarity is a measure of historical relatedness. In other words, the greater the physical resemblance between two objects, the closer in time they were manufactured and deposited. Based on the assumptions that spatial association and similarity measure relatedness, typological cross-dating is used to assign an age to an assemblage when the projectile points it contains are morphologically similar to points from other assemblages having association-based dates. Typological cross-dating is routinely used in this manner to assign 'approximate' ages to non-ceramic assemblages and surface assemblages that have no association with organic material (*e.g.*, Beck 1994; Bettinger *et al.* 1991; O'Connell & Inoway 1994). Often the spatial distance separating these assemblages is on the order of tens or hundreds of miles.

15.3 Causes of spatial proximity and morphological similarity

When individual artefacts have the same transport agent and environment of deposition there is generally a high probability that spatial proximity will result. When this is the case, it may be justifiable to conclude that the historical relatedness of such artefacts is demonstrated by their spatial proximity. Yet, there are other causes of spatial proximity. Artefacts are often clustered on the landscape because they share analogous depositional histories, even though unrelated individuals widely separated in time may have deposited them. In braided alluvial systems, for example, prehistoric settlement is often constrained to topographically high interfluves (Guccione *et al.* 1988). These geomorphic features often have archaeological deposits containing 'mixed occupations' or 'palimpsests deposits' containing artefacts from unrelated individuals or occupations (Rossignal & Wandsnider 1992; Schiffer 1987). Even in regions where settlement is relatively unconstrained, the deposition of individual artefacts in a particular area will often influence the probability for subsequent deposition of historically unrelated artefacts in the same area — a phenomenon Schiffer (1987, 62) referred to as the 'Arlo Guthrie trash-magnet effect'. Finally, the probability of random association must be considered when assessing whether or not artefacts are historically related. The probability of random association is a function of the configuration of populations across space and the nature of land use by lineages over time. Consequently, there are a number of processes that can result in the spatial proximity of artefacts having unrelated histories in terms of their 'systemic context' (Schiffer 1972).

When the cultural transmission of technological information takes place within a single lineage, all members of a given tool class are generally the products of a similar manufacturing trajectory, resulting in a high probability of similarity among tool-class members. Under these circumstances, it may be safe to assume that the morphological similarity of artefacts demonstrates their historical relatedness. However, bridging arguments based solely on morphological similarity conflate historical relatedness with other causes of similarity. When the similarity of two artefacts is the result of divergence from a common lineage they may in fact be close in age, or very different in age, depending upon when divergence occurred and the unique historical trajectories of each artefact (Hagerstrand 1967). Divergence can result in differential spatial trends in the ages of projectile points as a result of the unique histories of individual points post-dating divergence. Beck (1995), for example, demonstrates the existence of spatial/temporal patterns in Great Basin projectile point types (defined by haft attributes) which may result from divergence. If, on the other hand, similarity among these Great Basin types is caused by convergence, the artefacts may be morphologically similar even though they are historically unrelated. For example, convergence can be demonstrated by the global distribution of barbed projectile points. Even in the absence of direct contact, barbs may evolve independently as an engineering-design solution to similar environmental contexts. Finally, the probability that artefact similarity is the result of chance must also be taken into consideration. Projectile point types or any artefact type for that matter defined by a large number of attributes will have a much smaller probability of recurring by chance than types defined by fewer attributes. Thus, there is an inverse relationship between the complexity of type definition and the probability that artefacts will satisfy the criteria for type membership by chance alone (Dunnell 1971).

Even a partial summary of causal mechanisms is enough to suggest the magnitude of chronological error that can result when spatial proximity and morphological similarity alone are assumed to indicate relatedness. Determining on a case by case basis the causal mechanisms underlying artefact proximity and similarity is critical when these properties are the basis for chronological inferences. Recent research by Dunnell (1978, 1989, 2000) and others (Lipo *et al.* 1997; Lyman *et al.* 1997; Neiman 1995; O'Brien 1996; Wilhelmsen 1997) points out the empirical and theoretical importance of this kind of work. As discussed above, three causes of similarity must be distinguished — homology, analogy, and chance. As important as they are to chronology building these distinctions do not address the error introduced by association-based dating. In the remainder of this paper I outline a new methodology for directly dating projectile points using the thermoluminescence dating technique.

15.4 Thermoluminescence dating of projectile points

Thermoluminescence (TL) provides calendrical dates for heating events in prehistory. When such heating events are archaeological in nature, as in manufacture or use, the dated and target events are isomorphic and the problem of demonstrating historical relatedness is not an issue. When this is the case, TL dates are likely to have greater accuracy than radiocarbon dates. Because heat treatment technology in North America has at least an 8,000 year history as a component of projectile point manufacture (Morse & Morse 1983; Purdy 1982; Rick 1978), there is substantial warrant for examining the potential of TL to provide manufacturing dates for projectile points. If successful, this application may become the cornerstone for more accurate chronologies of projectile point change, as well as a more accurate means of assigning ages to non-ceramic assemblages and surface sites that lack other datable materials.

The TL technique dates the most recent heating event during which the temperature of a crystalline material exceeded approximately 375°C (Aitken 1989; Feathers 1997; Mercier *et al.* 1995). For ceramics this typically corresponds to firing during manufacture. For heated lithics, the event is either heat treatment during manufacture or accidental heating subsequent to manufacture. Not all heat-treated projectile points can be TL dated, however, since successful heat treatment can occur at temperatures lower than 375°C (Luedtke 1992; McCutcheon 1997). The two problems that must be resolved in dating heated points with TL are:

1. distinguishing points that have been heated in excess of 375°C, and
2. distinguishing between heating during manufacture and subsequent heating.

A resolution to the first problem is provided by the experimental heat treatment of cherts. Experimental heating has demonstrated that most cherts heated rapidly to between 350°C and 400°C either explode or exhibit features of catastrophic thermal alteration (Ahler 1983; McCutcheon 1997; Purdy 1974). Catastrophic thermal alteration implies that mechanical properties such as fracture toughness and fracture predictability are so adversely affected by the thermal event that the probability of subsequent manufacture by percussion is significantly reduced (Griffiths *et al.* 1987). The mechanism responsible for catastrophic thermal alteration is most likely rapid heating above the maximum coefficient of expansion for water, which is 373°C (Luedtke 1992). Beyond this temperature water under pressure expands rapidly through vaporization resulting in potlids, crazing, and crenulated fracture surfaces (McCutcheon 1997; Purdy 1974). Cherts can be heated and cooled rapidly without exploding or exhibiting any of these characteristics as long as the maximum temperature does not exceed 350°C (Luedtke 1992). Consequently, the appearance of potlids, crazing, and the crenulated fracture surfaces resulting from crazing, can often be used to identify material that has been sufficiently heated for TL dating.

The second problem, distinguishing between heating during manufacture and subsequent heating, has a probabilistic rather than empirical solution. When heat treatment is part of the manufacturing process for projectile points the associated debitage is likely to contain manufacturing rejects resulting from catastrophic heating. Rejection due to catastrophic heating may occur at various stages of manufacture depending on the amount and degree of crazing and/or potlids, though some points may be successfully manufactured in spite of excessive heating. Successfully manufactured points made from catastrophically heated chert will tend to exhibit potlids more often than crazing, since crazing results in more pervasive structure flaws. Catastrophic heating may also occur after manufacture when a projectile point falls into a hearth or a hearth is built on top of it, or when it is burned on a house floor or subjected to surface fire. In this case, there is no relation between the event of manufacture and the heating event. Catastrophic heating subsequent to manufacture is a chance occurrence, however, and its frequency is likely to be insignificant relative to the occurrence of catastrophic heating during manufacture. This is particularly true when heat treatment of chert is a routine step in projectile point manufacture, as it is in many regions of the United States. Furthermore, catastrophic heating from surface fires is rare since heat is generated for an insufficient amount of time to cause excessive internal heating (A. Skinner, *pers. comm.*, 1998).

By examining the distribution of dates for points of the same stylistic class it may also be possible to distinguish between points heated during and subsequent to manufacture. The chronological relationship between the thermal event dated by TL and the *Manufacturing Event* is illustrated in Figure 15.1 (a) for a single projectile point. The thermal event dated by TL occurs at a point in time relative to the *Manufacturing Event*, which encompasses primary shaping, heat treatment, and secondary shaping of a point. In general, the beginning and ending of the *Manufacturing Event* closely bracket the event dated by TL. The use history of a point covers a much longer period of time, beginning after the point is manufactured and ending at its final deposition. Within this range there may be multiple episodes of use, rejuvenation, recycling, and discard or loss. The use history of a point may be on the order of days or months, or when a point is recycled at a later date, the use history may encompass thousands of years. For a single projectile point, a TL date for a catastrophic thermal event can be a good estimate of when the *Manufacturing Event* occurred. However, a single date for a *Manufacturing Event* provides no information about how long a particular class of projectile point was in use. To identify how long a particular class of projectile was manufactured it is necessary to examine the range of *Manufacturing Events* for several points of the same class. As Figure 15.1 (b) shows, projectile points from the single class are not manufactured simultaneously but have a distribution of *Manufacturing Events* beginning and ending when the first and last members of the class were catastrophically heated. The range of this distribution defines the period of time over which projectile points in this class were manufactured. By identifying this range it should be possible to eliminate projectile points with problematic use histories. Problematic points are those in which the manufacturing dates are significantly later than the range defined by *Manufacturing Events*. These are most likely to be points that have been catastrophically heated subsequent to manufacture, that have been recycled at a later date, or that satisfy the class requirements due to convergent or chance similarity. Until recently, no attempts have been made to date projectile points using TL. The remainder of this paper discusses a pilot study designed to evaluate the use of TL in this context (Wilhelmsen & Miles 1998).

15.5 Sample preparation and experimental procedures

Six projectile points from surface sites in the Central Mississippi Valley were utilised for the pilot study (Fig. 15.2). All of the points exhibited potlids, crazing, and crenulated fracture surfaces. Due to the destructive nature of the technique, thorough metric and technological documentation was conducted prior to sample preparation. All sample preparation then took place in subdued red light so as not to bleach the natural TL signal. Several sample cores were removed from the centre of each projectile point using a variable speed drill equipped with a hollow $\frac{5}{8}$th inch diamond-impregnated bit. The drill was operated at low speed while the sample was cooled in a water bath to minimise heat. Two millimetres of material were removed from the light-exposed ends of the cores using a diamond wafering saw also operated at low speed and water cooled. One core was archived and the remaining cores crushed in a tool-steel mortar and pestle, and sieved to obtain a 90–125μm fraction and a $< 90\mu m$ fraction. Magnetic separation was conducted after sieving to remove steel tool fragments. Each fraction was then placed in dilute HCL (1:1) for 30 minutes to remove any carbonates that might contribute spurious TL, and rinsed in acetone to remove extremely small grains that might be affected by triboluminescence from crushing (Valladas 1978). The $< 90\mu m$ fraction was then settled in acetone to isolate the 1–8um fraction and this was settled on aluminum disks (Zimmerman 1971).

The 90–125μm grains were used for equivalent dose determination, and the fine grains for measurement of alpha efficiency. The coarse grains were dispersed on planchets and normalized by weight. TL was measured on a Daybreak glow oven in an N_2 atmosphere using a heating rate of 1°C/sec and a maximum temperature of 450°C. Emission was constrained by a Corning 7-59 (blue) filter and a Schott HA3 heat-absorbing filter. Irradiations were administered with a Sr-90 beta source and an Am-241 alpha source.

For equivalent dose, both additive and regeneration growth curves were constructed using 3–5 points at each dose increment. Regeneration followed heating to 450°C. Stable signal was determined by a plateau test. Dose intercepts for both additive dose and regeneration curves were plotted as a function of temperature, and the plateau region common to both was selected for analysis. Equivalent dose was estimated from the slide procedure (Prescott *et al.* 1993) where the dose shift required to bring the two curves into coincidence, corrected for any sensitivity change, was taken as the equivalent dose estimate. The slide method avoids errors of extrapolation in additive dose and sensitivity change in regeneration by using the shape of the regeneration curve to define the extrapolated region in additive dose. Alpha efficiency used the b-value method by comparing the slopes of the additive curves using beta and alpha irradiation.

Radioactivity for both the lithic and associated sediment sample was measured by thick-source alpha counting for uranium and thorium contents and by flame photometry for potassium-40. Moisture content in the sediments was estimated from typical water contents for appropriate soil textures determined for the region in soil survey manuals (Butler 1985; Graves Jr. 1983).

15.6 Experimental results

Figure 15.3 shows the glow curves for four successfully dated projectile points and the increase in signal with dose. The additive dose curves and the regenerated dose curves exhibit a systematic increase in TL with dose. Figure 15.4 shows the plateau regions for these four points with the plateau regions used in the analysis shaded in black. Two additional points failed the plateau test and could not be dated. One of these points (Sample Number UW256) is shown in the bottom of Figure 15.4. Note the absence of a plateau suggesting that the point was insufficiently heated in prehistory and therefore cannot be dated by thermoluminescence.

The results of the slide method for the four dated points, and their derived equivalent doses (DE), are shown in Figure 15.5. The high degree of scatter indicated by the error bars is likely to result from low natural signal. The additive dose points are represented by circles and the regeneration dose points by squares. The stars represent the regeneration points after they have been shifted to coincide with the additive dose points. Scaling for sensitivity changes were not necessary for any of the points, and a linear fit was found to be satisfactory.

The radioactivity results are shown in Table 15.1,

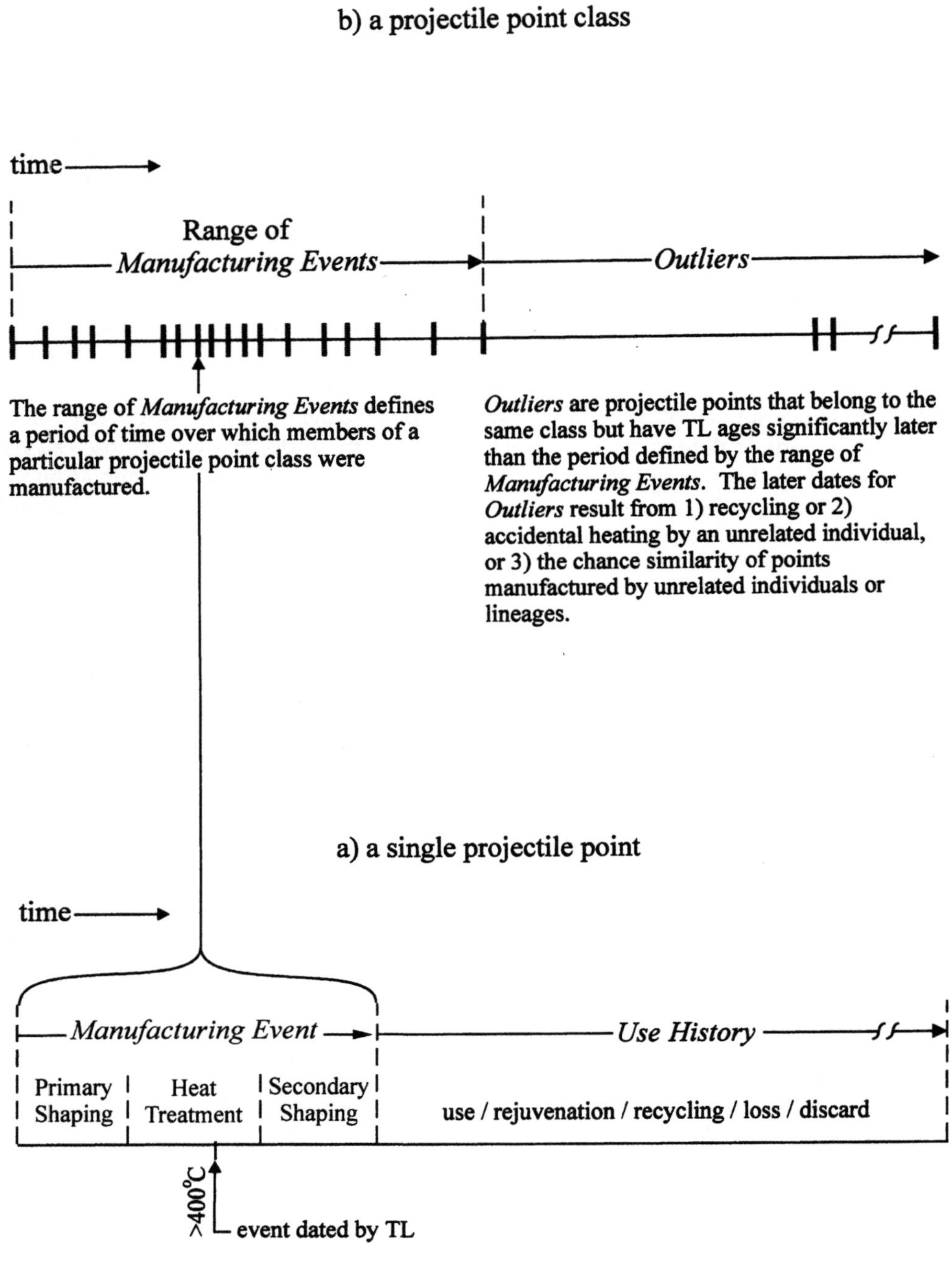

Figure 15.1: (a) The chronological relationship between a *Manufacturing Event*, the event dated by TL, and the *Use History* of a single projectile point. (b) The chronological relationship between a range of *Manufacturing Events* and outliers for a projectile point class containing several members.

Lab no.	Dose Rate Contributions				
	cosmic	α	β	γ	total dose
UW285	0.2495 ± 0.0501	0.0527 ± 0.0113	0.0939 ± 0.0097	0.4891 ± 0.0333	0.8852 ± 0.0620
UW288	0.2493 ± 0.0514	1.6041 ± 0.2039	0.7299 ± 0.0395	0.6386 ± 0.0450	3.2219 ± 0.2186
UW317	0.2497 ± 0.0154	0.0236 ± 0.0085	0.0999 ± 0.0146	0.6177 ± 0.0467	0.9909 ± 0.0715
UW321	0.2499 ± 0.0515	0.0423 ± 0.0123	0.0673 ± 0.0082	0.5583 ± 0.0429	0.9178 ± 0.0686

Table 15.1: Radioactivity results for all dated projectile points after conversion to Grays and adjustment for alpha efficiency and moisture content.

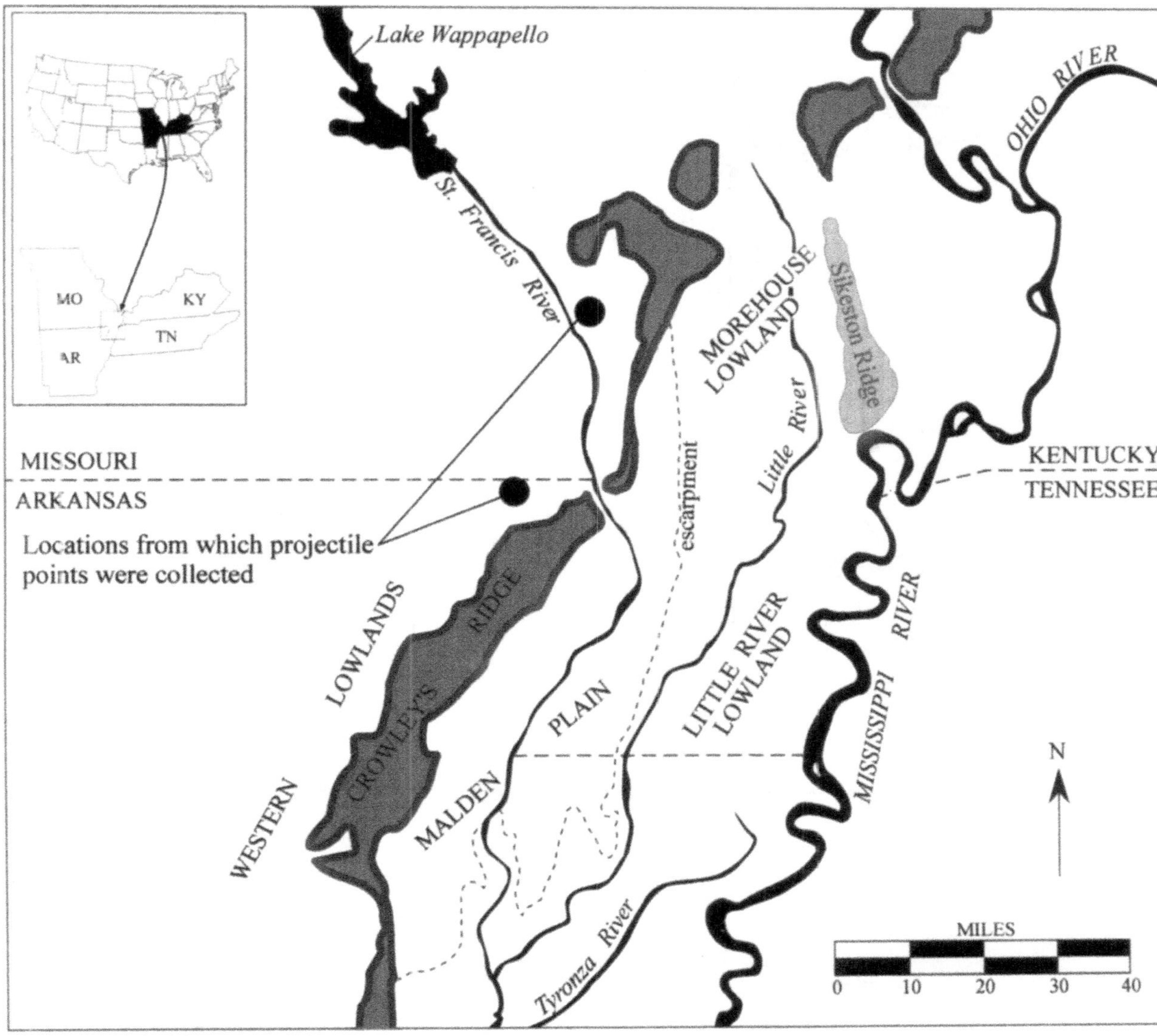

Figure 15.2: Map of the Central Mississippi Valley showing the location from which projectile points were collected and dated.

after conversion to Grays and adjustment for alpha efficiency and moisture content. As can be seen, the gammas provide the bulk of the dose rate, most of them originating from the external sediment. Point UW288 is the exception due to high uranium content. An accurate estimate of the external dose rate is therefore essential for dating these points. Previous studies have shown that the radioactive composition in sediments from site areas in the region does not vary greatly, due to widespread homogeneity in the composition of the Mississippi Valley sediments. For example, on 37 samples for eight different sites the gamma dose rates varied by only 10% (J. K. Feathers, *pers. comm.*, 1998). This low variance is beneficial for minimizing error in calculating external dose rates. It is noteworthy that the cosmic dose rate, calculated from global values (Prescott & Hutton 1988), and enhanced by the surface location of the samples, is more significant here than normal.

The derived ages for these points are shown in Table 15.2. The TL date for point UW285 places its manufacture at the early end of the Late Archaic period; the date for UW288 places its manufacture in the Early to Middle Woodland period; and the date for UW317 places its manufacture in the Middle Woodland period. As this table shows, there is good agreement between the TL dates and the culture historical ranges for morphologically similar points based on typological cross-dating. Although there is no reason to suspect that these dates should agree (since TL and ^{14}C date different events) the TL dates generally support the age ranges suggested by the association-based dates. In contrast, the TL date for point UW321 places its manufacture at AD 516 ± 489, which postdates the culture historical range for morphologically similar points by about 7,000 years. Resolving this disagreement requires more dates, as well as data on the depositional history of projectile points, settlement history

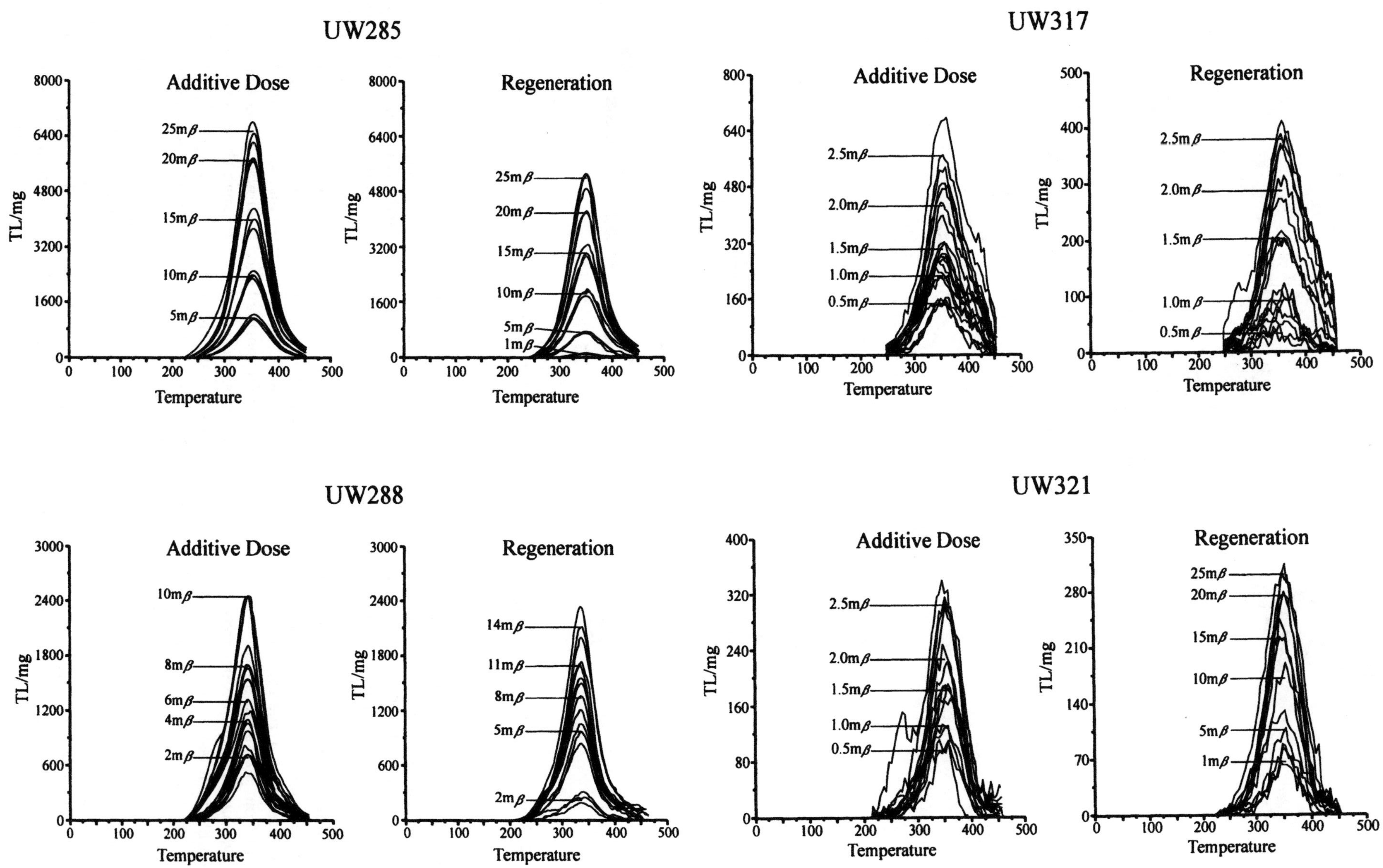

Figure 15.3: Glow curves for four successfully dated projectile points (Sample Numbers UW285, UW288, UW317, and UW321). The additive dose curves and the regenerated dose curves exhibit a systematic increase in TL with dose.

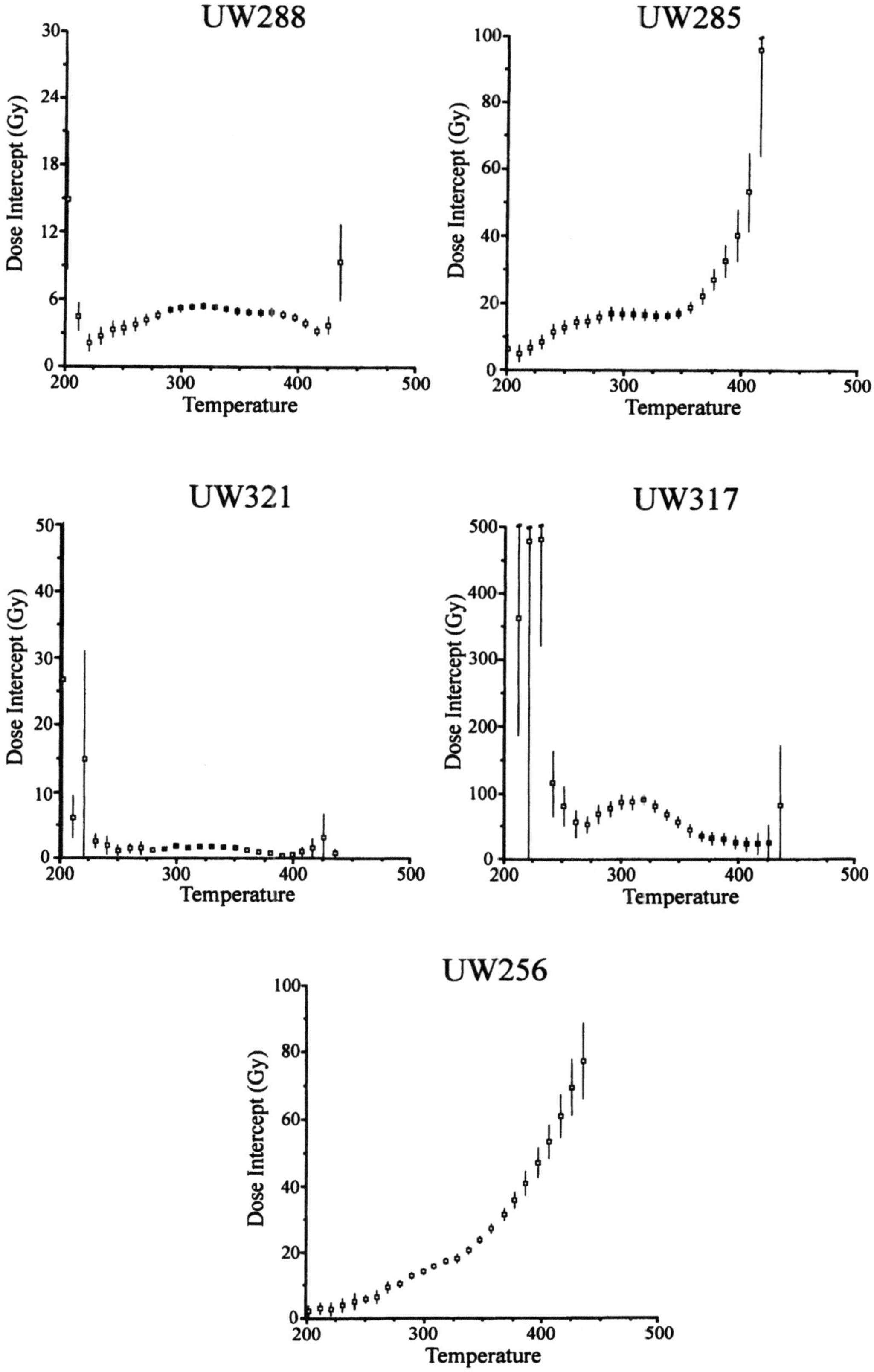

Figure 15.4: Scatterplots showing the dose intercepts for all projectile points that passed the plateau test and one that failed (Sample Number UW256) probably due to insufficient heating in prehistory. The broad plateau regions on the successfully dated points (infilled) indicate sufficient heating for TL and are used to calculate the equivalent dose (DE). Note the absence of a plateau region for Sample Number UW256.

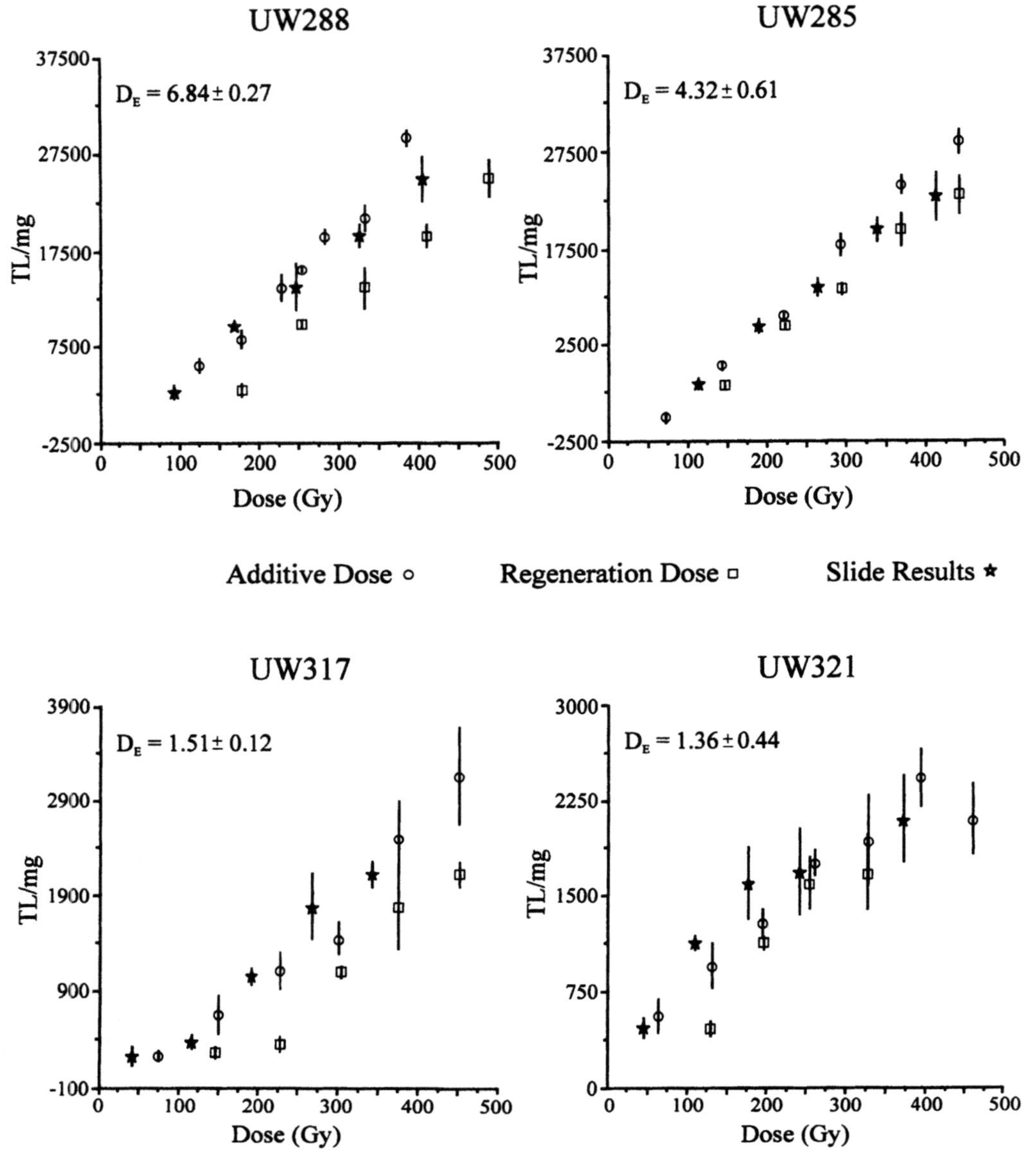

Figure 15.5: Scatterplots showing the results of the slide method for calculating the equivalent dose (DE) used in the dating equation.

of the area, and raw material availability. In stone-poor areas, for example, the use of a point may post-date the range of *Manufacturing Events* for the type by several millennia because recycling is a least-cost method of resource acquisition. The probability of recycling also increases in areas that are stone poor and repeatedly occupied. Since point UW321 comes from just such an environment, the TL date for its manufacture may be indicative of recycling. More dates may tell us something about the temporal structure of recycling as a means of resource acquisition.

15.7 Summary

Two directions for future work are suggested by these results:

1. reducing the identification error for catastrophically heated points, and
2. reducing sample size requirements.

Experimental studies demonstrate that colour and luster changes indicative of heat treatment do not guarantee a point was heated in excess of 375°C, but that potlids and crazing (which typically occur at temperatures above 350°C) may. In addition to examining the mechanical properties of unheated and heat-treated cherts, experimental studies should focus more closely on identifying the characteristics of catastrophic heating and the temperature ranges over which these occur (*e.g.*, McCutcheon 1997). Electron spin resonance may also be used to identify prehistoric heating temperatures and may in fact yield lower identification error (*e.g.*, Toyoda *et al.* 1998, 1993). This technique should be considered in future dating experiments. Minimizing the destructiveness of sample

Lab No.	TL Age	age by association/ similarity	culture historical type
UW285	2883 BC ± 765	2500–1000 BC	Ledbetter Stemmed
UW288	126 BC ± 166	1500 BC–AD 200	Burkett/Delhi
UW317	AD 474 ± 163	200BC–AD 200	Snyders
UW321	AD 516 ± 489	7500–6900 BC	Kirk Corner-notched

Table 15.2: Derived ages for all projectile points that passed the plateau test and were successfully dated.

preparation should also be an area of concern. Use of single aliquot analyses, such as the SARA method (Mejdhal & Botter-Jensen 1997), may reduce sample size requirements to the extent that smaller cores may be used, leaving the majority of the point intact.

Our ability to explain the evolution of projectile point technology is in part dependent on obtaining direct dates for manufacturing events. All things being equal, TL dates will always be more accurate than radiocarbon dates, if not as precise, when they are used to date archaeologically relevant target events. Because the target event is dated directly there is no need for bridging arguments that conflate the causes of spatial proximity and similarity, or that may be difficult or impossible to quantify. The results of this pilot study demonstrate that catastrophic heating is sufficient to zero the natural TL signal in projectile points during manufacture. Since catastrophically heated points often exhibit stylistic traits, TL can be used to evaluate and refine existing chronologies. Additionally, understanding the timing of manufacture and the use history of projectile points is within our grasp if we are not limited to ^{14}C dates, or other indirect dating methods. Understanding these characteristics of projectile points is critical for explaining the evolution of projectile technology.

Acknowledgements

The laboratory research for this paper was conducted in the Thermoluminescence Dating Laboratory at the University of Washington, which is directed by Dr Feathers. The first author would like to cite several individuals who assisted him directly and indirectly. Dr Norbert Mercier of the Laboratoire des Sciences du Climat et de l'Environnement, Gif sur Yvette, France, provided helpful comments on laboratory procedures. Michael Shott and Kim Kornbacher commented extensively on an earlier draft of this manuscript. While the contributions of these individuals have improved the final manuscript, the first author remains solely responsible for errors or omissions. A special thanks is extended to Mr and Mrs Taylor Miles of Kennett, Missouri who provided the samples for this research, and who generously supported the first author while working in the field. Most importantly, he would like to thank Dr Robert C. Dunnell, 'The Boss', for being such a demanding teacher and for inspiring my research.

References

AHLER, S. A. 1971. *Projectile Point Form and Function at Rodgers Rockshelter, Missouri.* Research Series, No. 8. Missouri Archaeological Society, Columbia.

AHLER, S. A. 1983. 'Heat treatment of knife river flint.' *Lithic Technology* 12(1): 1–8.

AITKEN, M. J. 1989. 'Luminescence dating: a guide for non-specialists.' *Archaeometry* 31: 147–59.

BECK, C. 1994. *Dating in Exposed and Surface Contexts.* University of New Mexico Press, Albuquerque.

BECK, C. 1995. 'Functional attributes and the differential persistence of Great Basin dart points.' *Journal of California and Great Basin Anthropology* 17: 222–43.

BERGMAN, C. A. & E. MCEWEN 1997. 'Sinew-reinforced and composite bows: Technology, function, and social implications.' *In* Knecht (1997), pp. 143–60.

BETTINGER, R. L., J. F. O'CONNELL & D. H. THOMAS 1991. 'Projectile points as time markers in the Great Basin.' *American Anthropologist* 93: 166–72.

BLEED, P. 1986. 'The optimal design of hunting weapons: Maintainability or reliability.' *American Antiquity* 51: 737–47.

BROYLES, B. J. 1966. 'Preliminary report: The St Albans site (46KA27), Kanawha County, West Virginia.' *The West Virginia Archaeologist* 19: 1–43.

BUTLER, R. E. 1985. *Soil Survey of Stoddard County, Missouri.* U.S Department of Agriculture, Soil Conservation Service.

CHAPMAN, C. H. 1947. 'A preliminary survey of Missouri archaeology Part II: Middle Mississippi and Hopewellian cultures.' *Missouri Archaeologist* 10(2): 57–94.

CHAPMAN, C. H. 1948a. 'A preliminary survey of Missouri archaeology Part III: Woodland Cultures and the Ozark Bluff Dwellers.' *Missouri Archaeologist* 10(3): 95–132.

CHAPMAN, C. H. 1948b. 'A preliminary survey of Missouri archaeology Part IV: Ancient cultures and sequence.' *Missouri Archaeologist* 10(4): 133–64.

CHRISTENSON, A. L. 1997. 'Side-notched and unnotched arrowpoints: Assessing functional differences.' *In* Knecht (1997), pp. 131–42.

COE, J. L. 1964. 'The formative cultures of the Carolina Piedmont.' *Transactions of the American Philosophical Society* 54(5): 1–130.

DEAN, J. S. 1978. 'Independent dating in archaeological analysis.' *Advances in Archaeological Method and Theory* 1: 223–55.

DUNNELL, R. C. 1971. *Systematics in Prehistory.* Free Press, New York.

DUNNELL, R. C. 1978. 'Style and function: A fundamental dichotomy.' *American Antiquity* 43: 192–202.

DUNNELL, R. C. 1989. 'Aspects of the application of evolutionary theory in archaeology.' *In* C. C. Lamberg-Karlovsky (ed.), *Archaeological Thought in America*, pp. 35–49. Cambridge University Press, Cambridge.

DUNNELL, R. C. 2000. 'Dating.' *In* L. Ellis (ed.), *Archaeological Method and Theory: An Encyclopedia*, pp. 143–50. Garland Publishing, New York and London.

FAWCETT, W. B. 1998. 'Chronology and projectile point neck-width: An Idaho example.' *North American Archaeologist* 19: 59–85.

FEATHERS, J. K. 1997. 'The application of luminescence dating in American archaeology.' *Journal of Archaeological Method and Theory* 4: 1–66.

FINLAYSON, B. & S. MITHEN 1997. 'The microwear and morphology of microliths from Gleann Mor.' *In* Knecht (1997), pp. 107–29.

FLENNIKEN, J. J. & A. W. RAYMOND 1986. 'Morphological projectile point typology: Replication, experimentation, and technological analysis.' *American Antiquity* 51: 603–14.

GRAVES JR., L. H. 1983. *Soil Survey of Butler County and Part of Ripley County Missouri.* U.S. Department of Agriculture, Soil Conservation Service.

GRAYSON, D. K. 1993. *The Desert's Past: A Natural History of the Great Basin.* Smithsonian Institution Press, Washington D.C.

GRIFFITHS, D. R., C. BERGMAN, C. CLAYTON, K. OHNUMA, G. V. ROBINS & N. J. SEELEY 1987. 'Experimental investigation of the heat treatment of flint.' *In* G. de G. Sieveking & M. H. Newcomer (eds.), *The Human Uses of Flint and Chert*, pp. 43–52. Cambridge University Press, Cambridge.

GUCCIONE, M. J., R. H. LAFFERTY III & L. S. CUMMINGS 1988. 'Environmental constraints of human settlement in an evolving Holocene alluvial system, the Lower mississippi Valley.' *Geoarchaeology* 3: 65–84.

HAGERSTRAND, T. 1967. *Innovation Diffusion as a Spatial Process.* University of Chicago Press, Chicago.

JUSTICE, N. D. 1987. *Stone Age Spear and Arrow Points of the Midcontinental and Eastern United States: A Modern Survey and Reference.* Indiana University Press, Bloomington and Indianapolis.

KNECHT, H. (ed.) 1997. *Projectile Technology.* Plenum Press, New York.

LIPO, C. P., M. E. MADSEN, R. C. DUNNELL & T. HUNT 1997. 'Population structure, cultural transmission, and frequency seriation.' *Journal of Anthropological Archaeology* 16: 301–34.

LITFIN, J. C., P. C. JACKSON & K. D. VICKERY 1993. *Chronological Seriation Approach to Fort Ancient Triangular Arrowpoints in the Central Ohio Valley.* Paper presented at the 58th Annual Meeting of the Society for American Archaeology, St. Louis.

LUEDTKE, B. E. 1992. *An Archaeologist's Guide to Chert and Flint.* Archaeological Research Tools No. 7. Institute of Archaeology, University of California, Los Angeles.

LYMAN, R. L., J. O'BRIEN & R. C. DUNNELL 1997. *The Rise and Fall of Culture History.* Plenum Press, New York.

MCCUTCHEON, P. T. 1997. *Archaeological Investigations of Stone Tool Heat-Treatment Technology in Southeast Missouri: An Experimental Approach.* Ph.D. thesis, Department of Anthropology, University of Washington, Seattle.

MEJDHAL, V. & L. BOTTER-JENSEN 1997. 'Experience with the SARA method.' *Radiation Measurements* 27(2): 291–4.

MERCIER, N., H. VALLADAS & G. VALLADAS 1995. 'Flint thermoluminescence dates from the CFR laboratory at Gif: Contributions to the study of the chronology of the Middle Palaeolithic.' *Quaternary Science Reviews* 14: 351–64.

MORSE, D. F. & P. MORSE 1983. *Archaeology of the Central Mississippi Valley.* Academic Press, Orlando.

NEIMAN, F. 1995. 'Stylistic variation in evolutionary perspective: Inferences from decorative diversity and interassemblage distance in Illinois Woodland ceramic assemblages.' *American Antiquity* 60: 7–36.

O'BRIEN, M. J. (ed.) 1996. *Evolutionary Archaeology: Theory and Application.* University of Utah Press, Salt Lake City.

O'CONNELL, J. F. & C. M. INOWAY 1994. 'Surprise valley projectile points and their chronological significance.' *Journal of Californian and Great Basin Anthropology* 16: 162–98.

PRESCOTT, J. R., D. J. HUNTLEY & J. T. HUTTON 1993 'Estimation of equivalent dose in thermoluminescence dating — the Australian Slide Method.' *Ancient TL* 11: 1–5.

PRESCOTT, J. R. & J. T. HUTTON 1988. 'Cosmic ray and gamma ray dosimetry for TL and ESR.' *Nuclear Tracks and Radiation Measurements* 14: 223–7.

PURDY, B. A. 1974. 'Investigations concerning the thermal alteration of silica minerals: An archaeological approach.' *Tebiwa* 17: 37–66.

PURDY, B. A. 1982. 'Pyrotechnology, prehistoric application to chert materials in North America.' *In* T. A. Wertime & S. F. Wertime (eds.), *Early Pyrotechnology: The Evolution of the First Fire-Using Industries*, pp. 31–44. Smithsonian Institution Press, Washington D.C.

RAU, C. 1876. 'The archaeological collections of the United States National Museum in charge of the Smithsonian.' *Smithsonian Contributions to Knowledge* 22(4).

RICK, J. W. 1978. *Heat-Altered Cherts of the Lower Ilinois Valley: An Experimental Study in Prehistoric Technology.* Prehistoric Records 2. Northwestern University Archaeological Program, Evanston.

ROSSIGNAL, J. & L. WANDSNIDER (eds.) 1992. *Space, Time, and Archaeological Landscapes.* Plenum Press, New York.

SCHIFFER, M. B. 1972. 'Archaeological context and systemic context.' *American Antiquity* 37(2): 156–65.

SCHIFFER, M. B. 1987. *Formation Processes of the Archaeological Record.* Albuquerque, University of New Mexico Press.

SHOTT, M. J. 1996. 'Innovation and selection in prehistory: A case study from the American Bottom.' *In* G. H. Odell (ed.), *Stone Tools: Theoretical Insights into Human Prehistory*, pp. 279–309. Plenum Press, New York.

THOMAS, D. H. 1986. 'Points on points: A reply to Flenniken and Raymond.' *American Antiquity* 51(3): 619–27.

TOYODA, S., R. C. DUNNELL & M. IKEYA 1998. *Determining the Temperature of the Prehistoric Lithic Heat Treatment by Using ESR (Electron Spin Resonance) Signals.* Paper presented at the 63rd Annual Meeting of the Society for American Archaeology, Seattle.

TOYODA, S., M. IKEYA, R. C. DUNNELL & P. T. MCCUTCHEON 1993. 'The use of Electron Spin Resonance (ESR) for the determination of prehistoric lithic heat treatment.' *Applied Radiation and Isotopes* 44: 227–31.

VALLADAS, H. 1978. 'Thermoluminescence dating of burnt stones from prehistoric sites.' *In* M. J. Aitken & V. Mejdahl (eds.), *A Specialist Seminar on Thermoluminescence Dating. Part One*, PACT 2, pp. 180–4. Council of Europe, Strasbourg.

WIANT, M. D. & H. HASSAN 1985. 'The role of lithic resource availability and accessibility in the organization of technology.' *In* S. Vehick (ed.), *Lithic Resource Procurement: Proceedings from the Second Conference on Prehistoric Chert Exploitation*, pp. 101–14. SIU Center for Archaeological Investigations, Occasional Paper 4, Carbondale.

WILHELMSEN, K. H. 1997. *Events of Manufacture Ordered by Attribute-Scale Occurrence Seriation.* Paper presented at the 62nd Annual Meeting of the Society for American Archaeology, Nashville.

WILHELMSEN, K. H. & T. MILES 1998. *Thermoluminescence Dating of Catastrophically Heated Projectile Points.* Paper presented at the 63rd Annual Meeting of the Society for American Archaeology, Seattle.

ZIMMERMAN, D. W. 1971. 'Thermoluminescent dating using fine grains from pottery.' *Archaeometry* 13: 29–52.

Kris H. Wilhelmsen
InterNAP Network Services
Two Union Square
601 Union Street, Suite 1000
Seattle, WA 98101
USA kris@internap.com
kwilhelm@u.washington.edu

James K. Feathers

Department of Anthropology
University of Washington
Box 353100
Seattle, WA 98195–3100
USA
jimf@u.washington.edu

16

A Petrological Study of the Emerald Mines in the Egyptian Eastern Desert

I. Shaw[1] and J. Bunbury[2]

[1]*Department of Archaeology, University of Liverpool;* [2]*Department of Earth Sciences, University of Cambridge*

Abstract

This study of five of the major emerald mining localities within the Sikait-Zubara region of Egypt's Eastern Desert investigates the potential for provenancing emeralds both from specific countries and from specific ancient Egyptian mines. The gemmological examination of several samples from Sikait-Zubara and elsewhere indicates that much of the geochemical variation in world emeralds is mirrored within single grains of emerald tailings; there are, however, two sources of emerald that can be distinguished from those of Egypt: the Urals and Colombia. These observations provide the potential for investigations of the world emerald market during the time of operation of the Egyptian mines and also the detection of forgeries manufactured from New World emeralds. The analyses also contribute to the general debate concerning the validity of archaeological provenancing, suggesting firstly that analysis must study variation within single grains as well as between grains sampled from different regions, and secondly that purely geochemical definitions of 'sources' or 'provenances' do not necessarily provide ready-made solutions to archaeological problems concerning the spatial movement of raw materials and artefacts.

16.1 Introduction

Emeralds are a green form of the mineral beryl ($Be_3Al_2[Si_6O_{18}]$) whose formation requires the juxtaposition of granite pegmatites (a source of beryllium) and a source of reduced iron oxide. Hence emeralds are rare and, being of great hardness, were prized as gem stones both before and after the development of the technologies required for polishing and facetting the stones. Apart from Egypt, the other known sources are Austria, Brazil, Colombia, India and the Urals.

In this paper we describe several emerald-mining sites in the Egyptian Eastern Desert (in Wadi Gimal, Wadi Nuqrus, Wadi Sikait and Gebel Zubara — Fig. 16.1) and present the results of an electron microprobe analysis of samples of emerald from each of the mines. We then discuss the implications of this evidence in terms of the petrological sourcing of beryl from ancient Egyptian mines. Such provenancing of emeralds has the potential to elucidate the history of emerald procurement and exchange, firstly by allowing the relative importance of the Egyptian emeralds to be assessed in terms of the changing world emerald market, and secondly by clarifying the basis on which the mines were worked within the Egyptian Eastern Desert.

On a more methodological level, however, our case-study also contributes to the ongoing debate regarding the nature and validity of provenancing studies in archaeology. As Hughes (1998, 113) notes, in the context of obsidian sourcing, 'chemistry-based analyses, by themselves, help us apprehend only a part of the variability we are interested in as archaeologists'. Our analyses of Egyptian beryl suggest that geochemistry, far from being the key to all our provenancing problems (as Grubessi *et al.* 1989, 1990, tend to suggest), actually represents only a small part of a more complex picture, comprising archaeological, anthropological and geological factors of equal importance.

16.2 A brief history of Egyptian emerald mining

In the 1930s, Alfred Lucas analysed a number of supposed Egyptian beryls in items of jewellery of the pharaonic period (such as Middle Kingdom items from the Dahshur necropolis) and concluded that in all cases the stones were actually green feldspar or olivine. Although an emerald was supposedly identified in a

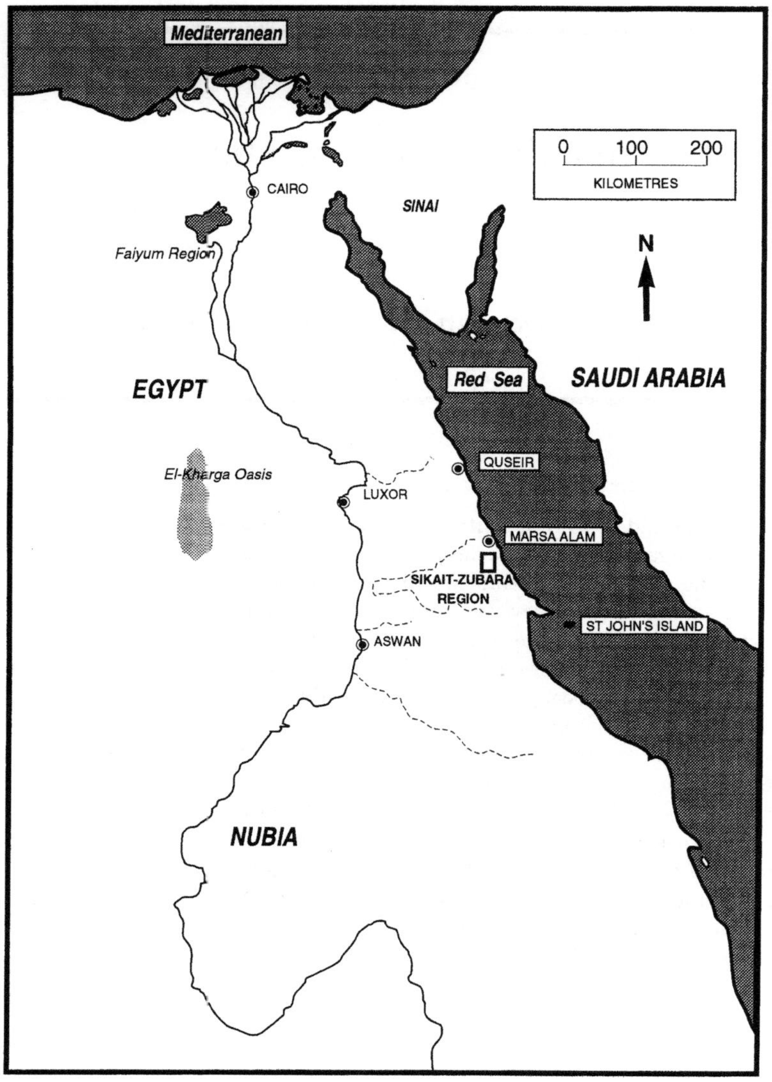

Figure 16.1: Map of Egypt showing the location of the Sikait-Zubara region.

necklace from the Predynastic site of el-Kubaniya, immediately to the north of Aswan (Lucas 1962, 390), there have not yet been any definite scientific identifications of beryl in the Pharaonic period or earlier, and the stone does not appear to have been used regularly in Egyptian jewellery until the Roman period (*c.* 30 BC to AD 311) at the earliest.

The emerald mines in the Sikait-Zubara region of the Eastern Desert are thought to have been the only source of emeralds for Europe, Asia and Africa in the Greco-Roman period (with the possible exception of the Habachtal emerald mines in Austria), and they continued to be exploited until at least the Middle Ages, when Arab writers document the appearance of larger, heavier stones from the Indian subcontinent. Sinkankas (1986, 184–210), however, states that emeralds were not mined in India until the 1930s, therefore the Indian 'emeralds' at this date may actually have been green sapphires. In the post-medieval period, the importance of Egyptian emeralds was even further reduced by the arrival of large quantities of high quality emeralds from the New World.

The principal emerald-mining sites in the Eastern Desert have not been scientifically studied since the expeditions of Donald MacAlister (1900), E. S. Thomas (1909) and G. W. Murray (1925), and the Sikait-Zubara region as a whole has not yet been examined in terms of the changing strategies of emerald mining over time (although see Shaw *et al.* 1999, for a preliminary attempt).

16.3 Summary of the fieldwork

In March 1994, we undertook geoarchaeolological survey in the Sikait-Zubara region of Egypt, firstly in order to explore the possibility of sourcing ancient emeralds to specific Egyptian mines and secondly in order to obtain a better understanding of the ways in which emerald mining techniques were influenced by geological conditions and technological advances.

A preliminary *in situ* examination of potsherds in 1994 indicated that the five sites included in our survey covered an extremely wide chronological range, extending at the very least from the late Roman period at the Wadi Gimal A site to the 16th century at Gebel Zubara. Since the rock-cut shrines at Gebel Sikait are usually assigned to the Ptolemaic period, the full period of exploitation at Sikait-Zubara may have spanned more than 1500 years. It should be noted, however, that the proposed dates for each of the sites described below are based almost entirely on preliminary studies of a small sample of the ceramics (for more detailed discussion of the archaeological remains at Sikait-Zubara, and the logistics of Roman and Byzantine emerald mining, see Shaw 1999; Shaw *et al.* 1999).

16.3.1 Wadi Gimal A (*c.* 4th–5th centuries AD)

The Wadi Gimal A emerald mining site (34° 44.57' E, 24° 31.22' N), located at the southern side of the Wadi Gimal (Fig. 16.2), approximately 15km west of the Red Sea coast, had not been examined before the 1994 survey. This one site could serve as a good case-study for post-pharaonic emerald mining as a whole, as a result of the small and self-contained nature of the operations, as well as the relatively undisturbed condition of the archaeological remains.

The miners' settlement (consisting of many dry-stone buildings of various sizes, the largest being 6.4 × 8.5m in area) is located beside a rocky eminence in a small side wadi that empties into the southern edge of the broad Wadi Gimal. There are two main sources of emerald at Wadi Gimal A: pegmatitic deposits and schist lenses (see Table 16.1). The whole area is dissected by thin veins of quartz and granite pegmatite, measuring up to 30cm in width. The pegmatite veins have been thoroughly mined out, with the removal of blocks of metamorphosed granite wherever necessary. Mafic lenses — distributed evenly throughout the rock — have also been removed wholesale from the rock, in some cases by tunnelling.

The tip heaps at Wadi Gimal fall into two categories:

1. coarse-grained tips, generally immediately adjacent to an area of mining, and
2. fine-grained heaps distributed around the perimeter of the settlement itself, rising to a height of about 30 cm above the wadi floor and sometimes bunched to form a terrace pockmarked with depressions.

Fragments of beryl and chips of beryl-bearing rock are common within the fine-grained tips, but, with the exception of a single broken, bored bead, none of the beryl seems to have been worked *in situ*.

16.3.2 Wadi Gimal B (4th–5th centuries AD)

Further to the northeast in the Wadi Gimal, and close to the intersection with the north-south running Wadi Nuqrus (see fig. 2), are two small mining localities in close proximity to one another, one in a tributary wadi to the northwest of Wadi Gimal (34° 48.98' E, 24° 34.08' N) and the other in a winding diversion to the southeast. The most substantial settlement remains at Wadi Gimal B are located in the northwestern site, consisting of several large dry-stone buildings along the sides of the wadi. Evidence for the ancient quarrying of the contacts and pegmatites of granitic veins and other intrusions is widespread, but the volume of fine-grained quarry tips is very low.

Although the Wadi Gimal B settlement is somewhat smaller than that at Wadi Gimal A, the dozen or so principal buildings are much more impressive constructions, each of which would have probably required a similar number of man-hours for their construction as the main 'colonnaded' building at Wadi Gimal A. There is, however, a surprising dearth of both pottery and worked emerald material at Wadi Gimal B, particularly when compared with the large quantities of surface material at all of the other sites examined in 1994. It is difficult to explain this apparent conflict between the amount of effort expended on construction and quarrying at Wadi Gimal B and the relatively brief period of occupation suggested by the ceramic evidence.

16.3.3 Wadi Nuqrus (6th–9th centuries AD)

The Wadi Nuqrus mines and settlement (34° 48.98' E, 24° 37.27' N) are located on the east side of the Wadi Nuqrus, a few kilometres to the north of Wadi Gimal B. The mines in the Wadi Nuqrus were far more extensive and richer in emeralds than the two Wadi Gimal mines, and were therefore presumably exploited over a longer period. There would thus have been more incentive to invest in the settlement. On the northern side of the wadi, the buildings are situated over a band of metamorphosed sandy material that marks the upper limit of this part of the site.

Unfortunately, the geological context in which the emeralds formed at the Wadi Nuqrus mines has proved somewhat difficult to elucidate, given the fact that the location in which they formed (just below the main area of settlement) is now almost entirely covered by the miners' tip heaps, each consisting of a central depression surrounded by debris. It seems, however, that the lode formed an inclined slab parallel to the surface of the hill. This slab appears to have been worked from the bottom of the slope upwards, and the discarded material would thus have been tipped back onto the already exhausted workings. The dimensions of the tips - 3 to 4m wide and about 2m deep — are much greater than those at Wadi Gimal A and B, and there is no evidence for the two-tier tipping described at the Wadi Gimal sites. Grains of emerald and clasts containing emerald are common among the tips, and the great variation in the size of the clasts (ranging from pieces

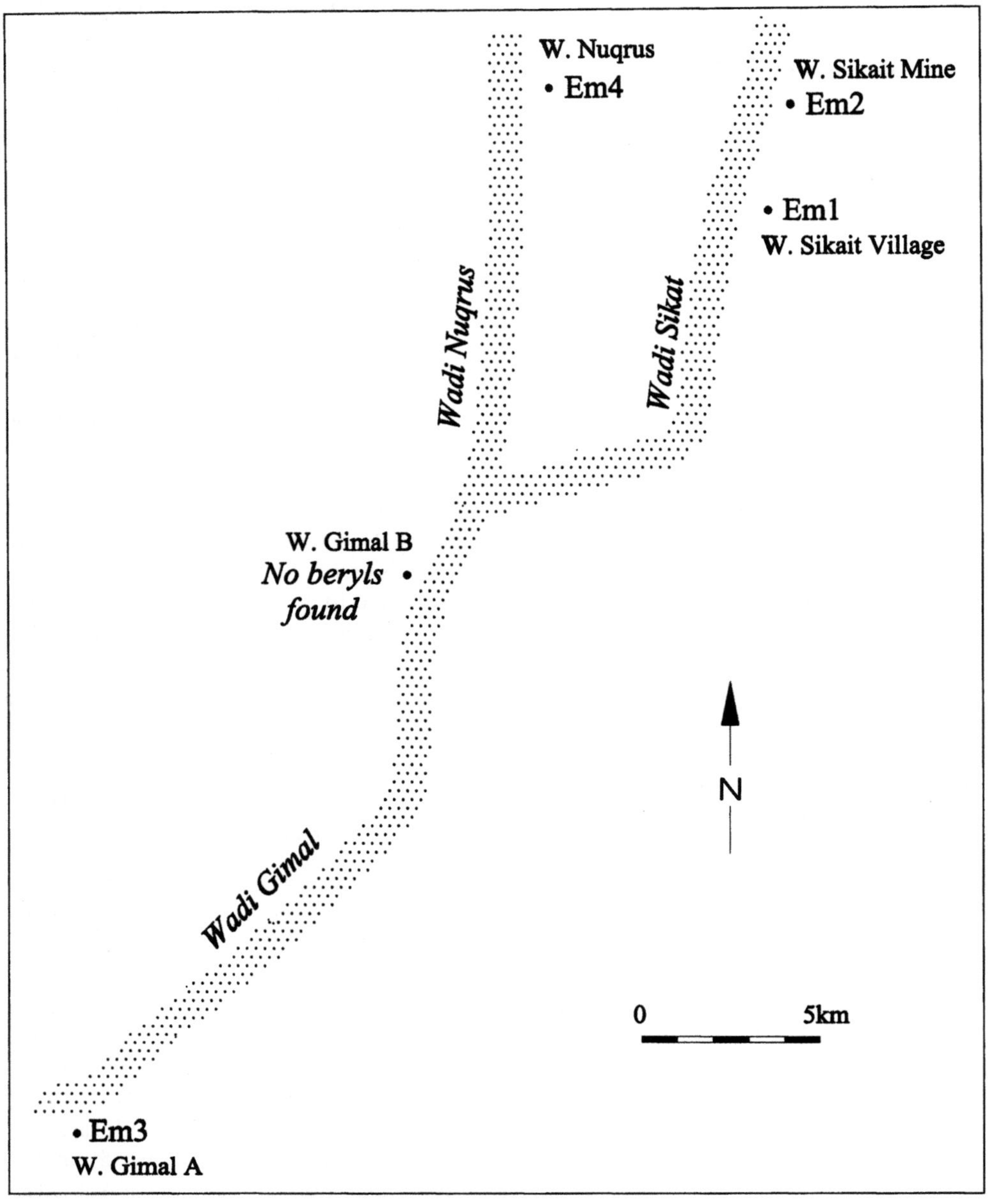

Figure 16.2: Sketch plan of the locations of the sites in Wadi Gimal, Wadi Nuqrus and Wadi Sikait.

Types of formation	Gimal A	Gimal B	Nuqrus	Sikait	Zubara
Boundaries of granite intrusions		✓			✓
Adjacent to quartz veins spawned by a granite pegmatite			✓	✓	✓
Within granite pegmatites	✓	?			
Mica schist lenses within a metamorphosed granite	✓				

Table 16.1: The different types of emerald formation at Wadi Gimal A and B, Wadi Nuqrus B, Wadi Sikait and Gebel Zubara.

with a width of 20cm to fine grains of sand) suggests that the minerals were extracted in one stage. On the south side of the wadi there are fewer tips and some indications of ancient excavation of the rock, consisting of a number of pits and a tunnel, focusing on the margins of a quartz band about one metre thick at the bottom of the hillside. A small rockfall shows that part of the hillside had been undermined, and a thick quartz band rises to the crest of the ridge northwest of the area of undermining. In the immediate proximity of the quartz band, two pits have been excavated, and to the east of these, large slabs of schist have been partly exposed.

16.3.4 Wadi Sikait (6th–9th centuries AD)

At the northern end of Wadi Sikait, on the slopes of Gebel Sikait (34° 47.48' E, 24° 39.85' N), is the largest of the late Roman mines. Although there is a small group of about 20 dry-stone huts close to the mines themselves, the major mining settlement, including a three-storey watch-tower and rock-cut shrines, lies several kilometres to the south (34° 47.80' E, 24° 37.94' N). The mines comprise about 15 separate irregular adits, which are much larger than those of either Gimal or Nuqrus. The report given by Frédéric Cailliaud (1821) suggests that they are also deeper than the latter.

The Sikait region is characterised by a sequence of folded schists and talc schists, again associated with granite pegmatites and quartz mineralization. As well as the main set of adits at the Sikait mine itself, there are also a few horizontal shafts cut into both sides of the wadi between the mine in the north and the settlement in the south. Even the rock-temples at the Sikait settlement are carved out of the soft part of a talc schist horizon. In our examination of the considerable tip heaps at the Sikait mine, we found only one fragment of emerald, consisting of three undamaged and intergrown hexagonal prisms of beryl. This comparative lack of fragments at the mine itself, however, was offset by the fact that very large quantities of fine processing heaps and broken emerald fragments were found at the Sikait settlement.

16.3.5 Gebel Zubara (12th–16th centuries AD)

The mines at Gebel Zubara (34° 48′ E, 24° 45′ N; see Fig. 16.3) are the most northerly of the five sites discussed in this paper, and the ancient importance of the site is indicated by its Classical name, *Mons Smaragdus* ('emerald mountain').

The miners' settlement at Zubara is spread out along a narrow wadi floor at the foot of the mountain. The relationship between landscape and settlement is therefore similar to that at Wadi Gimal A and B, and quite different to the hillside settlements of Wadi Nuqrus and Wadi Sikait. The Zubara buildings also have a distinctly curvilinear, vernacular character; and often the walls making up the huts or compounds are circular or spiral. It is also clear from the presence of pieces of modern sacking that, in addition to the medieval workings, relatively recent attempts have been made to exploit these mines.

All along the sides of the wadi are huge tip heaps primarily made up of fragments of schist. The gangue derives from numerous adits and pits, some shallow and others leading to labyrinthine systems of tunnels. The adits tend to follow the contact between the schists and the granite (and associated quartz mineralization). Whenever the emerald-bearing lodes gave out, new vertical shafts were sunk from the original galleries until a fresh lode was struck.

16.4 Summary and discussion of the ancient processing techniques employed at the Sikait-Zubara mines

One of the main problems with the study of mines — especially successful mines — is the fact that the mined material itself rarely survives *in situ*. It is therefore necessary to infer the nature of the mined material by determining which parts of the original geological context have been removed. In the case of the Egyptian emerald mines, this analysis is made easier by the fact that material was processed in the immediate vicinity of the mines, and therefore unusually small or poor quality fragments of the mined material have consistently been left behind. From these fragments it is possible to deduce firstly that the product of the mines was emerald, and secondly that, although the gems were associated with a variety of metamorphosed sedimentary rocks, they were always found in the vicinity of granite pegmatites (coarse grained material that forms when a water-rich melt crystallises). In the case of Egypt, the pegmatites are the last remnants of the crystallisation of a number of pre-Cambrian granites. These remnants of granite concentrate precious metals and other rare elements since they do not readily find a place in normal mineral structures. This situation may lead to the formation of deposits of metal (*e.g.*, the gold in the Wadi Hammamat) or other minerals, such as beryl, which are economically viable for mining.

The material extracted from the mines at Wadi Gimal, Wadi Nuqrus, Wadi Sikait and Wadi Zubara includes (1) pegmatitic veins, (2) the contact between pegmatites and schists, and (3) the granite as a whole (the latter only at the Wadi Gimal A mines, see Table 16.1). These pegmatites and schists occur over a wide area, and there is evidence for removal of veins from outcrops in an area that extends several kilometres from the main settlement (especially at Wadi

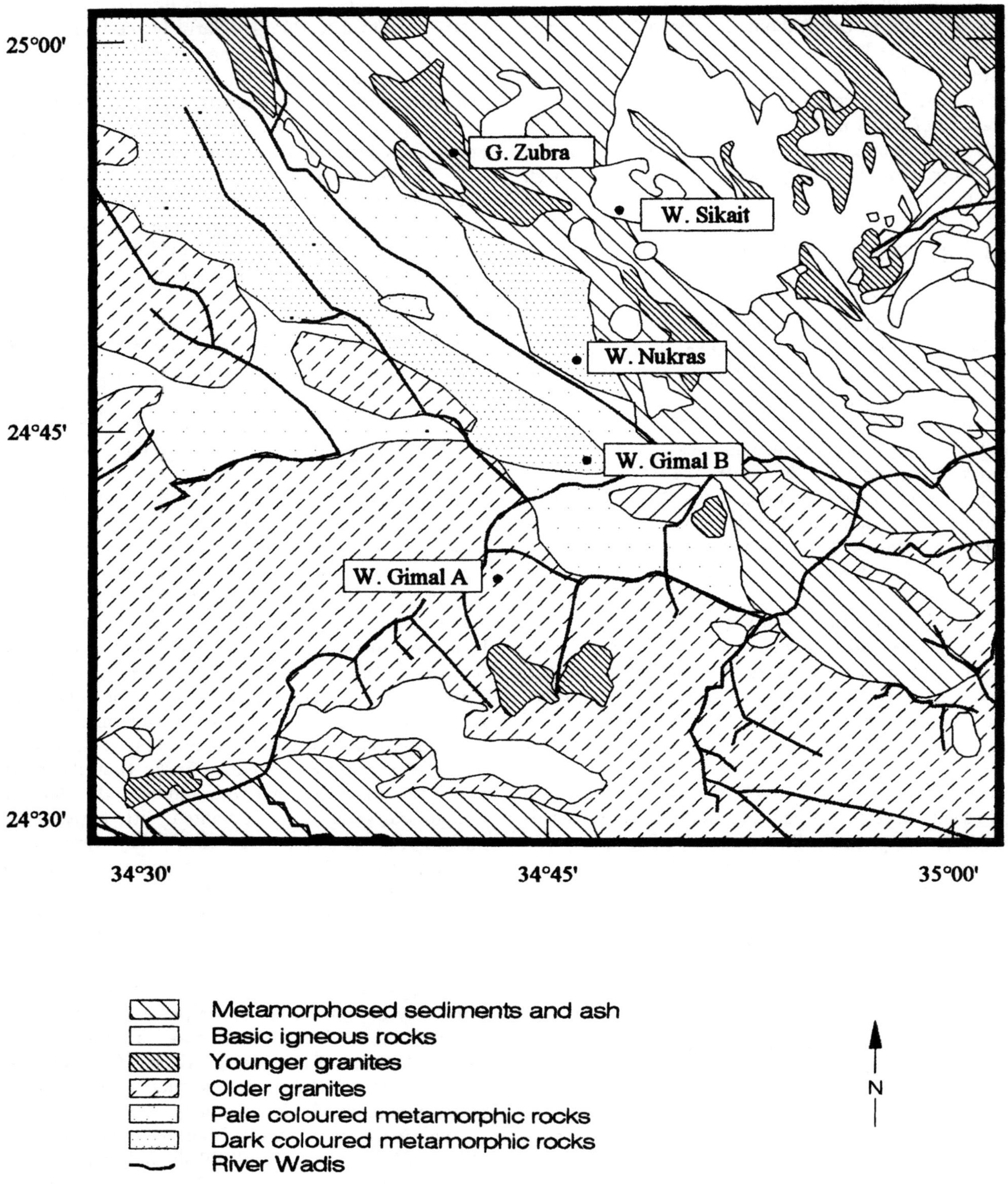

Figure 16.3: Geological map of the Sikait-Zubara region, showing at least two generations of granite

Gimal A). Rough selection of the material was performed at the outcrop, with material containing no emeralds being removed and discarded. Since, however, only coarse material is found at these workings, and no emerald fragments remain, we conclude that some form of fine processing was carried out at the settlements, where a distinctive type of fine-grained waste material can be found.

At all of the settlements, except Wadi Nuqrus, the fine-grained waste takes the form of terraces comprising groups of small circular heaps (with diameters of *c.* 50cm). Each heap includes emerald fragments and has a central circular depression. In some cases, the miners' settlements are constructed on top of the terraces formed from earlier workings, while against the sides of these buildings are piled the waste of subsequent activity. In the case of the Wadi Nuqrus mines, however, the emerald-bearing lode was just beneath or at the surface of a sloping piece of ground, and large circular depressions (*c.* 2m across) were constructed on the already exhausted seam. As the lode was worked out, the processing mechanism moved up the hill until no further material was yielded. The regularity of the depressions at Nuqrus suggests that some sort of machinery was used to process the emeralds. This machinery, which was operated at the settlement, was sufficiently portable to be moved around when the area around it was choked with waste, but it was evidently not taken out to the mines themselves to separate the emeralds at source. Modern beryl miners in Utah (western USA) still separate beryls from the waste material by hand, since mechanical devices break the beryl crystals along their basal cleavages. The 'machinery' used at the Nuqrus settlement may therefore simply have been some sort of light crushing device.

16.5 Electron microprobe analysis of the Sikait-Zubara samples

A number of geologists and archaeologists have published accounts of their visits to the Sikait-Zubara emerald-mining region of Egypt, particularly during the first three decades of the present century. In recent years, there have been two principal published reports documenting the analysis of Egyptian beryl (Grubessi *et al.* 1989, 1990; Jennings *et al.* 1993).

Jennings visited the Sikait-Zubara region in 1991, and his published account (Jennings *et al.* 1993) consists of geological descriptions of some of the mining sites as well as gemmological analyses of a number of specimens, including the measurement of their refractive indices and specific gravities, the determination of their pleochroic colours, and the study of their chemical composition using EDXRF. They state however, that, because 'none of the crystal fragments found by the authors on site was large enough or of high enough quality for gemmological testing' (Jennings *et al.* 1993, 112), the specimens of beryl were obtained from Egyptian gem dealers who were said to have either collected them personally from the mines or purchased them from local Bedouin. Jennings *et al.* were therefore analysing unprovenanced material, and consequently their results cannot be tied in to a provenancing study of the various separate mines in the Sikait-Zubara region (and it should also be noted that, given the circumstances of their collection, the gems cannot even definitely be said to derive from the Sikait-Zubara region at all).

Grubessi *et al.* (1989, 1990) used x-ray diffraction and infrared spectroscopy to compare Sikait-Zubara emeralds with those from mines elsewhere in the world. They reached the conclusion that emeralds from different provenances could be distinguished on the basis of geochemical criteria. Grubessi *et al.* demonstrated that the most effective way to distinguish the emeralds was by electron microprobe analysis (EPMA) of a suite of elements. So that a direct comparison could be made with their results, we also used EPMA in the new research described below. EPMA has the additional advantage that it can be used to analyse small patches of material *in situ*, thus allowing the characterisation of compositional variation within crystals (*e.g.,* due to zoning) and between crystals.

In 1994, we obtained samples from four sites in the Sikait-Zubara mining region (Gimal A, Nuqrus, Sikait mines and Sikait settlement, see Fig. 16.2) and used an electron microprobe to examine them, with the principal aim of reassessing the assumptions and results of Grubessi *et al.* with regard to the sourcing of emeralds.

It is obvious that any method of analysis which is used to study both emerald fragments recovered from mines and actual ancient gems surviving in the form of jewellery must be non-destructive. In addition, it is essential that the statistical populations of the emeralds from different geographical sources are shown to be distinct and that the fragments of discarded emerald at the mines are proven to be representative of the better quality emeralds that have been actually removed from the site.

The emerald fragments collected from each of the sites in the Sikait-Zubara region were thin-sectioned and examined under a petrological microscope before using an electron microprobe to undertake major element analysis of the polished section. Optical microscopy indicated that the fragments of beryl were polycrystalline and that individual crystals were also zoned (Figs. 16.4–16.5). Since a crystal of consistent structure is required in order to form a gem-quality emerald, it is assumed that gem-quality emeralds would demonstrate less variation in composition than poor quality fragments. However, there is no requirement for gem-quality beryl grains to have the same composition as each other. In order to repre-

Figure 16.4: Thin section of emerald grain (field of view 3mm) in cross polars, showing crystals of beryl, including a hexagonal basal section of a beryl which appears black in the micrograph.

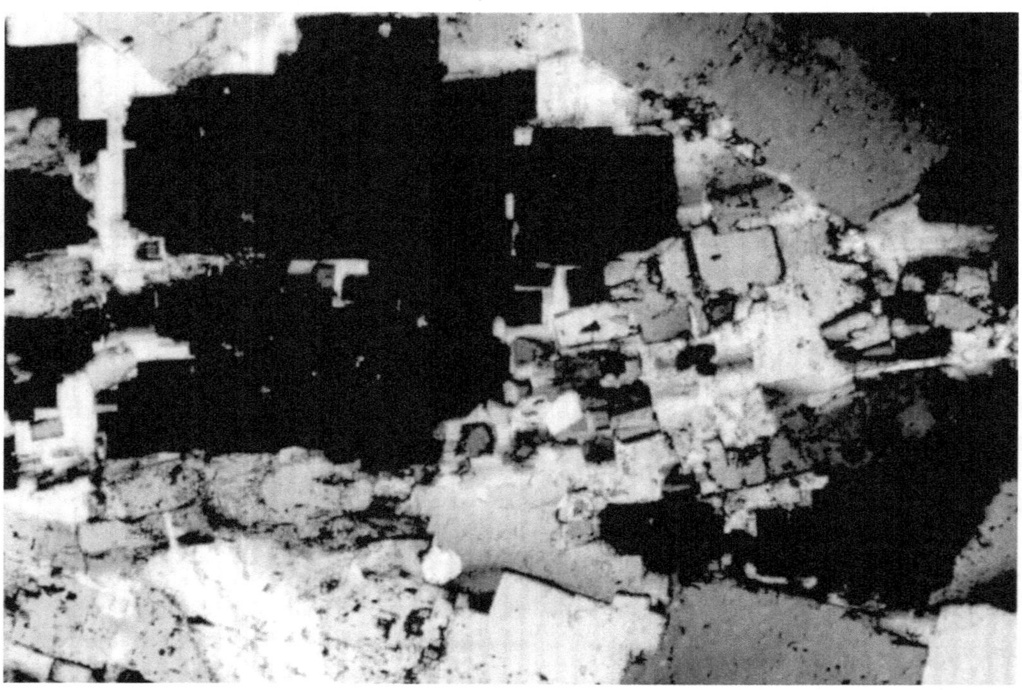

Figure 16.5: Thin section of emerald grain (field of view 3 mm) in cross polars, showing a polycrystalline area of beryl with a number of small grains in the centre of the micrograph.

sent the likely variation of a population of emeralds, we investigated the variation in a set of small grains, each in itself well-formed but as a whole forming what is described as 'emerald matrix'. The samples were analysed using the CAMECA SX50 electron microprobe in the Department of Earth Sciences at the University of Cambridge. The results of our analyses of each fragment are summarized in Table 16.2, while the results of the analyses of Grubessi *et al.* (1990) are presented in Table 16.3 (note that there are no analyses available for Indian emeralds, but true Indian emeralds, as opposed to green sapphires, were not discovered until earlier this century, see Sinkankas 1986).

The major data and trace element data were examined using element-element plots. Principal Component Analysis was used to identify sets of elements that showed most variation in composition for the elements analysed. The greatest separation of populations of emerald grains from different mines was registered in the co-variation of elements that occupy the two different cation sites in the beryl structure. In practice, Na_2O can be used as a proxy for the large ions that occupy the hexagonal channels in the silicate structure while MgO acts as a proxy for smaller and more highly charged ions that substitute into the silicate framework. This analysis indicated that, for all the elements analysed, the variation in composition of each fragment was as large as the variation between different fragments and indeed covered the range of most of the points identified by Grubessi *et al.* (1990). Thus, the range of results for one sample from Wadi Gimal (Em1) is just as great as the apparent differences between emeralds from Austria and Brazil. This situation is clearly illustrated by the scattergram in Figure 16.6 (based on data from our own analyses as well as those of Grubessi *et al.* 1990) showing the Na_2O and MgO contents of emeralds deriving from Egypt, Austria, Brazil, Colombia and the Urals. The inherent variability of beryl (and emerald) compositions can be ascribed to the flexibility of composition provided by large cation channels in the structure, and our analysis suggests that this inherent variability hinders efforts to distinguish emerald sources. Additional statistical processing of these data is unlikely to yield more conclusive results.

16.6 Conclusions

The analyses outlined above suggest that much of the geochemical variation in world emeralds is mirrored within single grains of emerald tailings. We therefore conclude that the work of Grubessi *et al.* (1990) on the chemical composition of emeralds is misleading. Thus, when statistical variation in the composition of emeralds from the Eastern Desert is taken into account, they not only do not vary from emeralds extracted from other mines in the Sikait-Zubara area, but they also show no consistent variation from emeralds mined in Austria or Brazil. On a more positive note, there are two types of emerald that *can* be distinguished from those of Egypt with the use of the electron microprobe: those from the Urals and Colombia. The study of the samples from the ancient Sikait-Zubara mines may therefore provide the potential for investigations of the world emerald market during the time of operation of the Egyptian mines and also the detection of forgeries manufactured from New World emeralds. Our future work, however, will require the use of techniques which are both more sensitive and non-destructive, such as XRF analysis (see Jennings *et al.* 1993; Jones *et al.* 1997).

Possible avenues of investigation include the use of the ratio of rubidium and strontium, elements that reside in the channels in the beryl structure. Although these elements are expected to partition variably into the channels, in the same way as other elements, the radiogenic strontium decays with time, therefore the Rb/Sr ratio gradually changes. The granites in the Eastern Desert (Fig. 16.3) are all part of the same orogenesis, and are therefore not expected to show much internal variation, but other world granites (and hence other emerald deposits) are of different age and should therefore yield distinctive Rb/Sr ratios. We also hope to test the hypothesis that the green colour of Egyptian emeralds is a result of the presence of reduced iron oxide (see Deer *et al.* 1992, 256–7).

Our re-analysis of the data presented by Grubessi *et al.* (1989) and our study of fresh samples from Sikait-Zubara suggest that researchers must take care to analyse variation at a number of different levels of magnitude. By concentrating on simple variation *between* samples from source regions on a global scale, researchers such as Grubessi *et al.* can reach conclusions that are not only simplistic but simply incorrect. On the other hand, the study of variation *within* individual sampled grains and *within* source regions can provide a much more complex and realistic view of the situation, thus allowing archaeologists to reformulate their research questions, taking account of limitations arising first from their chosen sampling methods (see Shackley 1998) and secondly from their chosen techniques of geochemical definition (*e.g.*, Na_2O and MgO contents versus Rb/Sr ratios).

Acknowledgements

We are grateful to the Institute of Archaeology, University College London and the Department of Earth Sciences, University of Cambridge, for their support during our research. We would like to thank Professor Mohammed Sharkawi of the Egyptian Educational and Cultural Bureau, London for reading and commenting on the first draft of this paper. We would also like to thank the McDonald Institute for Archaeological Research, University of Cambridge, which provided the funding for the preliminary geological survey of the Sikait-Zubara region in 1994.

Element	Sample													
	Em1A (9)	SD	Em1B (9)	SD	Em2 (14)	SD	Em3A (13)	SD	Em3B (15)	SD	Em4A (8)	SD	Em4B (12)	SD
SiO_2	64.74	0.60	64.52	0.46	65.12	0.49	65.22	0.28	65.03	0.41	65.27	0.25	64.88	0.90
TiO_2	0.02	0.037	0.01	0.04	0.02	0.05	0.01	0.04	0.02	0.04	0.02	0.03	0.01	0.03
Al_2O_3	14.99	0.22	13.90	0.90	15.06	0.37	14.65	0.15	14.61	0.28	15.06	0.25	14.78	0.27
Cr_2O_3	0.08	0.07	0.21	0.09	0.13	0.05	0.29	0.13	0.34	0.13	0.16	0.10	0.16	0.03
FeO	0.51	0.07	0.78	0.26	0.45	0.09	0.40	0.08	0.41	0.08	0.54	0.09	0.43	0.07
MnO	0.00	0.02	0.01	0.03	0.00	0.03	0.00	0.02	0.00	0.04	0.01	0.03	0.00	0.04
NiO	0.01	0.06	0.00	0.03	0.02	0.04	0.00	0.06	0.00	0.05	0.00	0.04	0.01	0.03
MgO	2.22	0.12	2.80	0.41	2.33	0.18	2.54	0.20	2.49	0.17	2.28	0.09	2.60	0.25
CaO	0.01	0.02	0.13	0.08	0.01	0.03	0.01	0.01	0.00	0.03	0.02	0.02	0.03	0.05
Na_2O	1.79	0.11	1.99	0.24	1.80	0.12	2.04	0.20	1.99	0.10	1.79	0.07	1.87	0.14
K_2O	0.01	0.03	0.04	0.04	0.00	0.02	0.00	0.02	0.00	0.03	0.02	0.03	0.01	0.02
Cl	0.00	0.02	0.02	0.02	0.00	0.02	0.00	0.02	0.00	0.02	0.00	0.02	0.01	0.02
Total	84.32		84.34		84.89		85.15		84.87		85.08		84.75	

Table 16.2: Major element concentrations of emerald samples from Wadi Sikait settlement (Em1), Wadi Sikait mines (Em2), Wadi Gimal A (Em3) and Wadi Nuqrus (Em4).

Element	Sample						
	St Teresina	Zubara	Austria	Colombia	Urals	mean	SD
SiO_2	64.36	64.78	63.60	65.34	65.14	64.64	0.69
TiO_2	0.22					0.22	
Al_2O_3	13.99	14.84	14.54	16.51	17.72	15.52	1.55
Cr_2O_3	0.32	0.11	0.35	0.37		0.29	0.12
FeO	1.11	0.72	0.38	0.31	0.24	0.55	0.36
MnO							
NiO							
MgO	2.68	2.36	2.27	0.87	0.44	1.72	1.00
CaO							
Na_2O	2.03	1.64	1.72	0.49	0.51	1.28	0.73
K_2O		0.07				0.07	
Cl							
BeO	13.03	13.11	13.01	13.37	13.46	13.20	0.21
Total	100.13	99.89	98.51	100.30	99.86	99.74	

Table 16.3: Major element concentrations of emerald samples from St Teresina (Brazil), Zubara (Egypt), Habachtal (Austria), Muzo (Colombia) and the Ural Mountains.

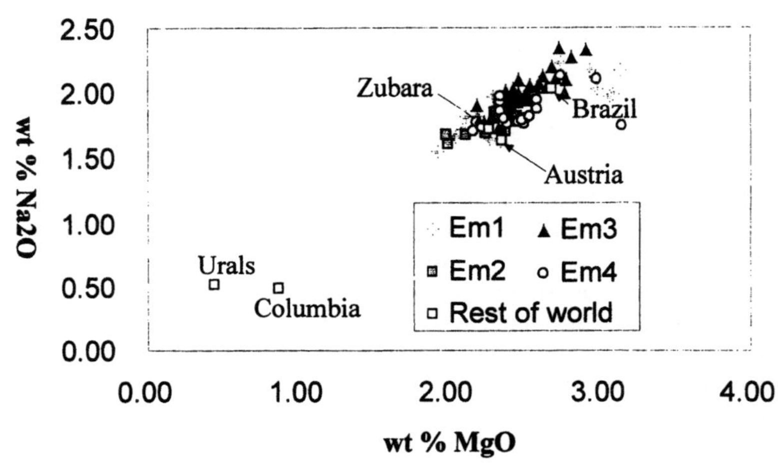

Figure 16.6: Scattergram showing the Na_2O/MgO contents for samples of emerald from Egypt, Austria, Brazil, Colombia and the Urals (data for rest of world from Grubessi *et al.* 1990).

References

CAILLAUD, F. 1821. *Voyage à l'oasis de Thèbes et dans les déserts situés à l'orient et à l'occident de la Thébaide, fait pendant les années 1815–1818.* Paris. 2 volumes.

DEER, W. A., R. A. HOWIE & J. ZUSSMAN 1992. *An Introduction to the Rock-Forming Minerals. Vol.1: Ortho- and Ring-Silicates.* Longman, London, second edition.

GRUBESSI, O., C. AURISICCHIO & A. CASTIGLIONI 1989. 'Lo smeraldo delle miniere dei faroni.' *La Gemmologia* 14: 7–21.

GRUBESSI, O., C. AURISICCHIO & A. CASTIGLIONI 1990. 'The pharaohs' forgotten emerald mines.' *Journal of Gemmology* 22(3): 164–77.

HUGHES, R. E. 1998. 'On reliability, validity and scale in obsidian sourcing research.' *In* A. F. Ramenofsky & A. Steffen (eds.), *Unit Issues in Archaeology; measuring time, space and material*, pp. 103–14. University of Utah Press, Salt Lake City.

JENNINGS, R. H., R. C. KAMMERLING, A. KOVALTCHOUK, G. P. CALDERON, M. K. EL BAZ & J. I. KOIVULA 1993. 'Emeralds and green beryls of Upper Egypt.' *Gems and Gemmology* 29(2): 100–15.

JONES, G. T., D. G. BAILEY & C. BECK 1997. 'Source provenance of andesite artefacts using non-destructive XRF analysis.' *Journal of Archaeological Science* 24: 929–43.

LUCAS, A. 1962. *Ancient Egyptian Materials and Industries.* E. Arnold, London, fourth edition. Revised by J. R. Harris.

MACALISTER, D. A. 1900. 'The emerald mines of northern Etbai.' *Geographical Journal, London* 16: 537–49.

MURRAY, G. W. 1925. 'The Roman roads and stations in the eastern desert of Egypt.' *Journal of Egyptian Archaeology* 11: 138–50.

SHACKLEY, M. S. 1998. 'Gamma rays, x-rays and stone tools: some recent advances in archaeological geochemistry.' *Journal of Archaeological Science* 25: 259–70.

SHAW, I. 1999. 'Sikait-Zubara.' *In* K. Bard (ed.), *Encyclopedia of the Archaeology of Ancient Egypt*, pp. 731–3. Routledge, London.

SHAW, I., J. BUNBURY & R. JAMESON 1999. 'Emerald mining in Roman and Byzantine Egypt.' *Journal of Roman Archaeology* 12: 1–13.

SINKANKAS, J. 1986. *Beryl.* Butterworth, London.

THOMAS, E. S. 1909. 'The mineral industry of Egypt: emeralds.' *Cairo Scientific Journal* 3(38): 267–72.

Ian Shaw
School of Archaeology, Classics and Oriental Studies
University of Liverpool
12 Abercromby Square
Liverpool L69 3BX
England
`ishaw@liv.ac.uk`

J. Bunbury
Department of Earth Sciences
University of Cambridge
Downing Street
Cambridge CB2 3DZ
England
`jmb21@esc.cam.ac.uk`

For Product Safety Concerns and Information please contact our EU representative GPSR@taylorandfrancis.com Taylor & Francis Verlag GmbH, Kaufingerstraße 24, 80331 München, Germany

Batch number: 10732722

Printed by Printforce, the Netherlands